建筑设计详解

Architectural Design Process Demystified

3

U0384135

中国林业出版社
China Forestry Publishing House

图书在版编目（CIP）数据

建筑设计详解 . 3 / 佳图文化主编 . -- 北京 : 中国林业出版社 , 2016.9

ISBN 978-7-5038-8577-8

Ⅰ . ①建… Ⅱ . ①佳… Ⅲ . ①建筑设计－研究 Ⅳ . ① TU2

中国版本图书馆 CIP 数据核字 (2016) 第 135565 号

中国林业出版社 · 建筑家居出版分社
责任编辑：李 顺 李 辰
出版咨询：（ 010 ）83143569

出 版：中国林业出版社（ 100009 北京西城区德内大街刘海胡同 7 号 ）
网 站：http://lycb.forestry.gov.cn/
印 刷：广州中天彩色印刷有限公司
发 行：中国林业出版社
电 话：（ 010 ）83143500
版 次：2016 年 9 月第 1 版
印 次：2016 年 9 月第 1 次
开 本：889mm × 1194mm 1 ／ 16
印 张：16
字 数：200 千字
定 价：290 .00 元

PREFACE 前言

From concept to building is not a process of simple duplication or deliberately piling up, but a transformation of abstract beauty in concept to beauty of building in form, an expression of conceptual connotation to concrete beauty of building. From concept to building is a process to realize our vision of a better city in line with the law of beauty.

According to the latest design concepts in current international construction industry and from the perspective of professional architectural design, this book is a collection of carefully selected cases mainly involved in shopping centers, commercial complexes, urban complexes, office buildings, hotel buildings, art and cultural buildings and transportation buildings and so on. In content layout, each case is analyzed in keywords, features and design concept in the company of a large variety of technical drawings such as plans, sections and analysis graphics. With rich content and full and accurate information, the book tries to give architects and readers in related industries visual enjoyment and design inspiration.

从概念到建筑，不是简单复制的过程，也不是刻意堆砌的过程，而是将概念的抽象美转化成建筑的形式美、将概念的抽象内涵转化成建筑的具象表达。从概念到建筑，是按照美的规律，实现“我们让城市更美好”的愿景的过程。

本套书依据国际现行建筑行业的最新设计概念，站在建筑设计的专业角度，精心挑选案例。本书案例主要涉及了购物中心、商业综合体、城市综合体、办公建筑、酒店建筑、文化艺术建筑以及交通建筑等建筑形态。内容编排上，分别从案例的关键点、亮点、设计概念等方面入手，并配合大量的各种技术图纸，如平面图、剖面图、分析图等。本书内容丰富、资料详实，希望能给建筑设计师及相关行业读者带来视觉享受和设计启迪。

CONTENTS 目录

Office Building 办公建筑

Transportation Building 交通建筑

Office Buliding

办公建筑

People Oriented
人本理念

Interaction
互动交流

Enterprise Logo
形象标识

Green & Eco-friendly
绿色环保

Xinglin Bay Business Operation Center 7# Building 杏林湾营运中心 7 号楼

Keywords 关键词

Super High-rise
超高层

Prismatic Shape
棱柱造型

Detailed Design
细部设计

Location: Xiamen, Fujian, China
Developer: XINGLIN JIANSHE
Architectural Design: Hordor Design Group
Total Land Area: 12,757.57 m^2
Base Area: 4,631 m^2
Overground Floor Area: 76,319.5 m^2 (commercial floor area 14,099.5 m^2, office floor area 62,220 m^2)
Underground Floor Area: 20,155.2 m^2

项目地点：中国福建省厦门市
开 发 商：厦门市杏林建设开发公司
建筑设计：厦门合道工程设计集团有限公司
总用地面积：12 757.57 m^2
建筑占地面积：4 631 m^2
地上建筑面积：76 319.5 m^2（其中商业建筑面积14 099.5 m^2，办公建筑面积62 220 m^2）
地下建筑面积：20 155.2 m^2

Features 项目亮点

The concept originates in glittering and translucent white crystal. Crystal is clean and transparent, and each crystal prism stands upright in the cluster of druse, which conform with the surrounding environment in this case.

主楼的形体概念来自于晶莹剔透的白水晶。水晶干净而透明，晶簇群上的每一个晶柱都昂然挺拔，极符合本地块项目营运次中心的地位和精神。

■ Overview

Xinglin Bay Business Operation Center is located on the south of BRT line and the northwest of sea water. Across the water is Garden Expo Park, boasting convenient traffic, rich landscape resources and open view. 7# Building is a 140.5m super high-rise office building which has two floors underground and 30 floors overground, located in the central position of the center and adjacent to large commercial center and greenbelt, enjoying the view of Guanren Mountain.

■ 项目概况

该项目地处杏林湾营运中心区，位于杏林园博园水域中心地带，北临BRT快速公交线路，东南全线临海，与园博园隔水相望。交通便利，景观资源丰富，视野开阔。7号楼地块项目处于该营运中心的中间位置，北有大型商业中心，西临营运中心绿化带，西南方向可远眺官任山景，为地上30层、地下2层、建筑高度148.5 m的超高层写字楼。

■ Overall Planning

Overall planning revolves around "two axis, two centers", which serve as a guidance for landscape design to strive for the largest landscape scale that ensures sea view for the east, mountain view for the west and park view for the south. Situated in the northwest of the base, the main building overlooks the lush greenery in the bordering greenbelt and faces the high-rise building group on the east side in a modest yet dauntless imposing stature.

■ 总体规划

本案以营运中心"两轴两心"的总体规划结构为景观设计的依据，做到东面观海，西面观山，南面观园，争取最大的景观面。将主楼置于西北面，下临中心绿化带，给建筑增添了一份绿意葱茏，而对东面的高层建筑群，则显示了一种谦恭退让的态度而不失区域次中心气度，很好地明确了自己在区域中的定位。

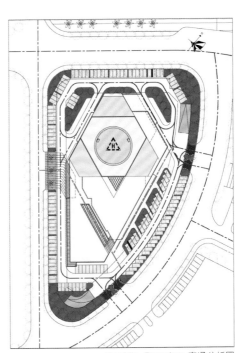

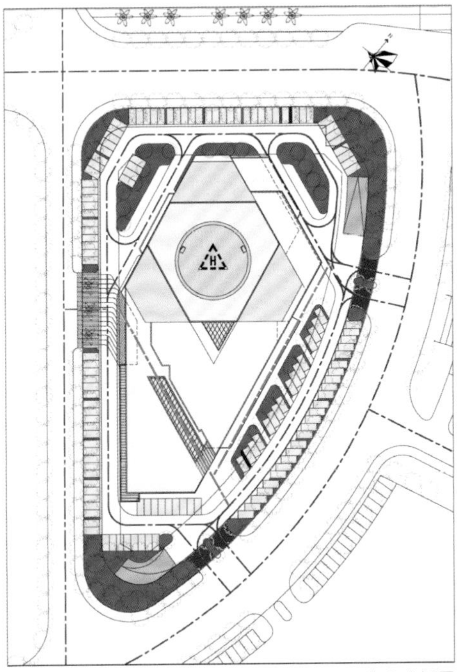

Traffic Drawing 交通分析图

- 基地主要出入口
- 基地人行出入口
- 车行流线
- 人行流线

作为150m超高层建筑项目，车流通畅的重要性可想而知。设计中以6m（部分地段5.5m）的环路贯通基地，连结东侧的两个主要出入口，结合绿地和停车的布置，做到商业和办公分流，楼宇入口引导明晰。

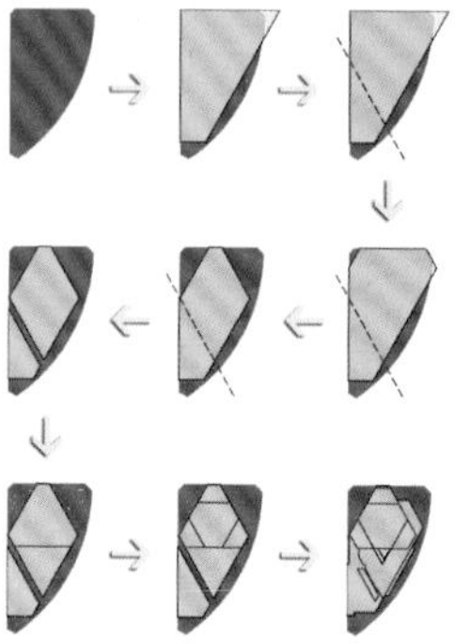

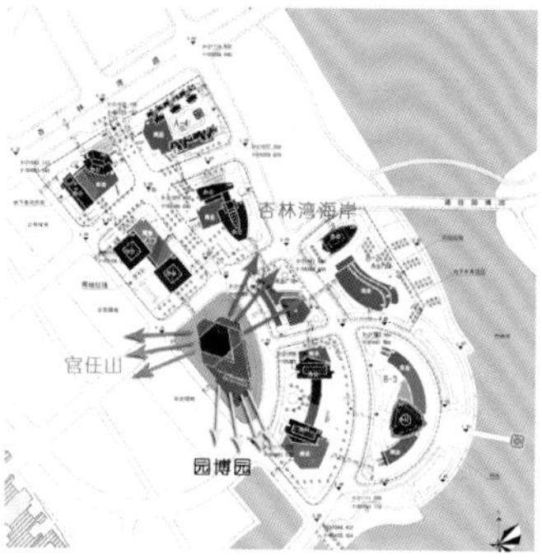

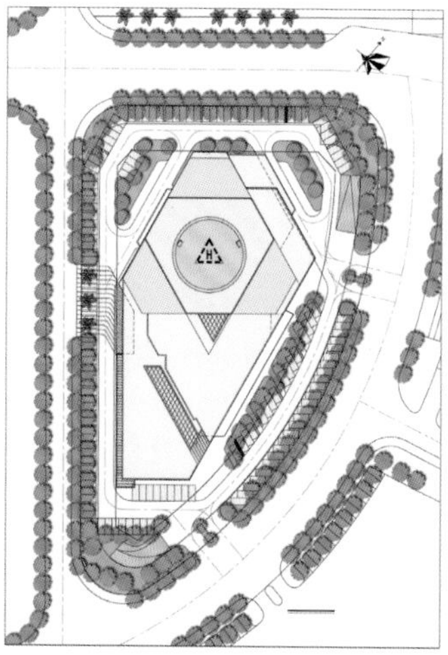

地块内的绿化设计，首先是围绕分组停车带布置的。此处的绿化客观上既成为基地内外的分界，也为人行道和停车带上的车辆提供了遮阳作用。
其次，在主楼西面和东面的两个出入口处，设计了部分结合停车的块状绿地，一面引导了车流，一面可结合长椅等等设施的布置，成为基地内的两处室外休闲场所。
最后，基地东北角和南面结合地下车库出入口布置的绿地，都为有限的基地面积争取了最大的绿化空间。

Landscape Analysis Diagram 景观分析图

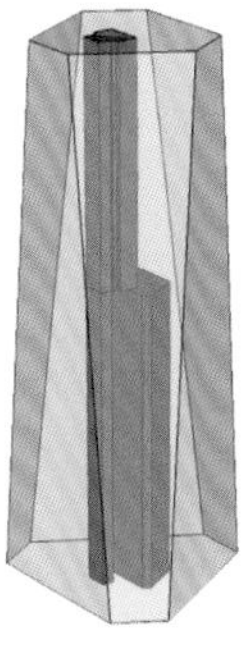

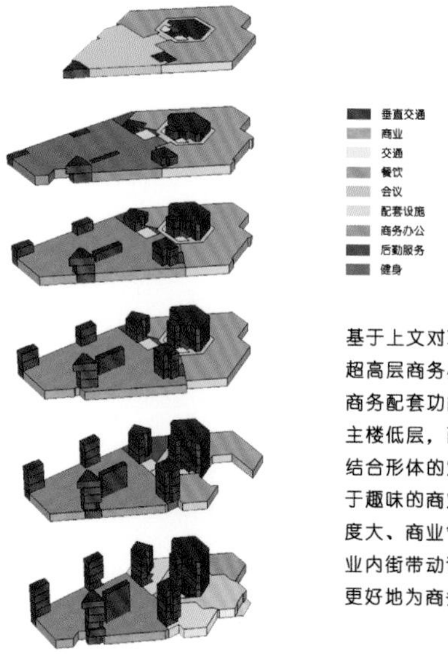

基于上文对项目的定位分析，设计功能以150m超高层商务办公建筑为主，辅以低层的商业和商务配套功能。布局上将商务配套功能多置于主楼低层，而将对外商业部分置于裙房。裙房结合形体的穿插、光线的设计，形成热闹并富于趣味的商业内街，弥补了该区与相邻地块跨度大、商业气氛不够浓郁的不足。更通过此商业内街带动该地块乃至周边的商业活力，以期更好地为商务功能服务。

Function Analysis Diagram 功能分析图

水晶晶柱＋底座 ⤳ 主楼＋裙房

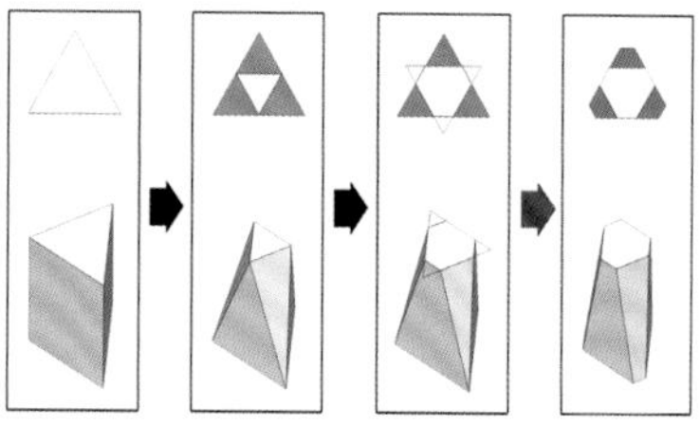

晶体＋不规则切面 ⤳ 玻璃幕墙＋有规律地削切

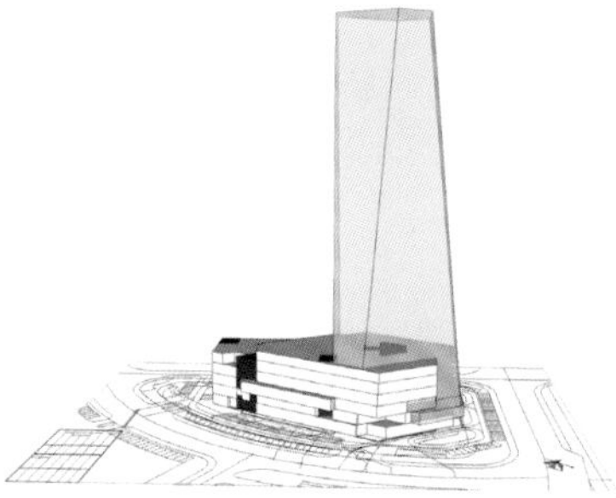

Shape Concept 形体概念分析图

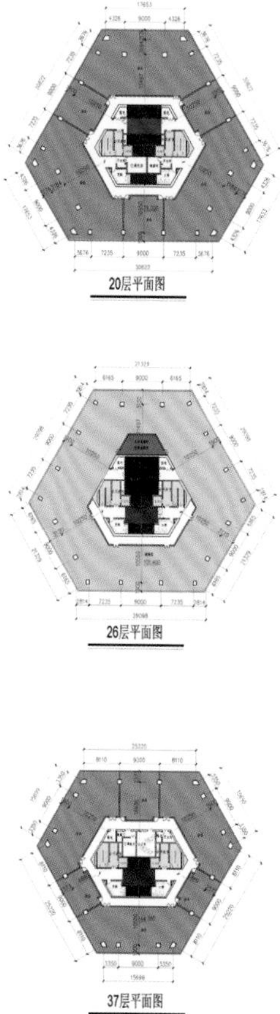

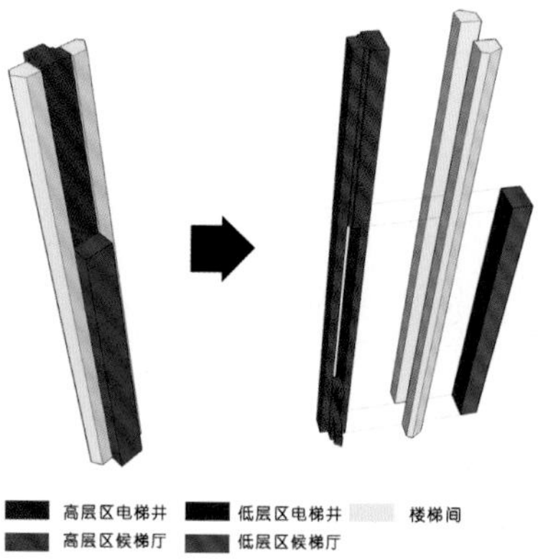

主楼核心筒的设计，使高层区与低层区电梯候梯厅分区明晰，合理利用分区空间布置配套服务功能，有效避免了交通面积的浪费。

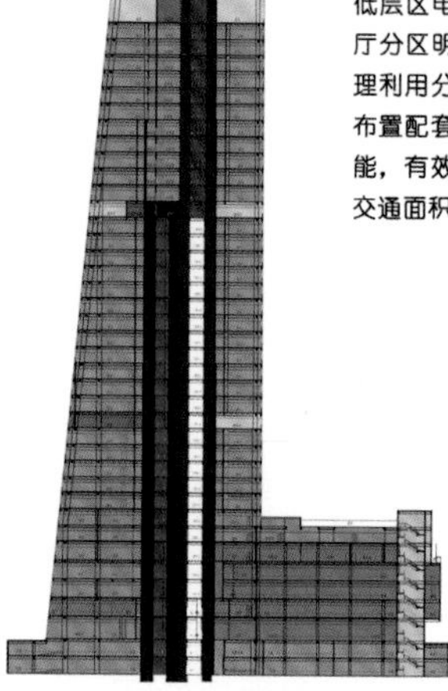

Main Core Tube Analysis Diagram 主楼核心筒分析图

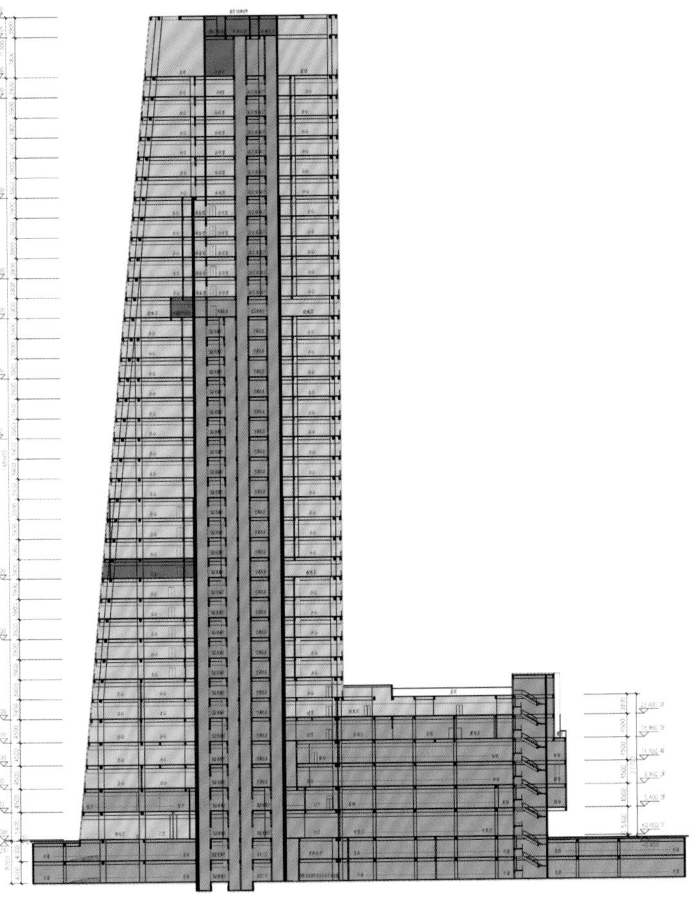

A-A Section A-A 剖面图

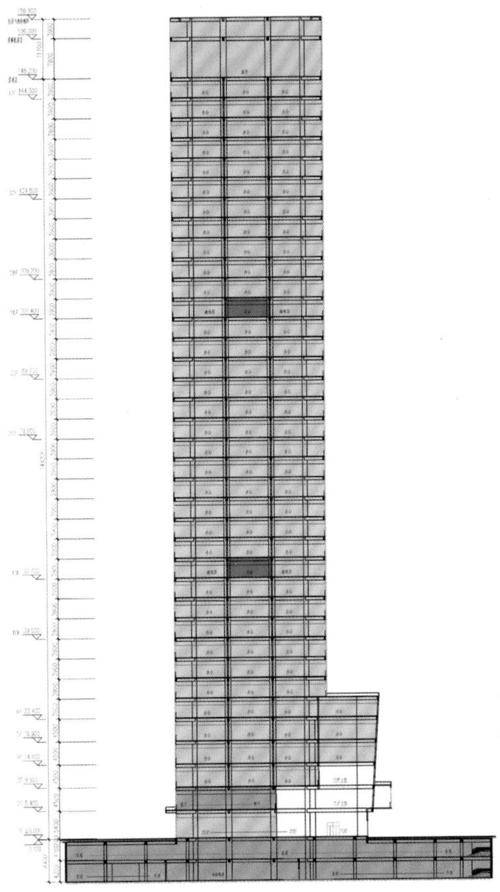

B-B Section B-B 剖面图

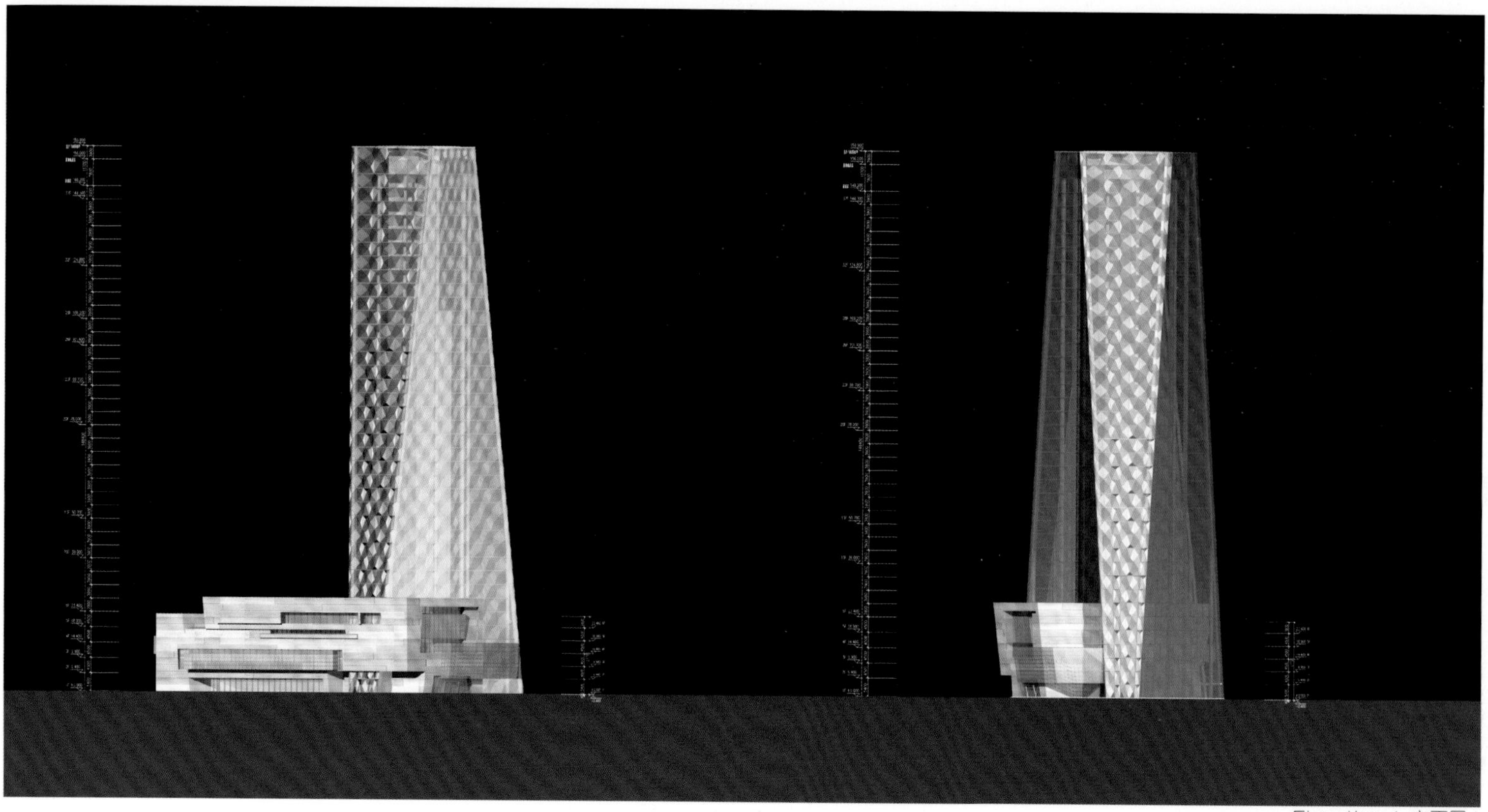

Elevation 1 立面图 1

■ **Shape Concept**

Through several rounds of scheme comparisons, designers finally used the shape comprised of crystal prism and triangle. The concept originates in glittering and translucent white crystal. Natural crystal is hexagonal drill-like that shaped in millions of years. Crystal is clean and transparent, and each crystal prism stands upright in the cluster of druse, which conform with the surrounding environment in this case. The main building transited from triangle to hexagon, upright, simple and modern, looks quite harmonious with the surroundings. Crystal clear tower shapes a contrast with the horizontally extended annex building in volume and material, just as an elegant and graceful crystal prism placing on an exquisite plate. Pyramids are rotated in a regular rule on the façade, creating different sparkling effects when the sun strikes the building in different angles. Reasonable building orientation provides a comfortable and relaxed working environment with favorable landscape view.

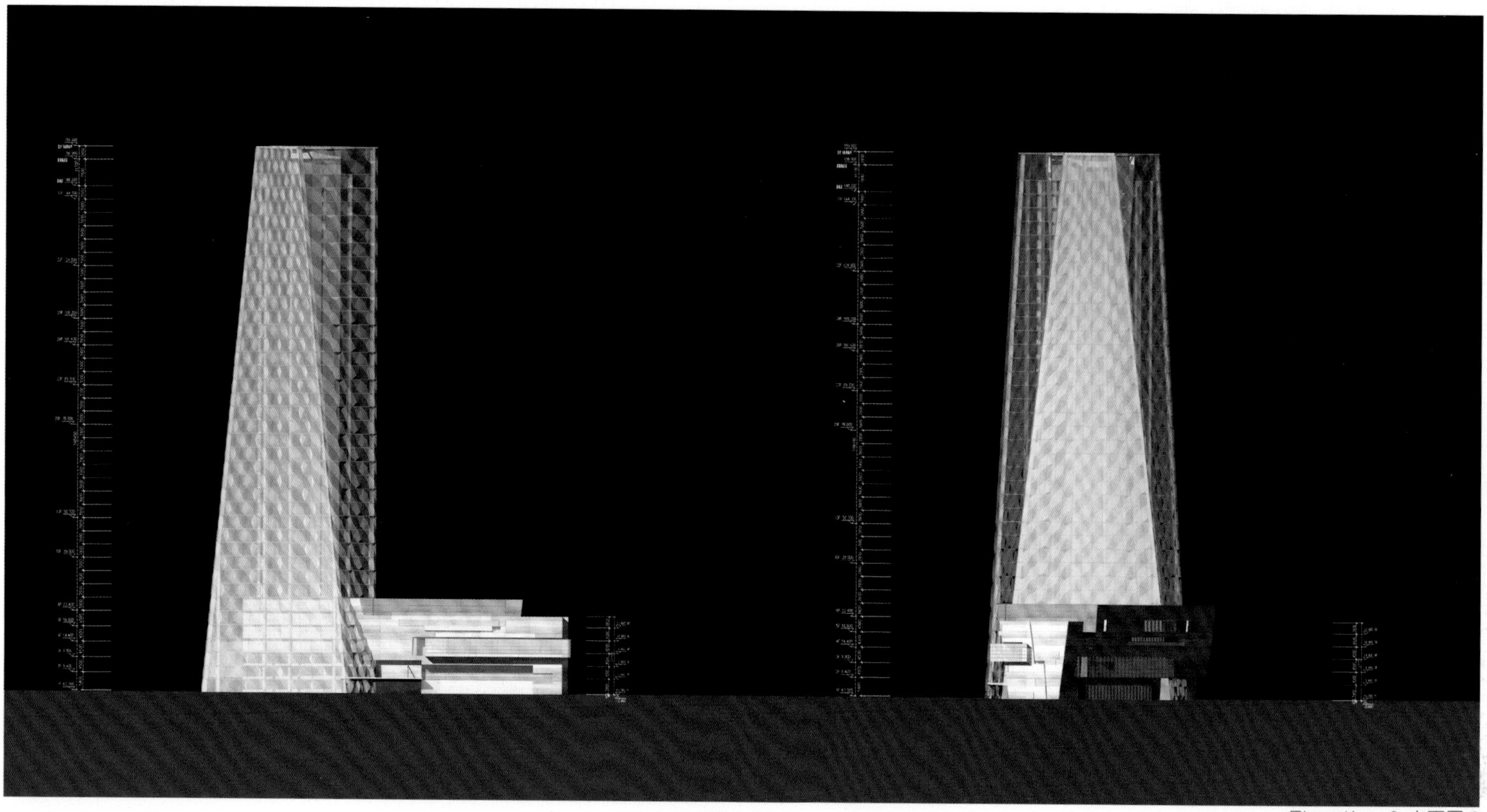

Elevation 2 立面图 2

■ 形体概念

通过多轮方案比选，设计团队确定了以水晶棱柱体结合三角形母体的形体方向。主楼的形体概念来自于晶莹剔透的白水晶。天然水晶呈六棱钻头状，是大自然经过亿万年形成的结晶。水晶干净而透明，晶簇群上的每一个晶柱都昂然挺拔，极符合本地块项目营运次中心的地位和精神。主楼形体由三角形过渡至切角六边形，形体挺拔、简洁、极富现代感，建筑造型和外观与周边环境协调一致。晶莹剔透的塔楼竖向形体与水平横向延伸的裙房在体量与材质上形成强烈对比，恰如一个放置于精美底盘之上的水晶棱柱，高雅华贵。立面细部由一个个偏转了中心的棱锥面经过多次旋转、镜像，以一定规律组合而成，在不同角度的阳光照射下形成如钻石切面般粼粼波光的效果。合理布局的建筑朝向为使用者提供了舒适、宽松、良好景观视线的工作环境。

Riverside International New Town Chuangye Building 滨河国际新城创业大厦

Keywords 关键词

Green Ecology
绿色生态

Spatial Organization
空间组织

Intelligentlzation
智能化

Location: Zhengzhou, Henan, China
Commissioned by: Zhengzhou Committee of Economic and Technological Development Zone
Architectural Design: KLP KELLER HU PARTNERSHIP LIMITED
Main Designers: Larry J. Keller, Sun Zhen
Total Land Area: 50,759m²
Floor Area: 127,000 m²
Date of Design: January, 2013

项目地点：中国河南省郑州市
委托单位：郑州经济技术开发区管委会
设计公司：美国KLP建筑设计有限公司
主要设计人员：拉里·凯勒（Larry J. Keller）、孙铮
用地面积：50 759m²
建筑面积：127 000 m²
设计时间：2013年1月

Features 项目亮点

Following the innovative concept of putting green ecology and the whole environment at the first place, it creates a symbol of mildness, open-mindedness and people-oriented services.

贯彻绿色生态和整体环境优先的创新理念，体现出该项目既平和严谨，又注重人性服务、全面开放的市民中心形象。

■ Overview

Zhengzhou Chuangye Building is located in the Binshui CBD of Zhengzhou Economic and Technological Development Zone. It is close to Jingnan 9th Road on the south and Jingkai 16th Street on the east, reaches Jingnan 8th Road on the north and borders Jingkai 15th Street on the west.

■ 项目概况

郑州创业大厦位于郑州经济技术开发区滨水商务中心区，南临经南九路，北至经南八路，东邻经开第十六大街，西至经开第十五大街。

■ Design Concept

The design follows the development idea of eco and services to create a governmental office building which will be in harmony with the surroundings, feature unique identity, promote local culture and serve the public. Based on the essence of Chinese culture — "harmonious, flexible and orderly" and the innovative ideas of "eco and green and environment first", it will build a symbol of mildness, open-mindedness and people-oriented services.

■ 设计理念

项目遵循"生态""服务"的核心发展理念，追求"融入环境、体现特性、传承文化、服务社会"的总体目标，设计立足于"中正平和，变通有则"的中原文化内涵，以绿色生态和整体环境优先的创新理念，体现了该项目既平和严谨，又注重人性服务、全面开放的市民中心形象。

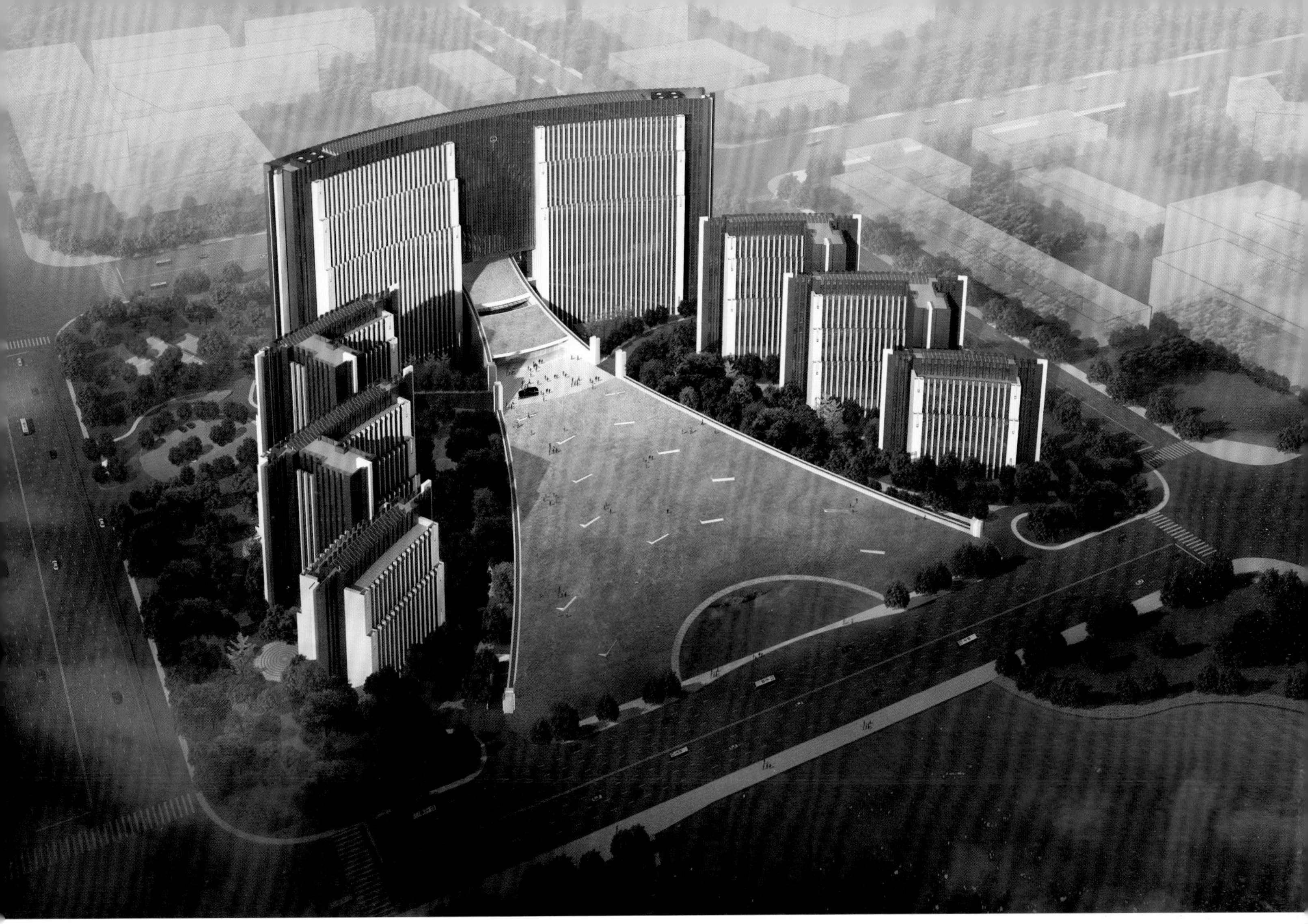

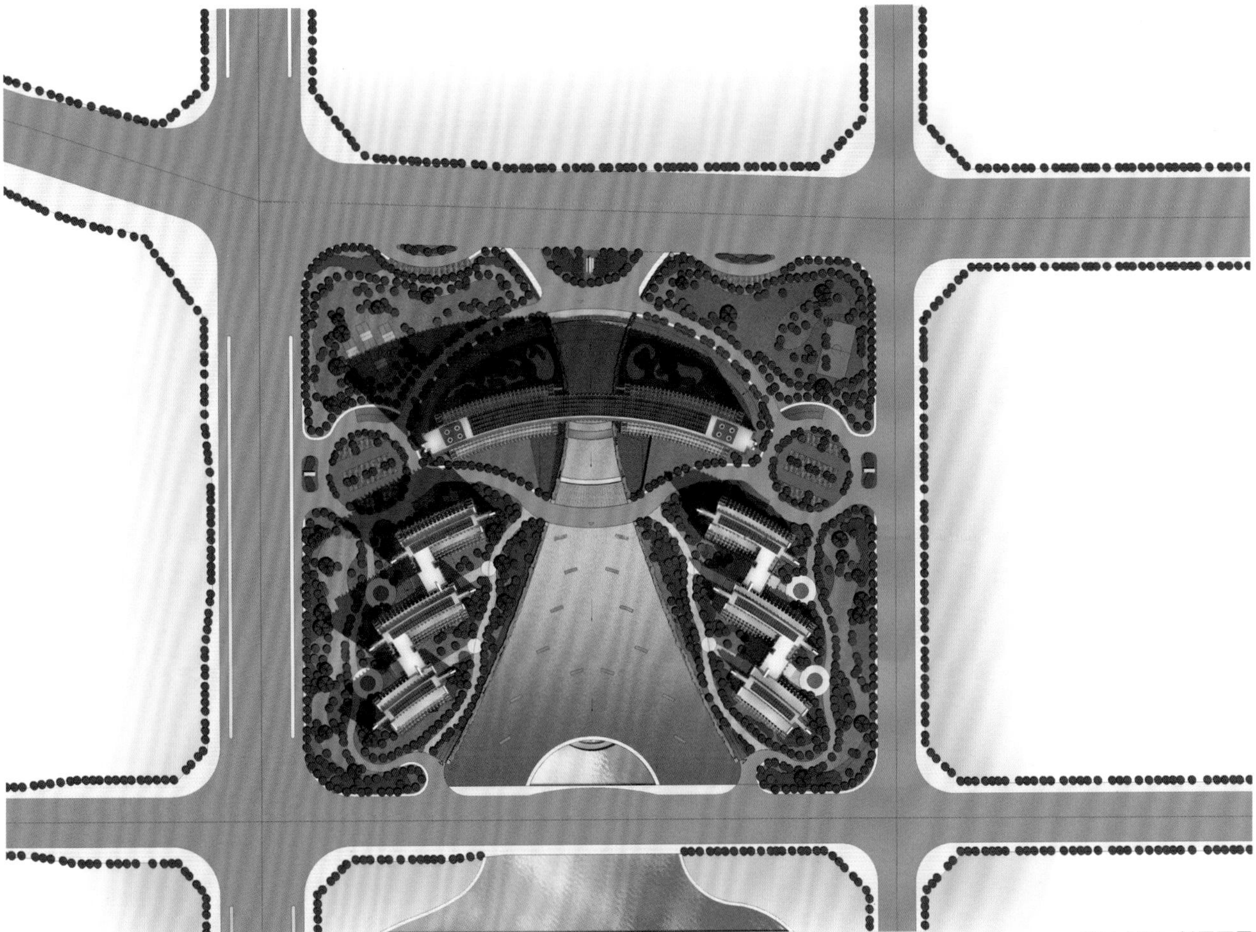

Site Plan 总平面图

■ **Design Features**

Landscape: making full use of the landscape resources;

Sunlight: taking advantage of its superior location to get plenty of sunlight;

Sharing: creating eco and sharing spaces inside;

Green: bringing outdoor green landscapes into the building;

Intelligence: adopting standard 5A intelligent systems;

Humanity: people oriented, offering pleasant working spaces for the staff with ecological, efficient, intelligent and humanized designs. It aims to create a simple, elegant, economical and practical office building with well organized spaces and transportation.

■ **设计特色**

景观办公：充分发挥景观优势并使之最大化；

阳光办公：利用地域优势、得到充足的阳光；

共享办公：创造内部的生态共享空间；

绿色办公：把室外绿化空间引入到建筑内部；

智能办公：采用5A标准智能化系统对大厦进行配置；

人性办公：以人为本，为办公人员提供舒适的交往空间，力求在建筑的生态化、高效智能化、人性化设计等方面突出个性，使建筑物造型庄严大方，结构简洁、经济实用，空间组织高效合理，交通流线顺畅便捷。

Alibaba Shenzhen Office Tower
阿里巴巴深圳大厦

Keywords 关键词

Glass Curtain Wall
玻璃幕墙

Architectural Expression
建筑表现

Landscape Environment
景观环境

Location: Shenzhen, Guangdong, China
Architectural Design: WEAVA Architects
Design: June, 2011

项目地点：中国广东省深圳市
建筑设计：法国韦瓦建筑设计公司
设计时间：2011年6月

Features 项目亮点

The striking iconic shape and the regular transparent glass on the curtain wall break the monotony of the facade. When night falls, spectacular lighting system for the building would impress people with a shocking visual impact which interplays with the daytime scene.

项目标志性的形状引人注目，幕墙随着透明玻璃规则变化，打破了墙面的单调；当夜幕降临，壮观的照明系统为建筑赋予新颖震撼的表现力，与白昼的景色竞相生辉。

■ Overview

Alibaba Tower is situated in a strategic point in Shenzhen—Nanshan New Town and will be a part of the new financial district together with other high-rise buildings.

■ 项目概况

阿里巴巴大厦位于深圳的战略性位置——深圳南山的新城建设区。在新的总体规划设计中，阿里巴巴大厦将同其他高层建筑一起，成为新的金融街区的一部分。

■ Urban Integration

The concept of the tower originates in the surroundings and restriction of urban planning. The requirements of urban planning, like plot ratio and height limit, challenge architects a lot. And the main challenge is to design the four volumes in a definite closed scope and break the dull dispersed urban pattern. Therefore, the first step is to connect the four branches and create a unique urban region, making the tower well-proportioned. And the next is about the direction, more specifically, how to adapt it to the surrounding environment. For example, the periphery faces city, the interior faces the sea, and whether the design could bring a sense of openness for the people walking in the building.

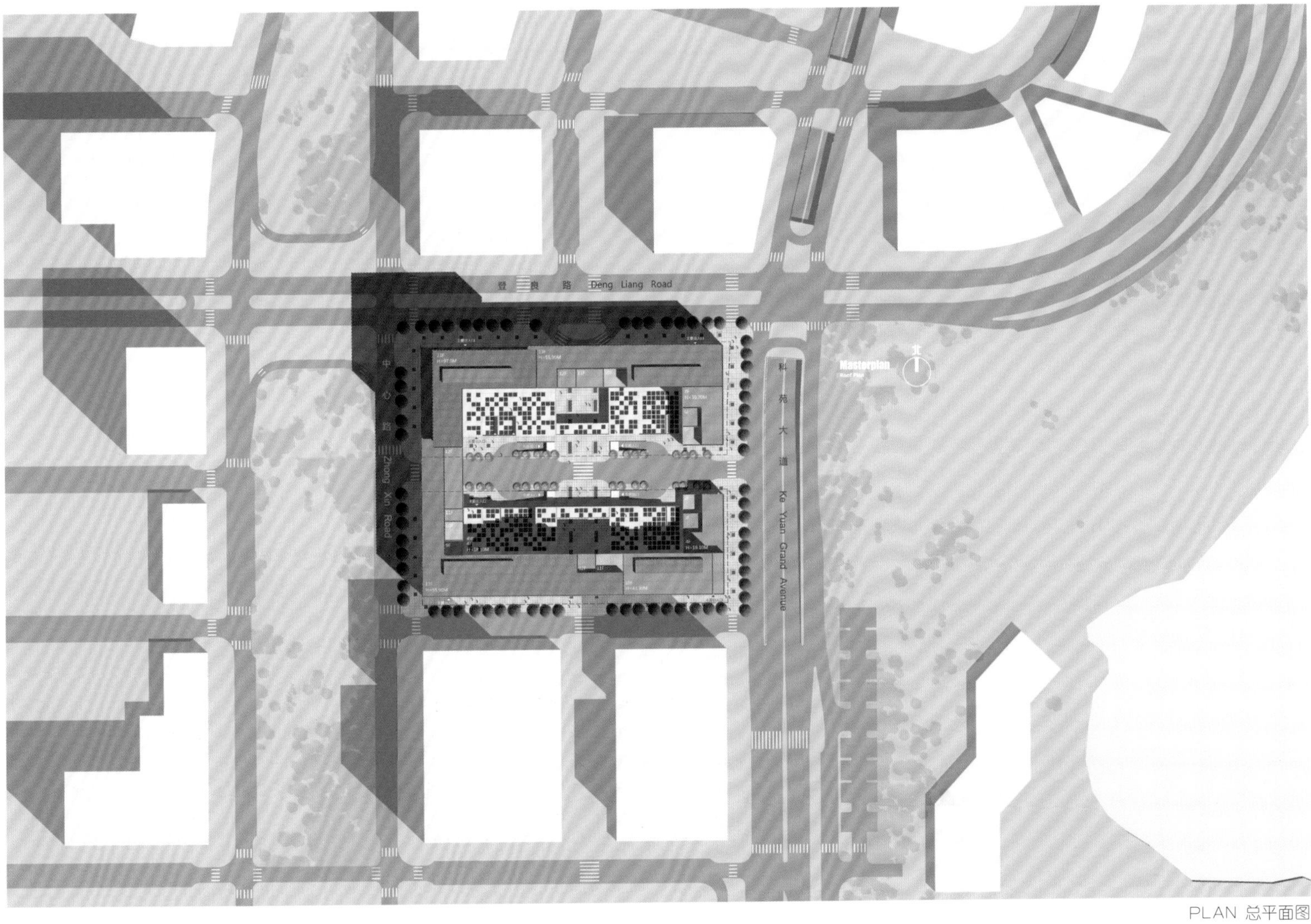
登良路 Deng Liang Road
中心路 Zhong Xin Road
科苑大道 Ke Yuan Grand Avenue
Masterplan
北

PLAN 总平面图

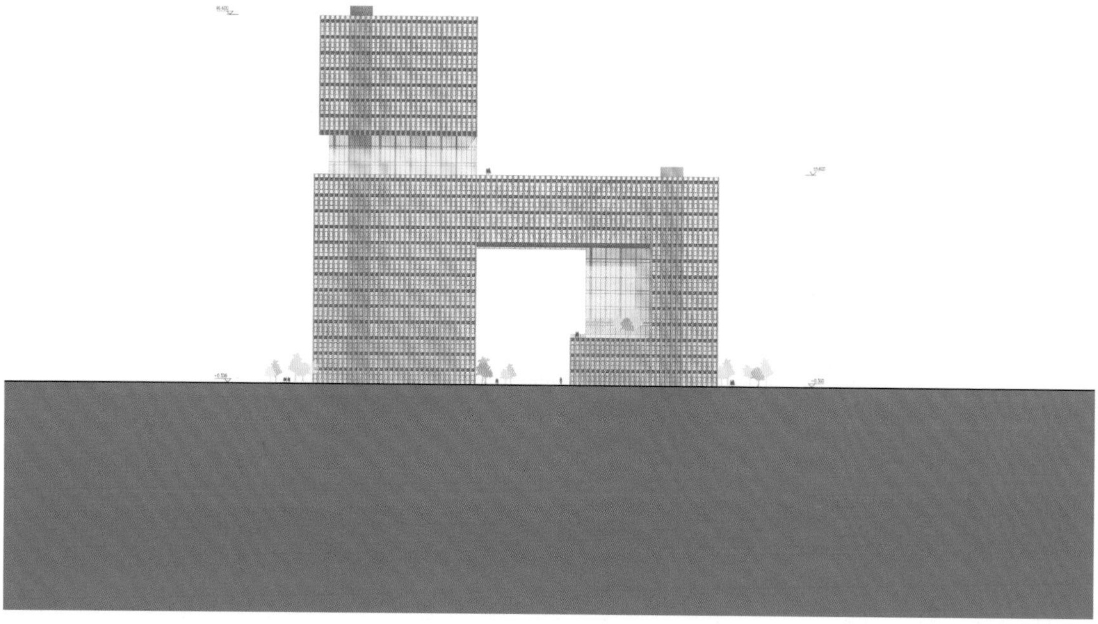

West Elevation 西立面图

■ 城市整合

阿里巴巴大厦的概念来自于它的周边环境和其在城市规划的束缚下所做的调整。城市规划的各项规定，如容积率、高度限制等等，给建筑设计带来了阻碍。因此设计的主要挑战就在于突破在一个固有的范围内的四栋楼体的封闭性，突破这单调而分散的城市格局。由此，设计的第一步就是要连接四个已有分支，创造出一个独特的城市区域，使得阿里巴巴大厦匀称而协调。接下来是它的方向，更确切地说是它对周遭环境的适应。比如，它的外围面对着城市，而内部面向大海，这一设计是否能给在这座大厦中步行的人们带来开阔之感。

East Elevation 东立面图

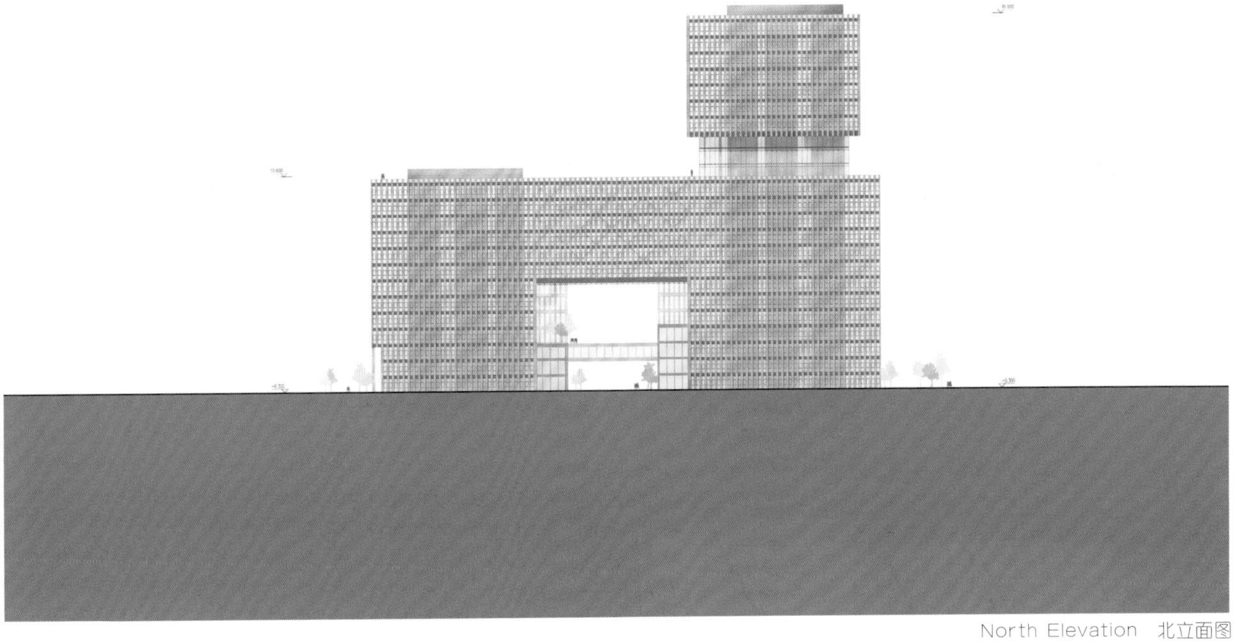
North Elevation 北立面图

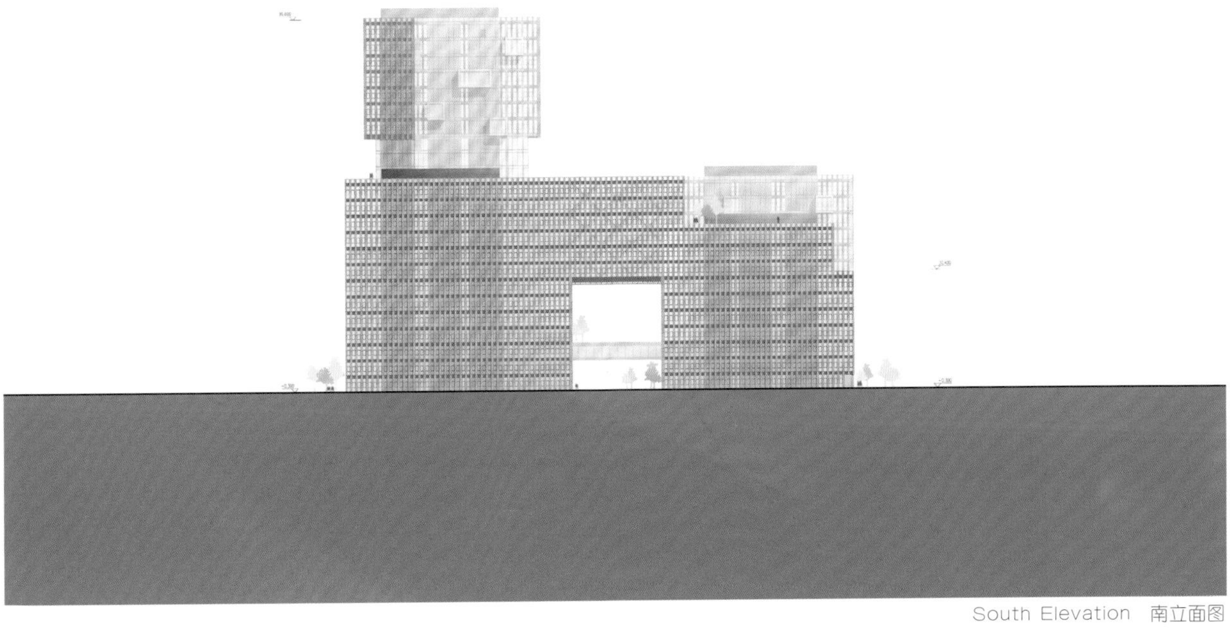
South Elevation 南立面图

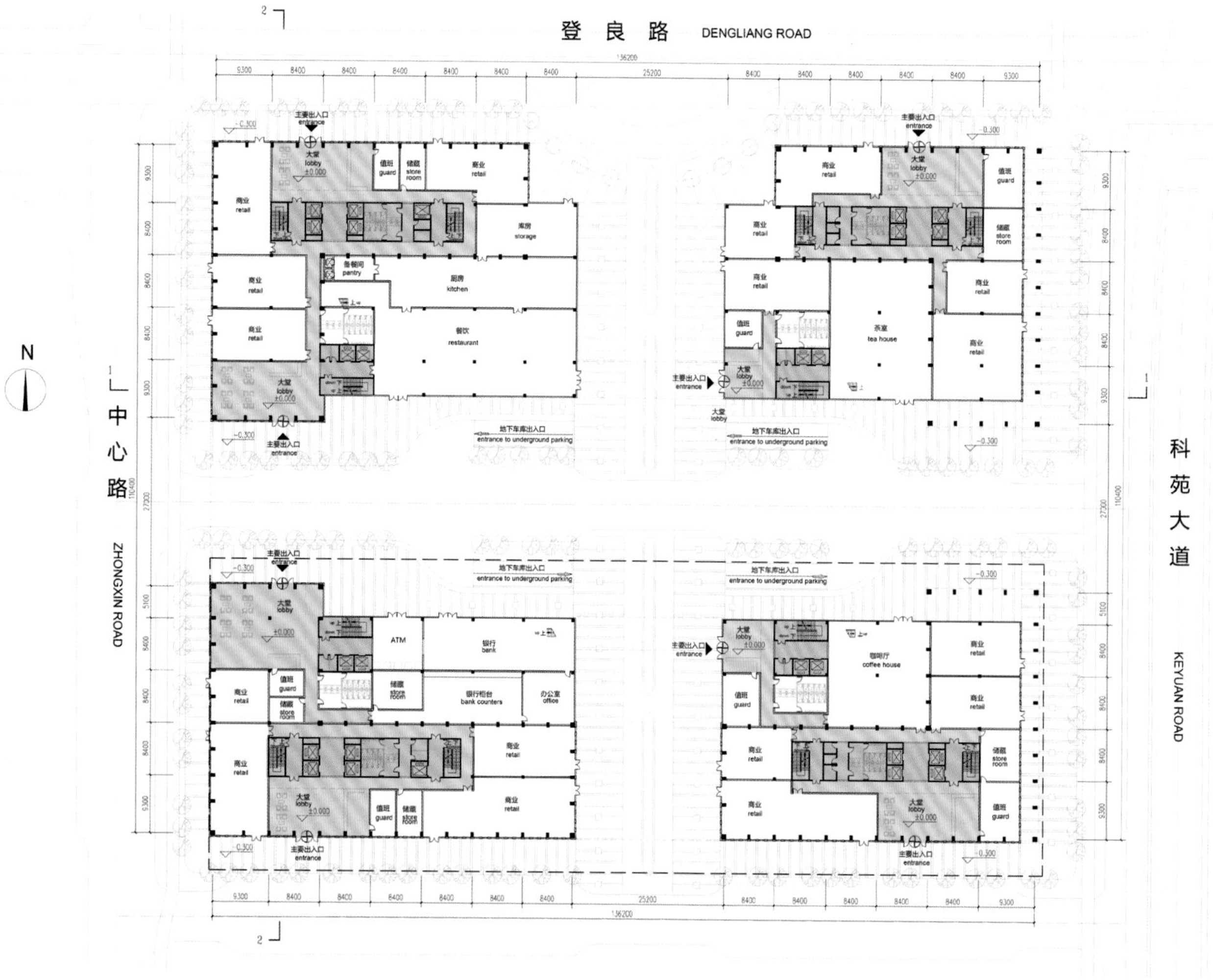

First Floor Plan 一层平面图

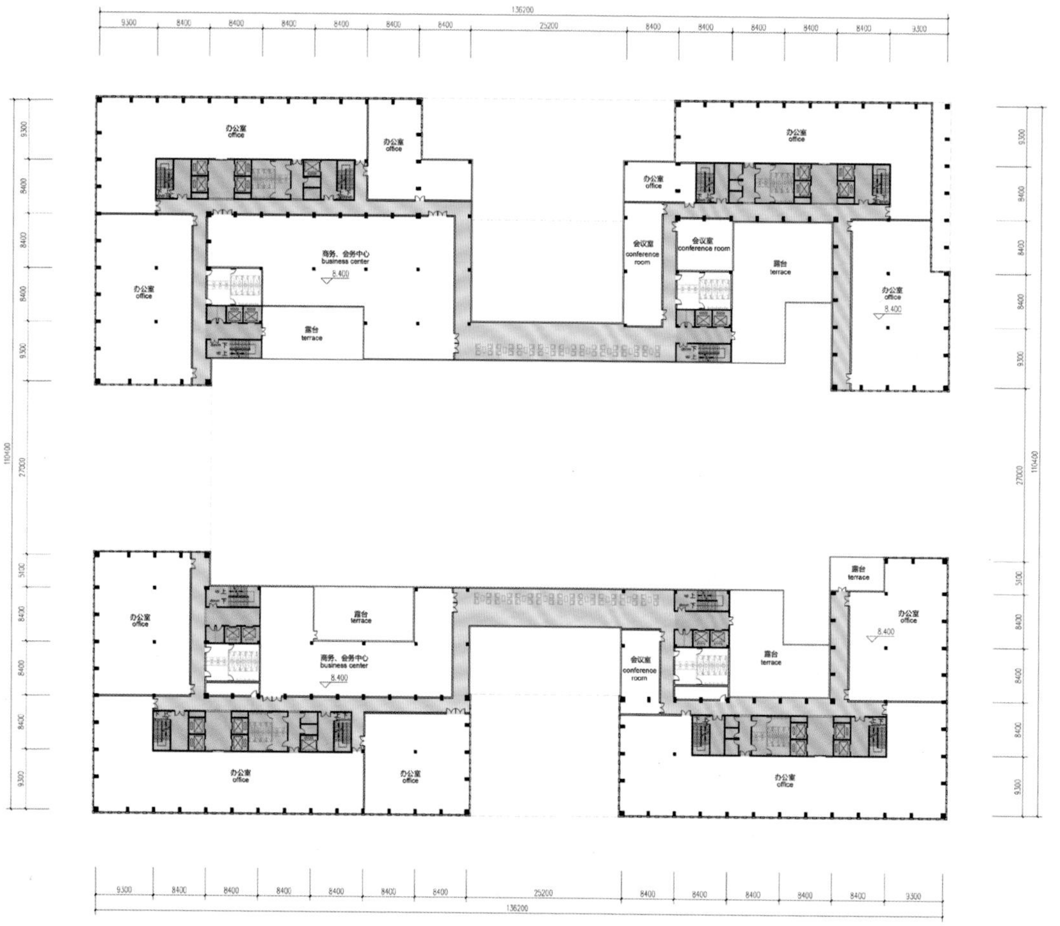

Third Floor Plan 三层平面图

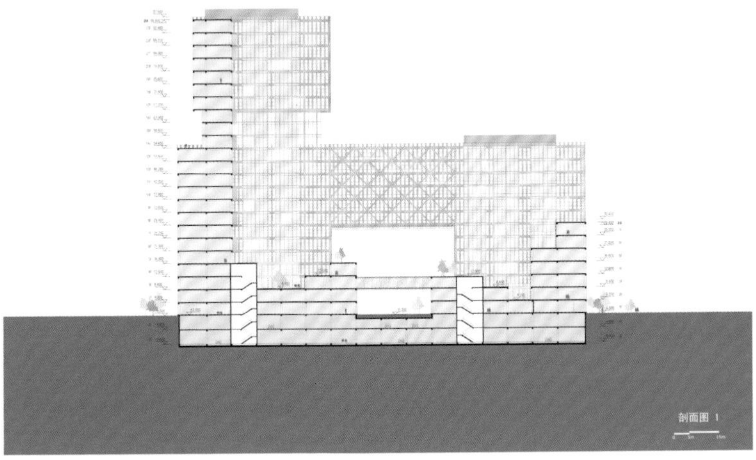

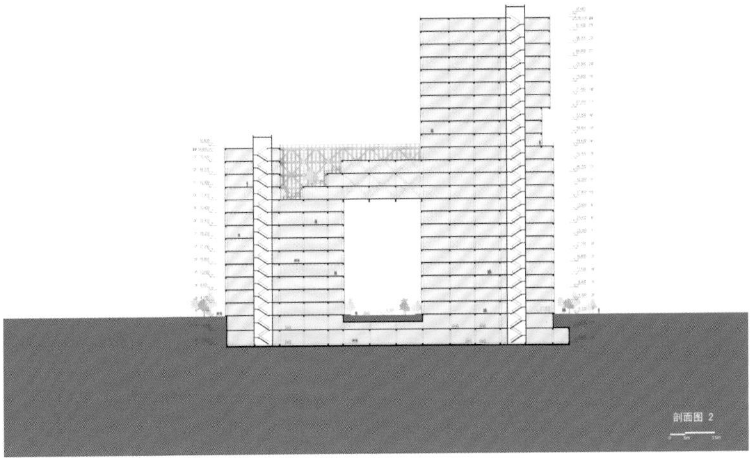

■ Architectural Expression & Changing in Scale

The architectural expression of USB is simple, and in coordination with the background and orientation of the building. In design , natural light is utilized in the most optimal way, such as on one hand the glass curtain towards the east seaside scenery, absorbing sunshine every morning, on the other hand, the more robust and opaque walls face the south and west to prevent direct strong sunlight. Two different designs for walls realize a natural and high-efficient balance between light requirement and sun protection. Seen from the interior courtyard, the scattered bay windows break the layout, comparing to the other side, which provides people with an open sea view and great landscape.

■ 建筑表达和尺度变化

“都市景块”的建筑表达简单，且与建筑的背景和方向协调。在设计方案中，日照的自然光被最优化利用，比如一方面玻璃幕墙朝向东面，每日早晨汲取阳光，面对着海边的自然景色，另一方面，更为坚固和不透明的幕墙则面向南面和西面，以防止直接的强日照。由此，使用两种不同的墙面设计的原因就很明显，在光线需求和光照防护平衡关系上也自然高效了。从内部的庭院来看，散落的凸窗又打破了格局，与面向城市的一边相对比，为人们提供了开阔的海景和优美的室外景观。

Chengdu Bochuan Logistics Park
成都博川物流基地

Keywords 关键词

Axis-Network System
轴网系统

Three-dimensional Space
立体空间

Clear Cut
棱角分明

Location: Chengdu, Sichuan, China
Client: Fujian Huaya Group
Architectural Design: PEDDLE THORP MELBOURNE PTY LTD.
Programme: Logistics Office
Land Area: 265,000 m^2
Floor Area: 750,000 m^2
Plot Ratio: 2.37
Status: Project Approval Stage

项目地点：中国四川省成都市
客　　户：福建华亚集团
建筑设计：澳大利亚柏涛（墨尔本）建筑设计有限公司
项目性质：物流办公
用地面积：265 000 m^2
建筑面积：750 000 m^2
容 积 率：2.37
状　　态：方案报批阶段

Features 项目亮点

The architects create sharing courtyards by designing double-height first floors, and establish the dialogue between buildings and courtyards with the landscape axis.

通过双首层的立体设计使院落共享化，通过景观轴的处理，达到建筑与建筑、建筑与院落、院落与院落的互动。

■ Architectural Design

Chengdu Bochuan Logistics Park plans to cover a land area of about 265,000 m^2 and a floor area of 750,000 m^2; its overall planning concepts are mainly embodied in the following aspects:

1.The pulsation — Through the stretching and dislocation of the original network, it will form scattered and organic linear radiation to create figure–ground relation of modern science and technology.

2. The Emphasis on the Interaction and Communication of Modern Office – The architects create sharing courtyards by designing double–height first floors, and establish the dialogue between buildings and courtyards with the landscape axis.

3. Three–dimensional Space — The landscape and buildings are formed by certain axis and network; landscape and buildings adopt the unified style, and the axis and network go on integrated design.

4. The Promotion of the Enterprise Image — The toughness of the steel, the efficiency of the logistics industry as well as the government support for the regional economy are as important as the blood for the human body. The architectural technique of acute angle and cutting can strengthen the volume feeling, making the overall building look clear–cut.

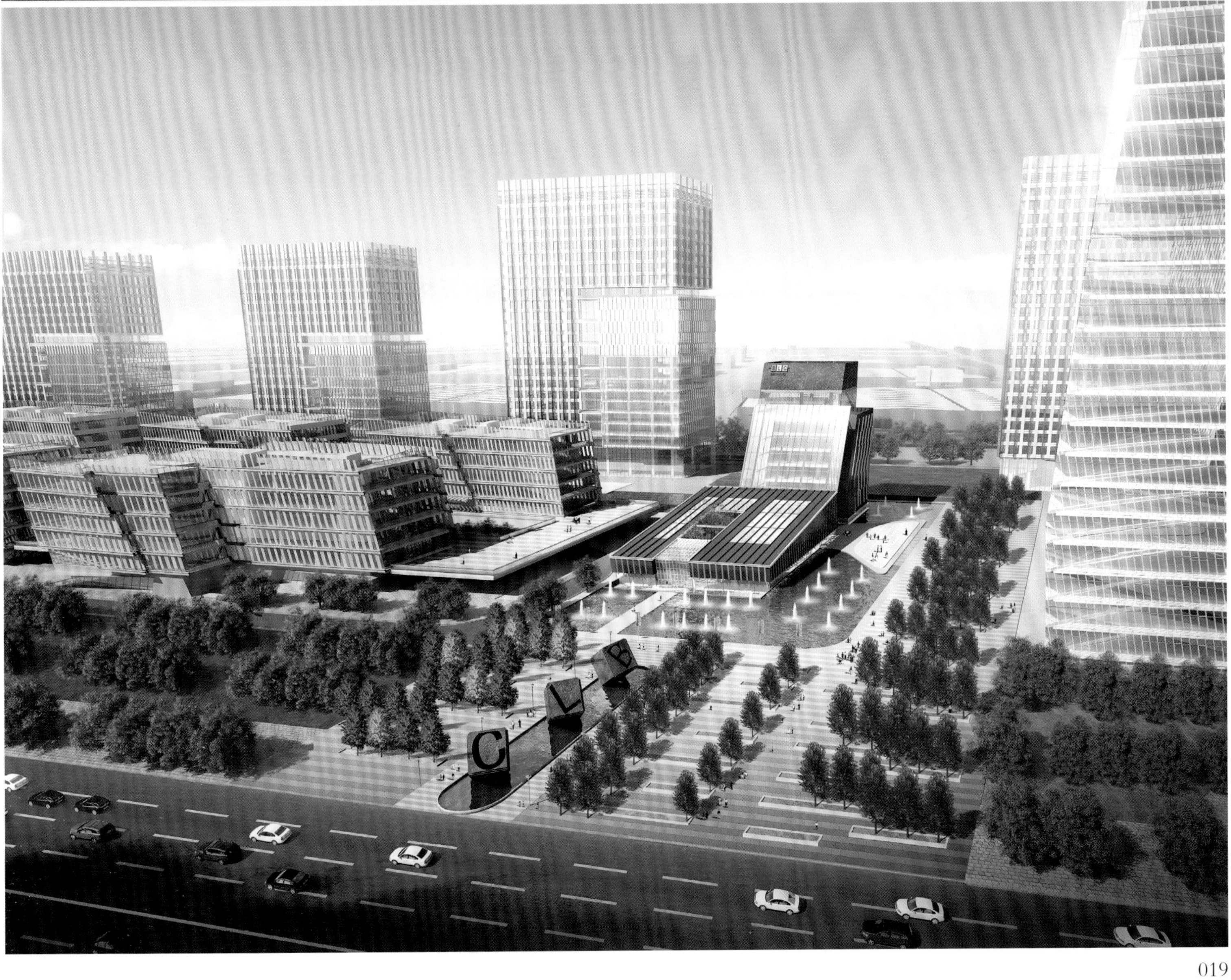
C
L
B

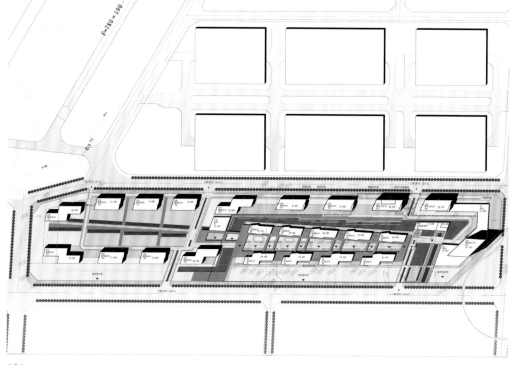

■ 建筑设计

成都博川物流基地规划用地面积约为265 000 m^2，建筑面积为750 000 m^2，其总体的规划概念主要体现在以下几个方面：

1.脉动——设计通过原始网格的拉伸与错动，形成错落有致的线性放射分布，打造现代科技的园区规划。

2.强调现代办公互动与交流——设计师通过双首层的立体设计使院落共享化，通过景观轴的处理，达到建筑与建筑、建筑与院落、院落与院落的互动。

3.立体空间思考——所有景观与建筑都由一定的轴网形成，景观与建筑统一风格，轴网统一进行一体化设计。

4.企业形象的提升——钢铁的硬朗和物流产业的效率以及国家的支持，如同血脉对人体的重要。通过锐角与切割的建筑手法加强型体感觉，使得整体看上去棱角分明。

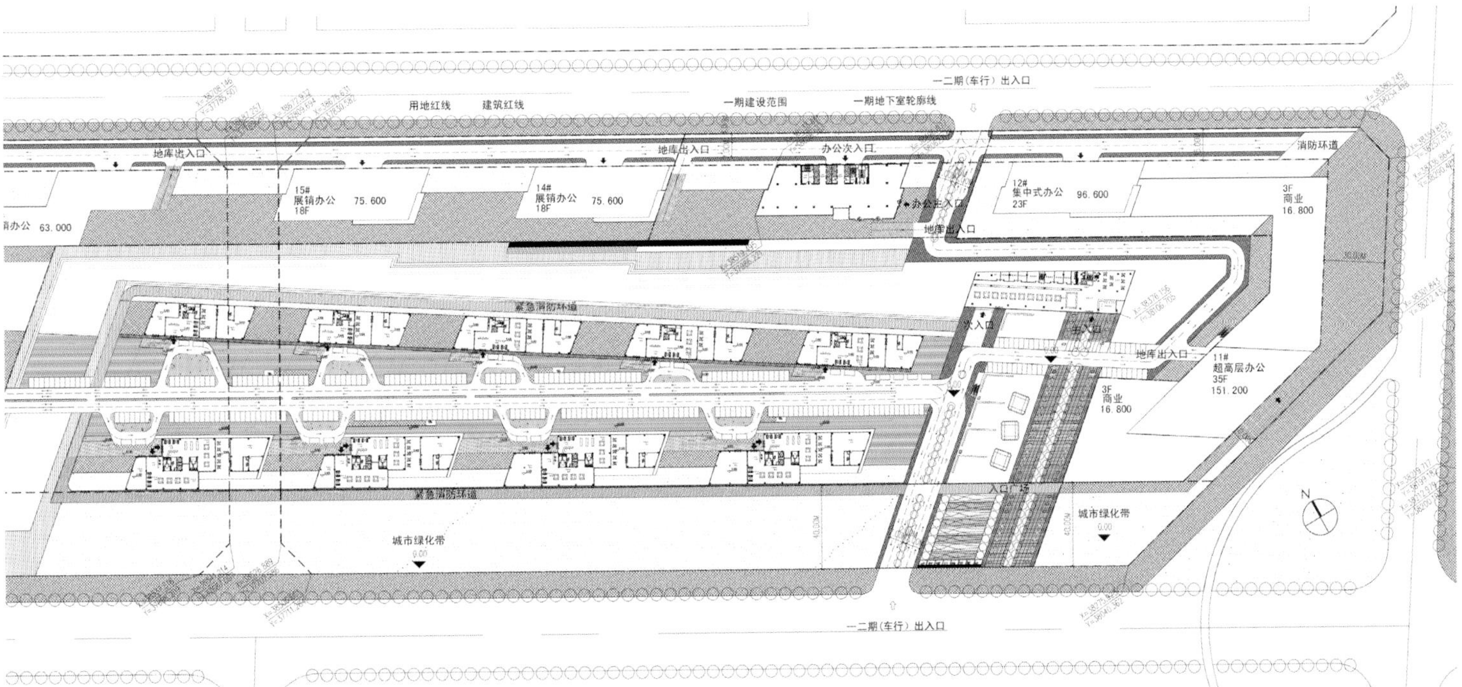

First Floor Plan 一层平面图

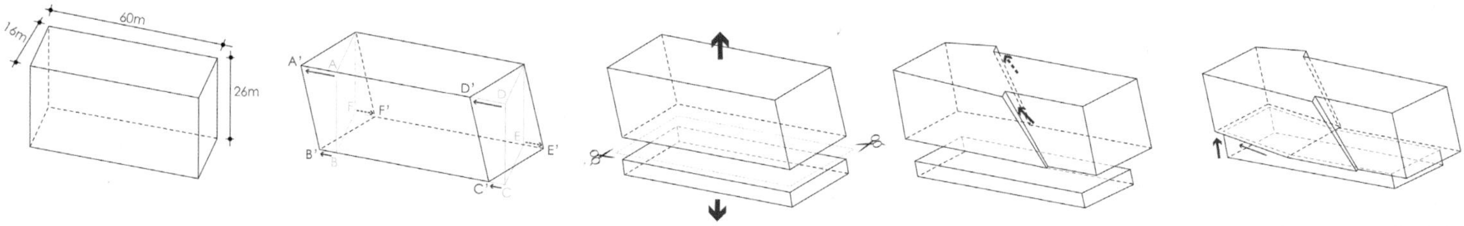

Development 形体演变

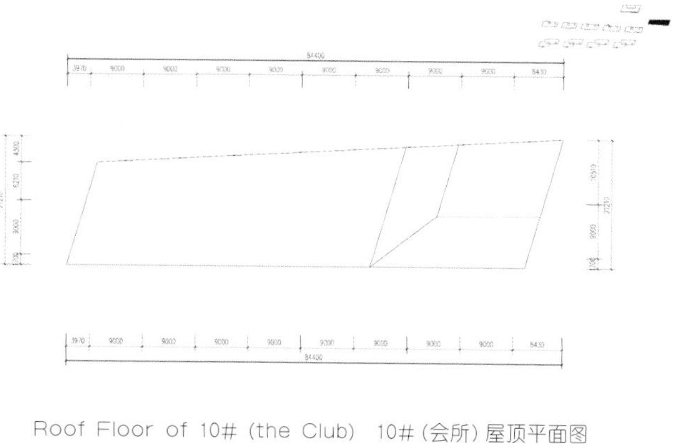

Roof Floor of 10# (the Club) 10# (会所) 屋顶平面图

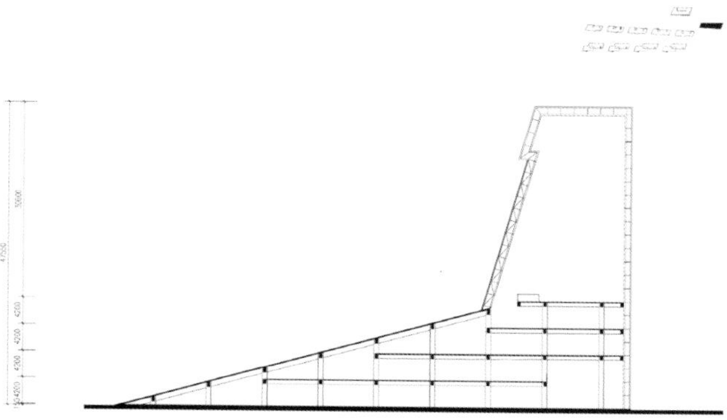

Section of 10# (the Club) 10# (会所) 剖面图

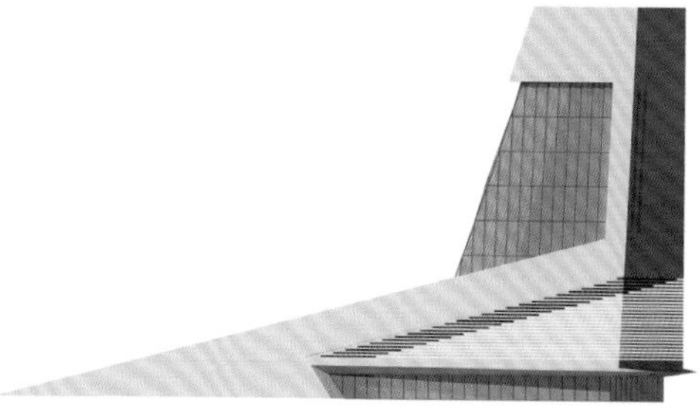

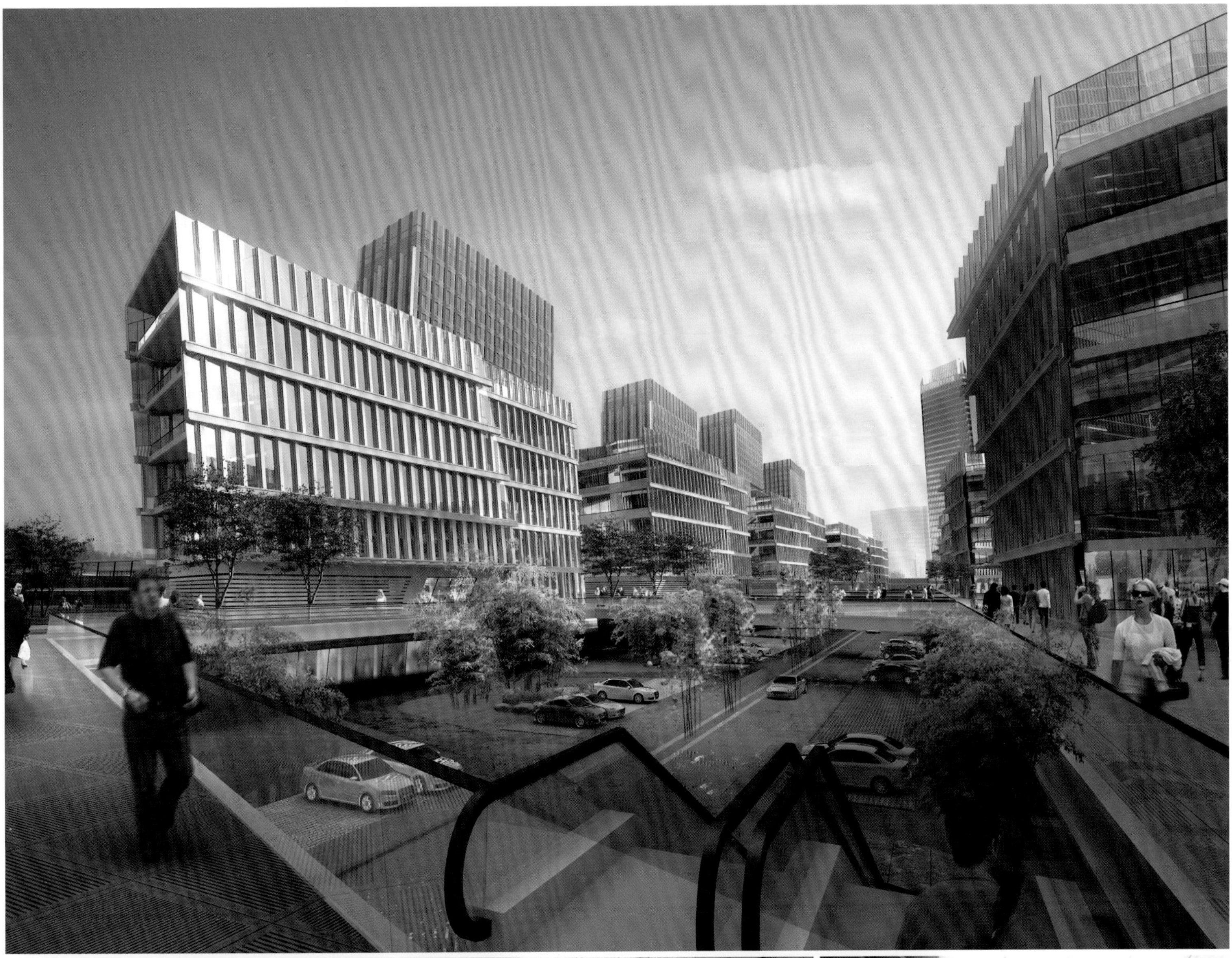

Dahai Guomao Tower
大海国贸大厦

Keywords 关键词

Multi-integration
多元整合

Three-dimensional Space
立体空间

Green Building
绿色建筑

Features 项目亮点

The function and space are typically organized horizontally and vertically based on the concept of "three-dimensional development, environment optimization".

采用"立体发展、环境优化"的方式进行功能布局和空间组织，在强调平面整合的同时，突出地下、地面、空中立体化的竖向规划组织特色。

Location: Dongying, Shandong, China
Developer: Dongying Dahai Real Estate Development Co., Ltd.
Architectural Design: HIC
Total Land Area: 25,921 m^2
Total Floor Area: 128,777 m^2
Building Density: 25.8 %
Plot Ratio: 3.5
Green Coverage Ratio: 54.6%

项目地点：中国山东省东营市
开 发 商：东营大海房地产开发有限公司
设计公司：上海翰创规划建筑设计有限公司/上海翰创建筑设计事务所
总用地面积：25 921 m^2
总建筑面积：128 777 m^2
建筑密度：25.8%
容 积 率：3.5
绿 化 率：54.6%

■ Overview

This site is located on the west of culture media center, east of hotel land and south of the business district, facing court and procuratorate across Fuqian Street on the north, and most of the surrounding plots are for residential area.

■ 项目概况

项目位于府前大街以北，东临文化传媒中心，西接酒店用地，北侧是商务办公区，南面与法院、检察院隔路相望，周边地块大部分为居住区。

■ Overall Layout

In overall layout, a south-north landscape axis is set up to connect city administration cultural axis in the south and business district & residential area in the north. The headquarters is in the central position of the axis, overlooking the building groups in the business district and organizing the open space and greening system organically. The project consists of three parts: twin towers, annex building and underground. The 22 overground floors are dedicated to office (east tower) and hotel (west tower), which are connected by landscape atrium, conference center and roof garden. The annex building is used for office and hotel supporting services. And the 2 underground floors are for subsidiary rooms, equipment rooms and parking lot.

■ 总体布局

总体布局上，规划形成南北景观轴线，南连城市行政文化轴线，北接商务办公区、居住区。大海总部经济大厦位于轴线中心位置，统领商务区建筑群，有机组织整合开敞空间与绿化系统。本案规划一栋地上双子塔楼建筑，建筑分为双塔楼、裙房、地下三部分。地上22层，主楼东塔主要功能为办公，西塔为酒店，两栋塔楼通过景观中庭、会议中心及屋顶花园相连接。裙房为办公酒店配套服务用房。地下2层，主要功能为酒店及办公后勤辅助用房、设备用房及停车场。

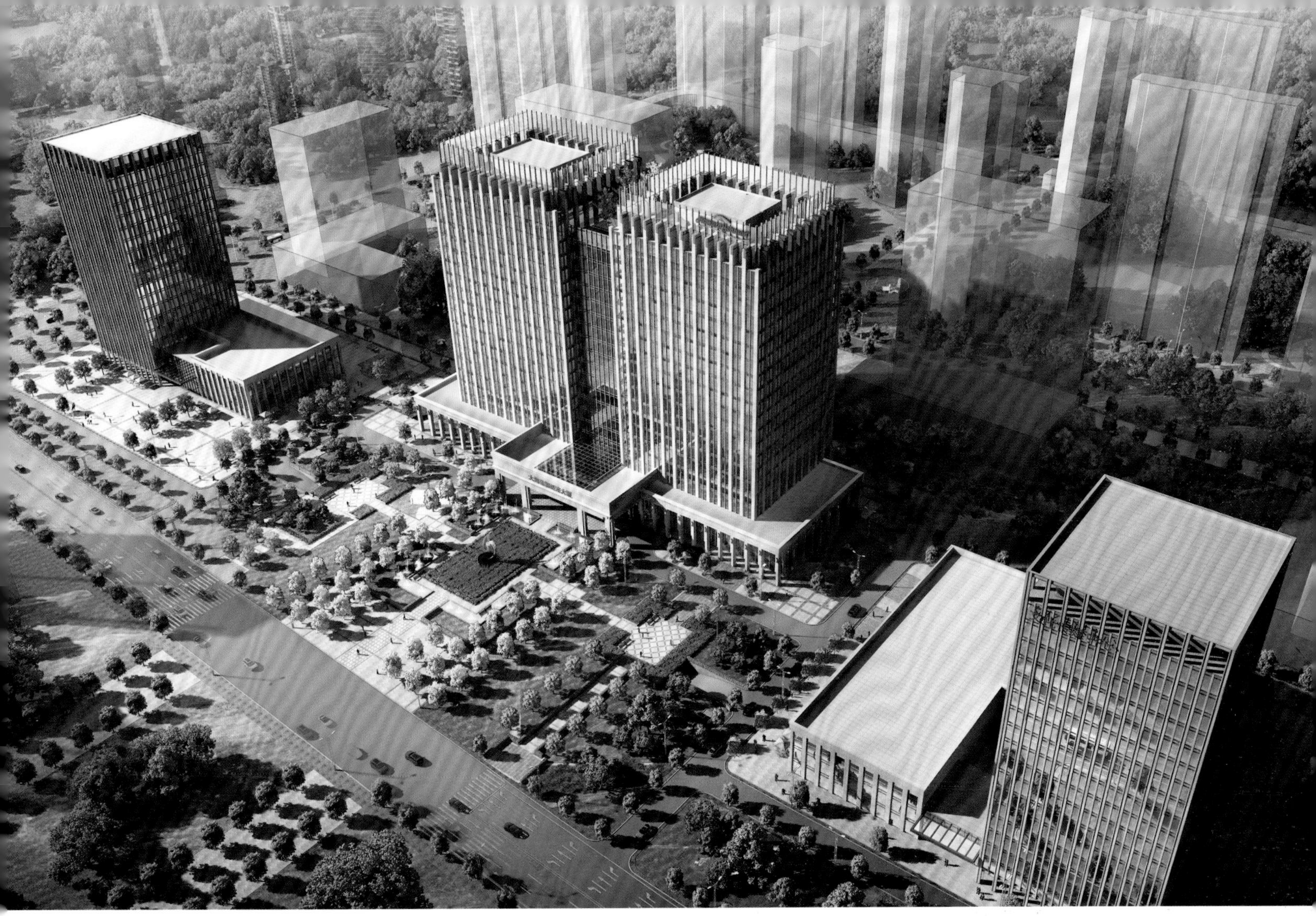

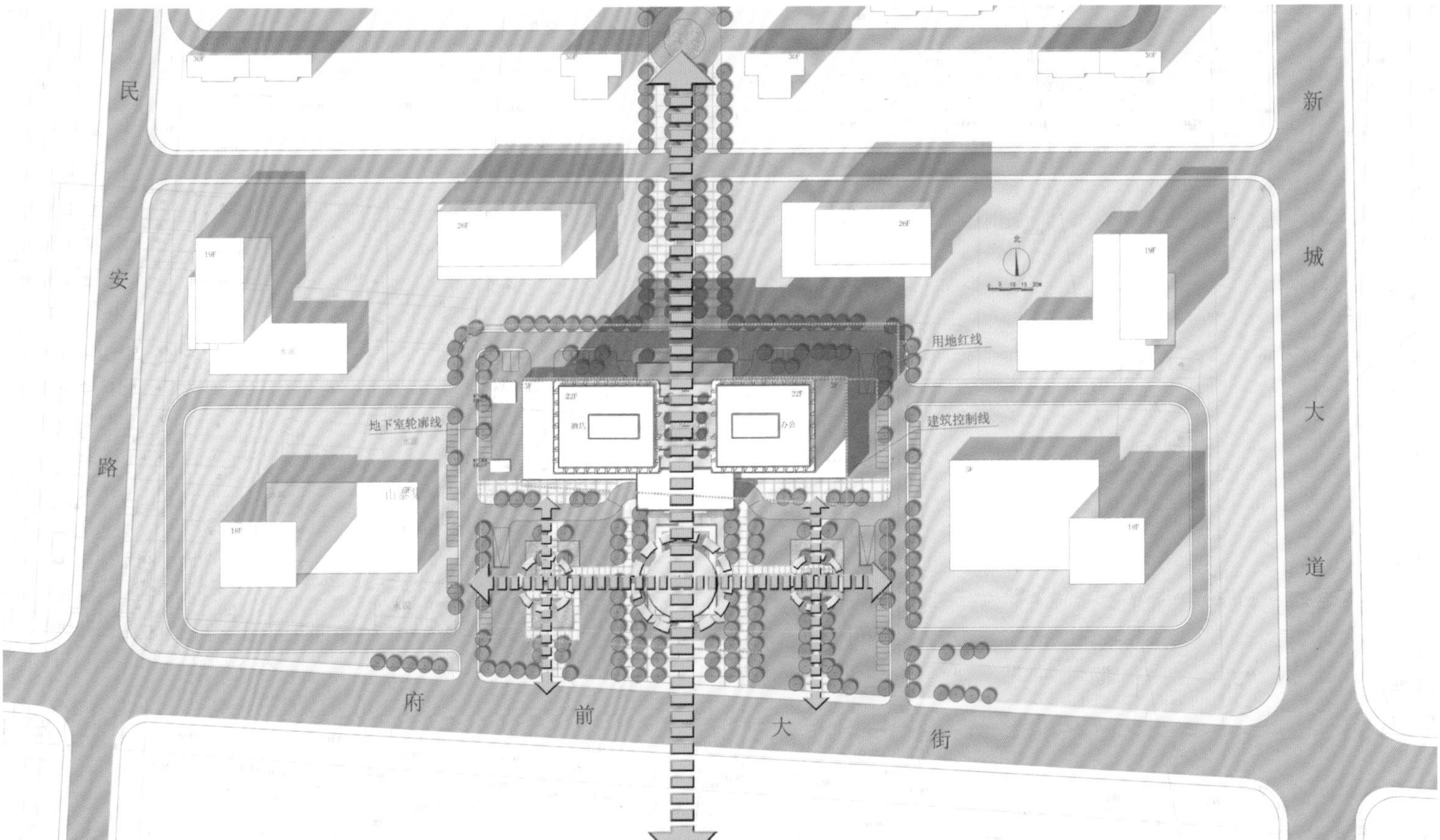

本案景观结构为一主轴、三次轴、一中心、两节点。通过南北主轴与南面城市行政文化轴线相联系，与北面居住区轴线相连接，形成新区清晰的城市生态绿轴。基地内主、次轴线，景观节点形成主次分明、脉络清晰、层次丰富的空间景观环境，通过景观核心统一成有机整体。

景观轴线

景观节点

Landscape Analysis Diagram 景观分析图

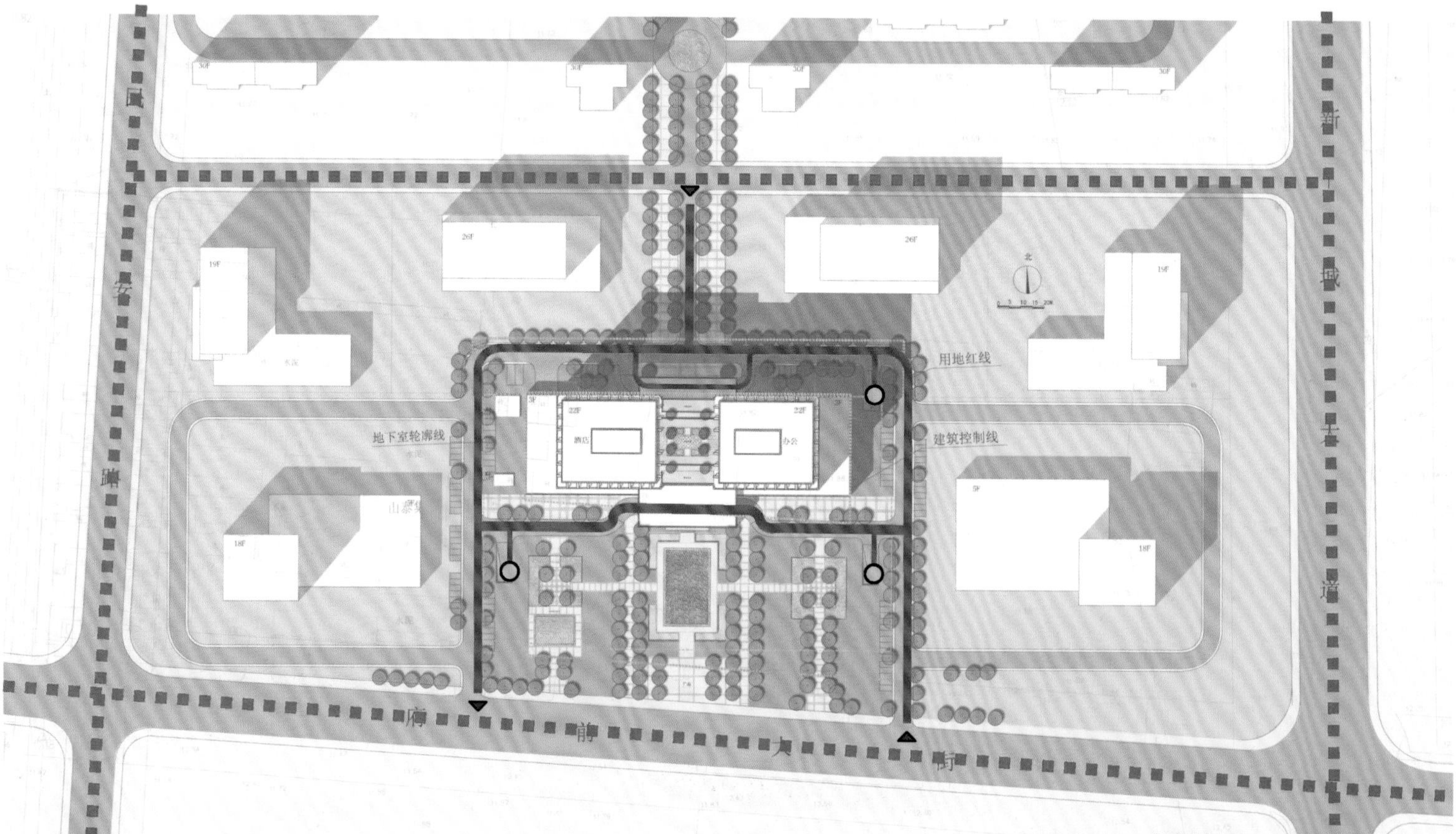

本案于用地南侧府前大街上设置车行主入口，北侧规划道路上设置车行次入口，两个出入口通过环通道路形成有机整体。基地设置三个地下车库出入口。基地不同性质流线避免交叉干扰，使得交通便捷、流畅、高效。

外部交通
内部交通
▲ 用地对外出入口
〇 地下车库出入口

Traffic Stream Diagram 车行流线图

—— 主要步行流线
主要步行节点
▲ 酒店、办公出入口

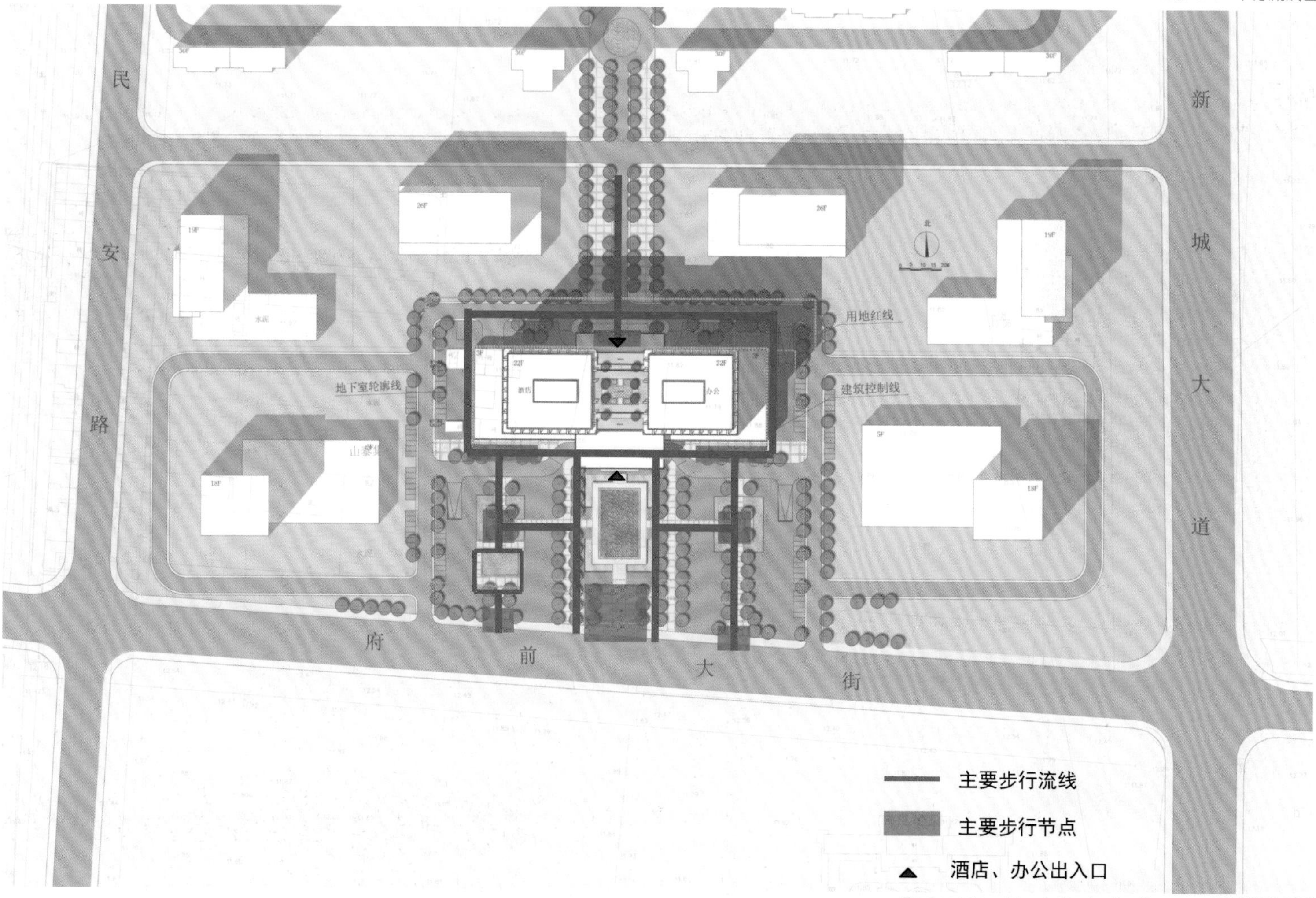

Pedestrian Flow Analysis Chart 人行流线分析图

■ Architectural Design

The modest and dignified building not only reflects the cultural deposit of Guangrao and corporate image of Dahai Group, but impresses people with modernity. It accommodates office, hotel, business and parking, forming a highly efficient, three-dimensional functional and spatial organization, emphasizing the organic integration of urban function, creating complex building with various development modes. The function and space are typically organized horizontally and vertically based on the concept of "three-dimensional development, environment optimization". Elevated entrance hall and landscape atrium shape a multi-dimensional spatial environment. Sky garden and conference center unite the two towers as one, providing a superior office environment. Supporting services such as banks, banquet hall, fitness center, recreational center, and catering provide a high-grade service platform for the high-end business office and financial center.

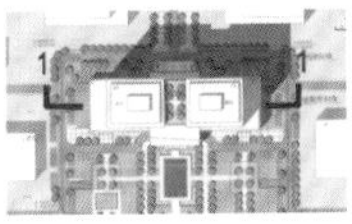

Section 剖面图

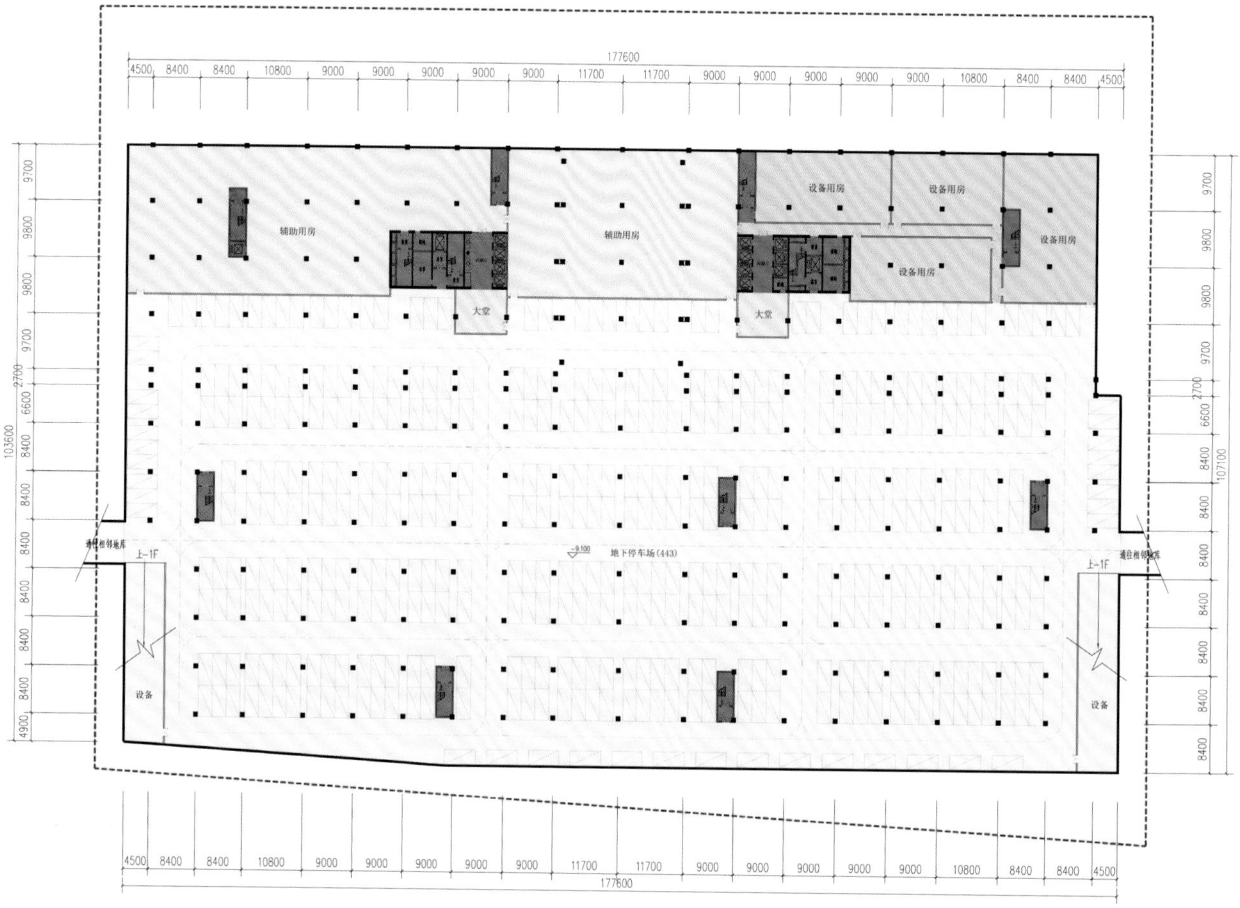

Plan of the 2nd Basement Floor 地下2层平面图

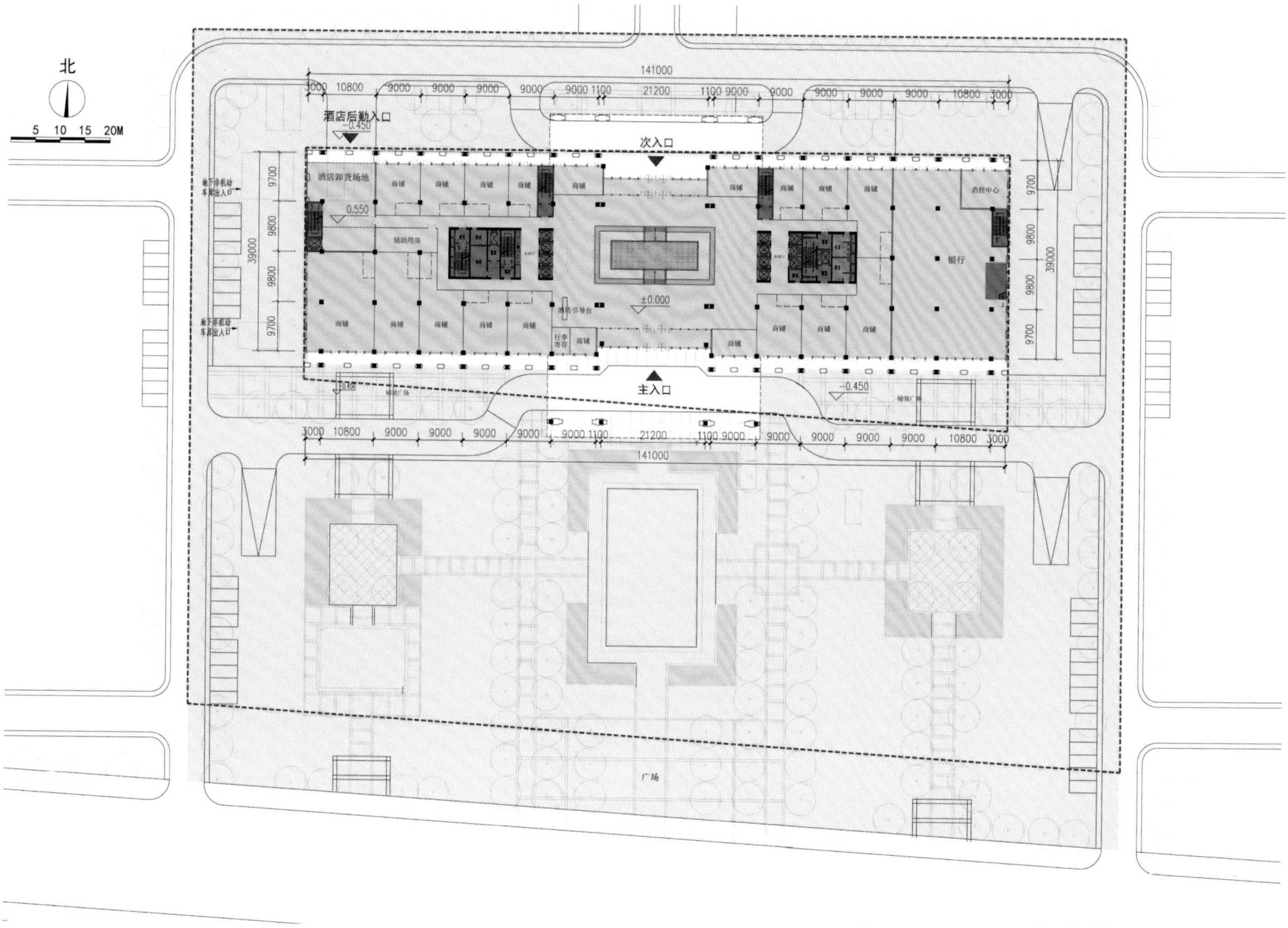

First Floor Plan 一层平面图

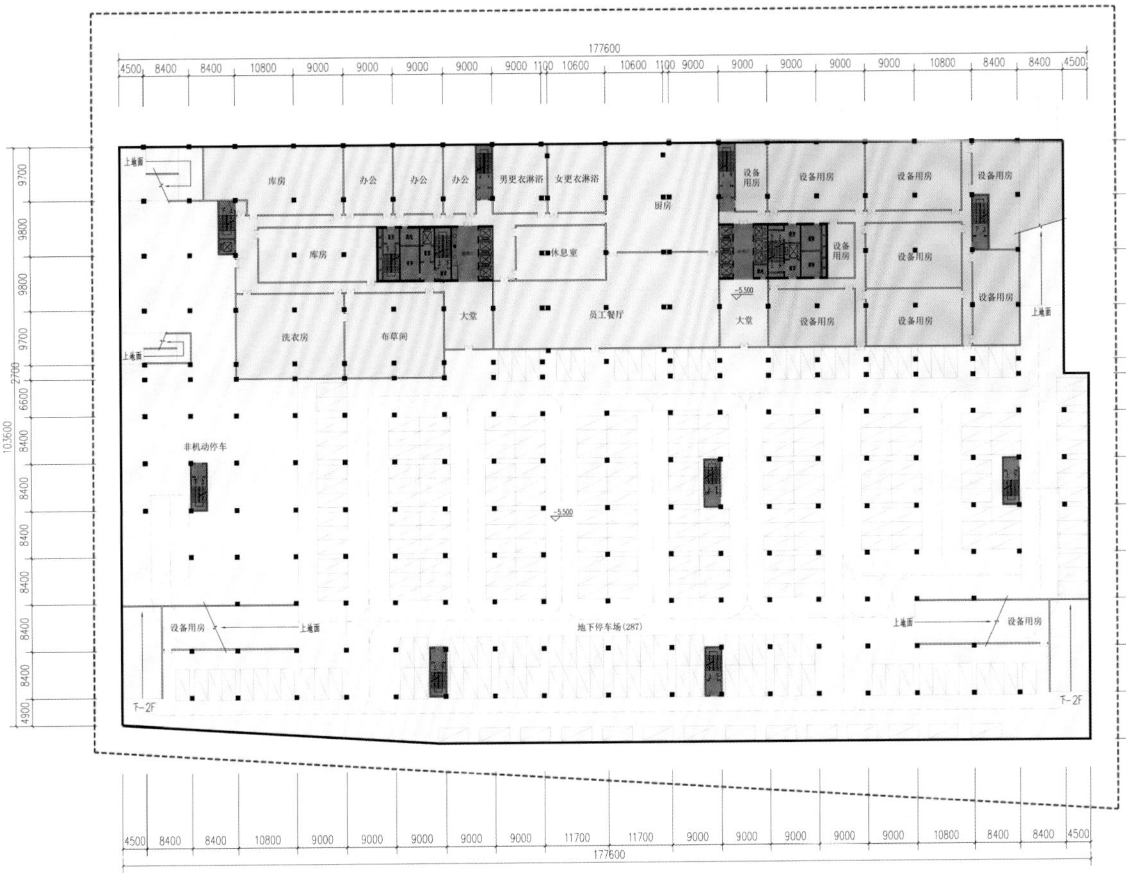

Plan of the 1st Basement Floor 地下一层平面图

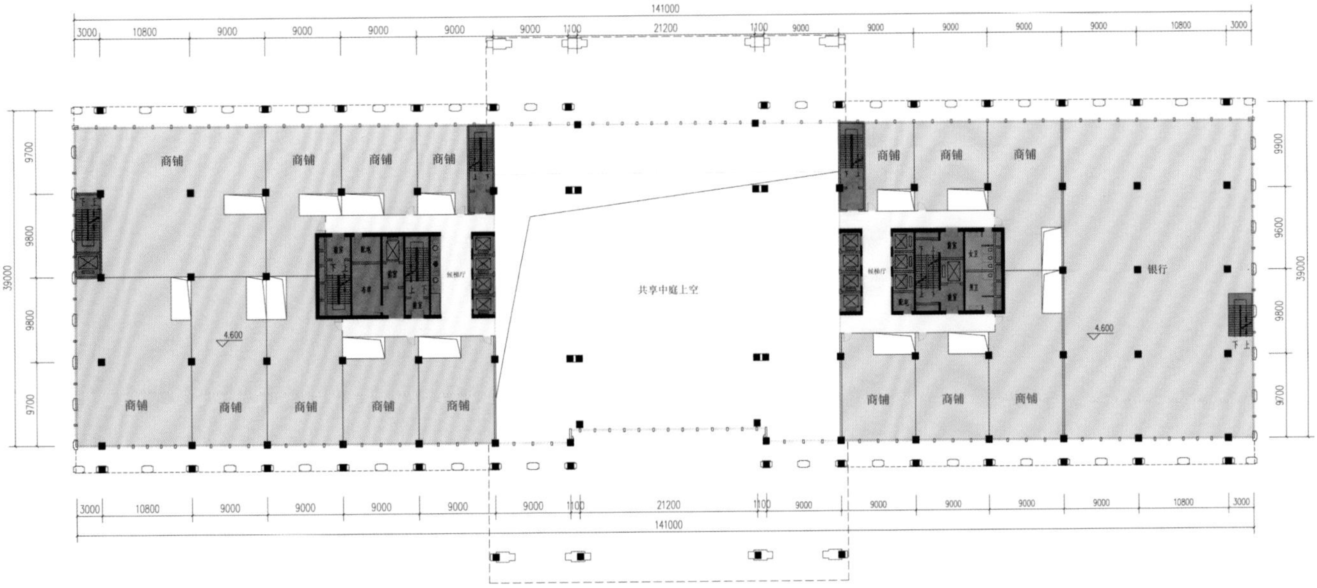

Second Floor Plan 二层平面图

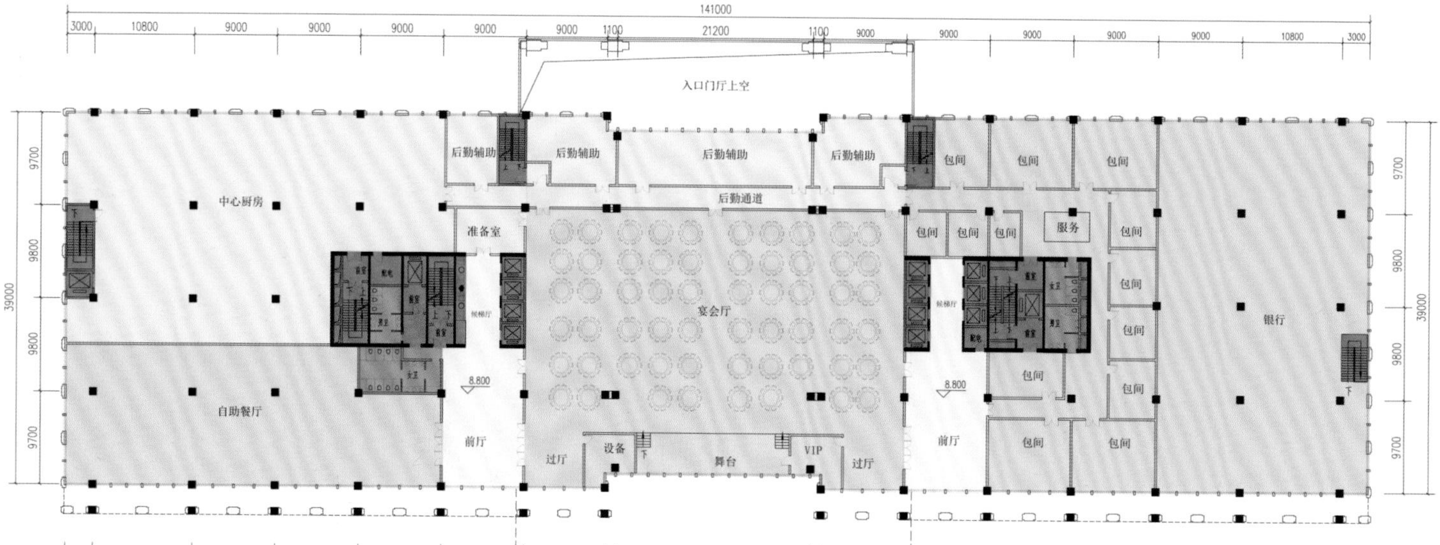

Third Floor Plan 三层平面图

■ 建筑设计

规划形成南北景观轴线，建筑坐北朝南，稳重大气，既体现广饶城市的文化底蕴、大海集团的企业形象，又具有现代建筑的时代气息。将办公、酒店、商业、停车四大需求整合，形成高效集约、立体多元的功能与空间组织，强调城市功能的有机整合，创造多种开发模式集聚的复合建筑。采用“立体发展、环境优化”的方式进行功能布局和空间组织。在强调平面整合的同时，突出地下、地面、空中立体化的竖向规划组织特色。建筑通过挑高的入口大堂、景观中庭形成立体多元的空间环境，并通过空中花园、行政酒廊会议中心使得办公、酒店两栋塔楼联系成整体，为高端商务办公提供优越的办公环境。银行、宴会厅、健身中心，娱乐中心、餐饮等完善的配套服务设施，为高端商务办公和金融中心提供高端的服务平台。

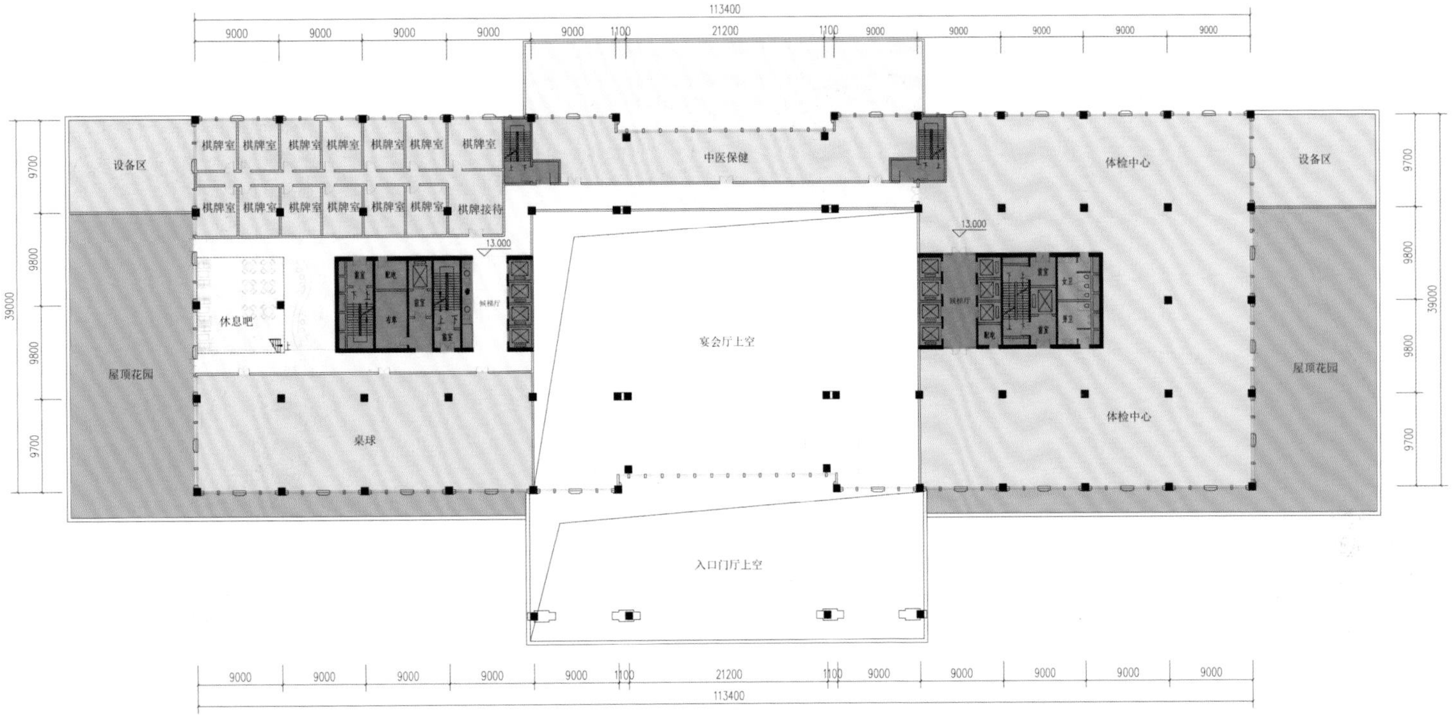

Fourth Floor Plan 四层平面图

■ Ecological & Green Building

Thanks to chimney effect, reasonably arranged green landscape platform and interior & exterior climate boundary, solar power generation, water recycling, the resources are saved, environment is protected and pollution reduced as much as possible, thus creating a healthy, comfortable and high-efficient convenient space.

■ 生态、绿色建筑

通过运用拔风效应的阳光中庭，绿色景观平台、室内外气候边界合理设置等建筑设计方法，以及太阳能发电、中水回收利用等先进技术方法的运用，最大限度地节约资源、保护环境和减少污染，为人们提供健康、舒适和高效的使用空间。

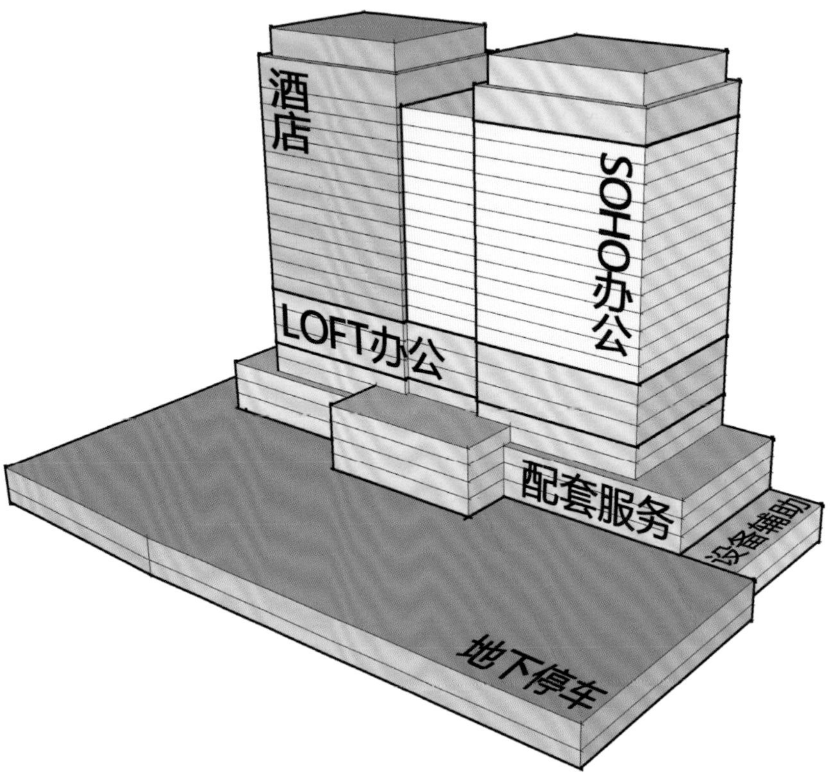

Function District Drawing 功能分区图

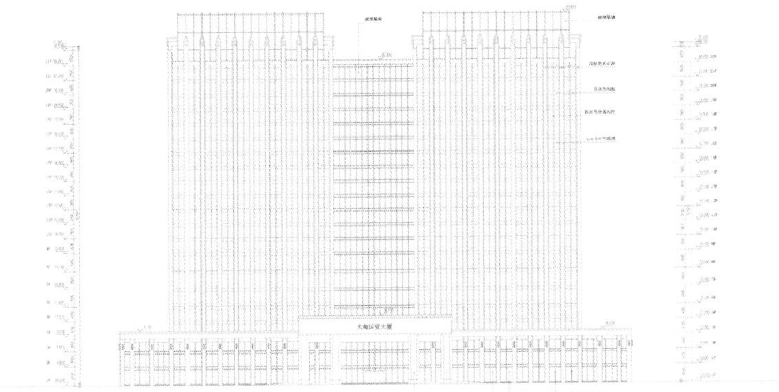

South Elevation 南立面图

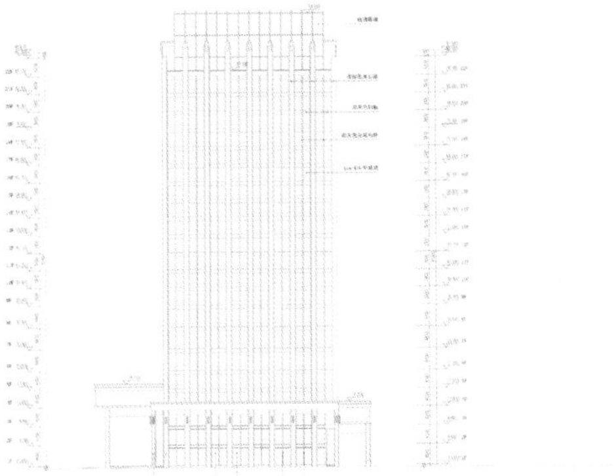

East Elevation 东立面图

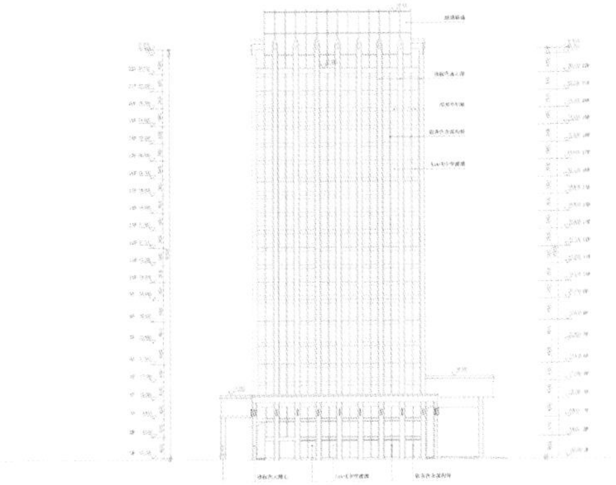

West Elevation 西立面图

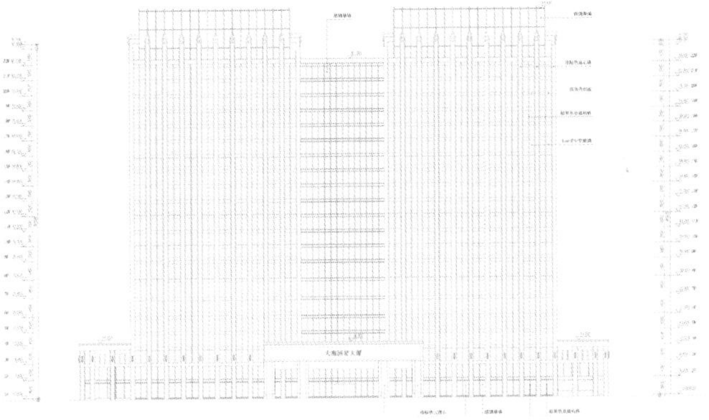

North Elevation 北立面图

Ningbo Digital
宁波数字媒体大楼

Keywords 关键词

Translucent Facade
半透明表皮

Multi-function Space
多功能空间

Mass Design
造型设计

Location: Ningbo, Zhejiang, China
Owner: Ningbo Daily
Architectural Design: Synarchitects
Total Floor Area: 60,000 m^2

项目地点：中国浙江省宁波市
业　　主：宁波日报
建筑设计：Synarchitects
总建筑面积：60 000 m^2

Features 项目亮点

The design is to build a solitaire reacting to the surrounding in its own way. The main idea in this concept is to express the power of the new and irresistible media. This building is like a polished and hidden gem ready to sparkle.

设计旨在建造一个与周围环境相呼应的独立建筑，表达新媒体不可阻挡的力量，整个大楼就像是一块光滑的即将绽放光芒的宝石。

■ Overview

The building performs the following functions: It contains editorial departments of different web-based companies, the newspaper *Ningbo Daily*, a national media platform and auxiliary areas of multi-function.

■ 项目概况

宁波数字媒体大楼主要功能包括一些网络公司的编辑部、宁波日报总部、一个国家级媒体平台和其他附属的多功能空间。

■ Architectural Design

After different approaches to find the right answer, the designers decided to build a solitaire reacting to the surrounding in its own way. The curve-shaped surfaces on the north-east and south-west facade are facing the new dense business areas neighboring the building site. It consists of a semi-transparent media-facade echoing the latest news. Main structure is given by two towers, connected to each other with bridges covering different functions. The main idea in this concept is to express the power of the new media one can't restrain. This building is like a polished and hidden gem ready to sparkle. It is a new favorite for the new Media Age.

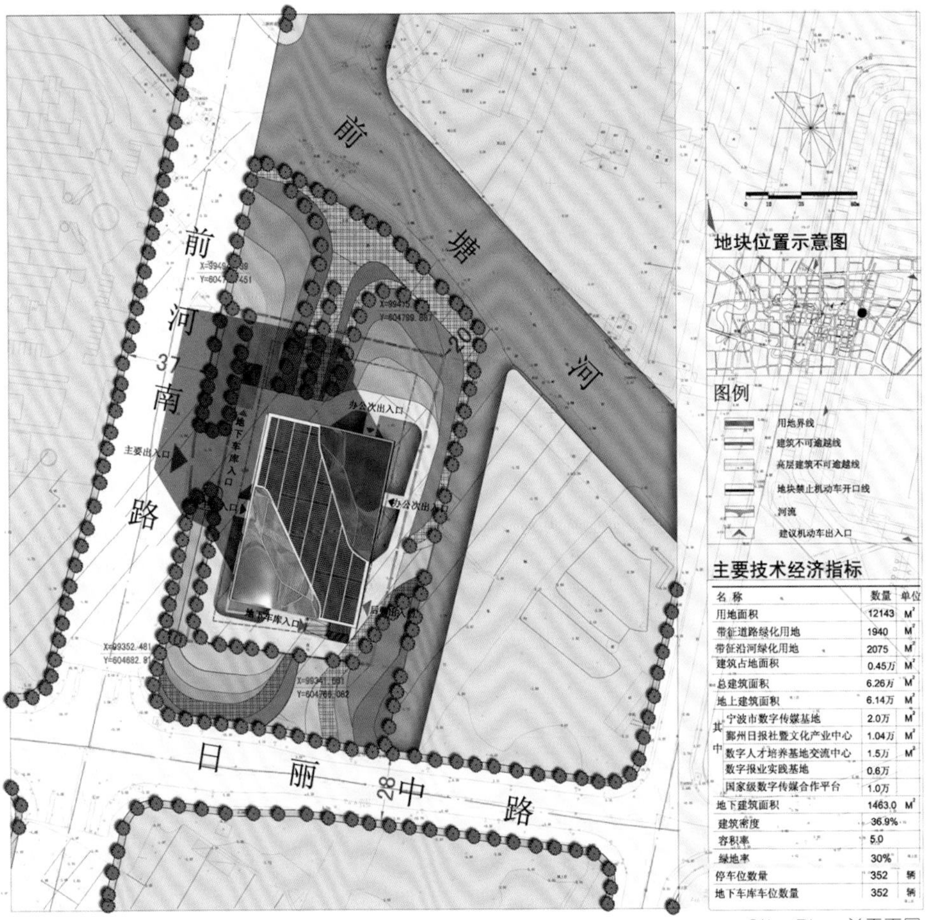

名称		数量	单位
用地面积		12143	M^2
带征道路绿化用地		1940	M^2
带征沿河绿化用地		2075	M^2
建筑占地面积		0.45万	M^2
总建筑面积		6.26万	M^2
地上建筑面积		6.14万	M^2
其中	宁波市数字传媒基地	2.0万	M^2
	鄞州日报社暨文化产业中心	1.04万	M^2
	数字人才培养基地交流中心	1.5万	M^2
	数字报业实践基地	0.6万	
	国家级数字传媒合作平台	1.0万	
地下建筑面积		1463.0	M^2
建筑密度		36.9%	
容积率		5.0	
绿地率		30%	
停车位数量		352	辆
地下车库车位数量		352	辆

Site Plan 总平面图

■ 建筑设计

设计师经过考虑决定建造一个与周围环境相呼应的独立建筑，东北面和西南面的弧形表面紧邻密集的商业区。它包括一个半透明的媒体表皮，这里呈现最近的新闻。主要结构是两座塔楼，它们由桥连接，包含不同的功能。设计的主要理念是表达新媒体不可阻挡的力量。大楼就像是一块光滑的即将绽放光芒的宝石，是这个新媒体时代的宠儿。

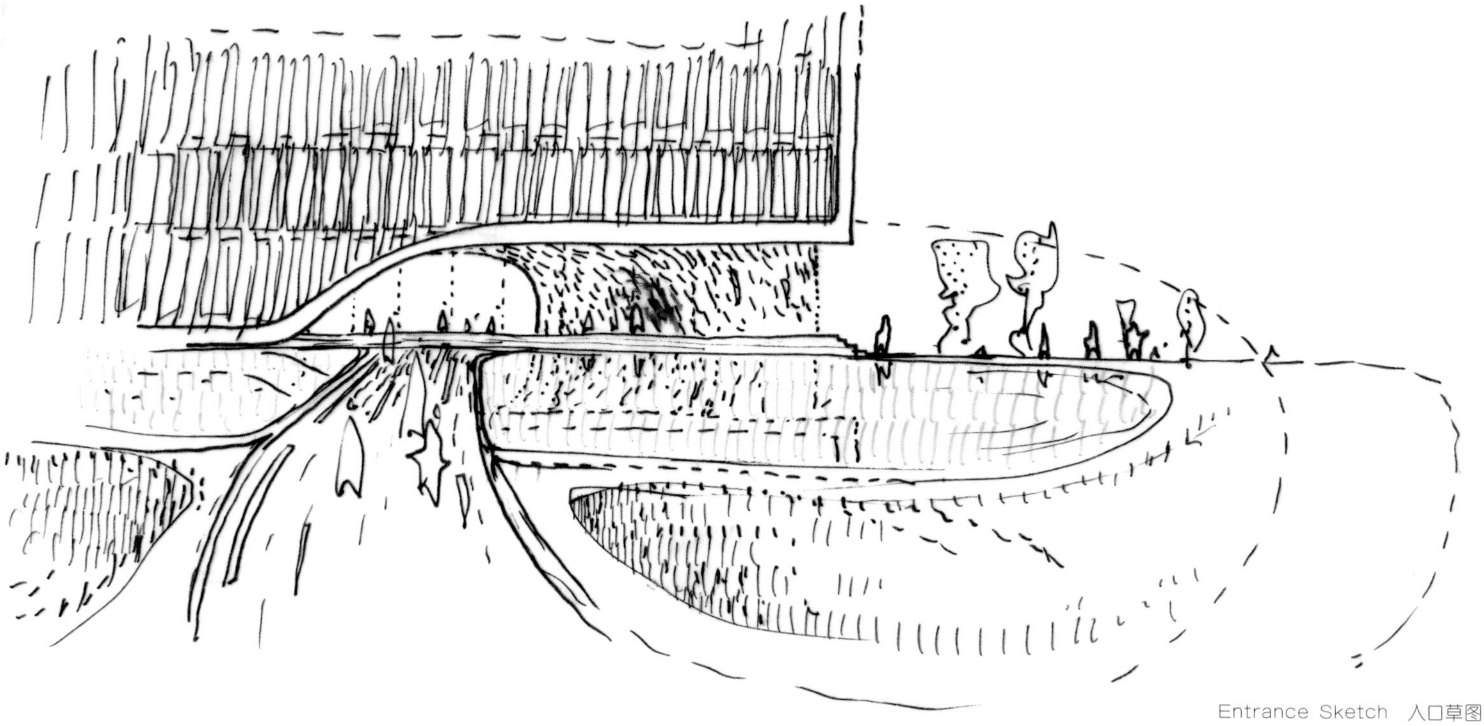

Entrance Sketch 入口草图

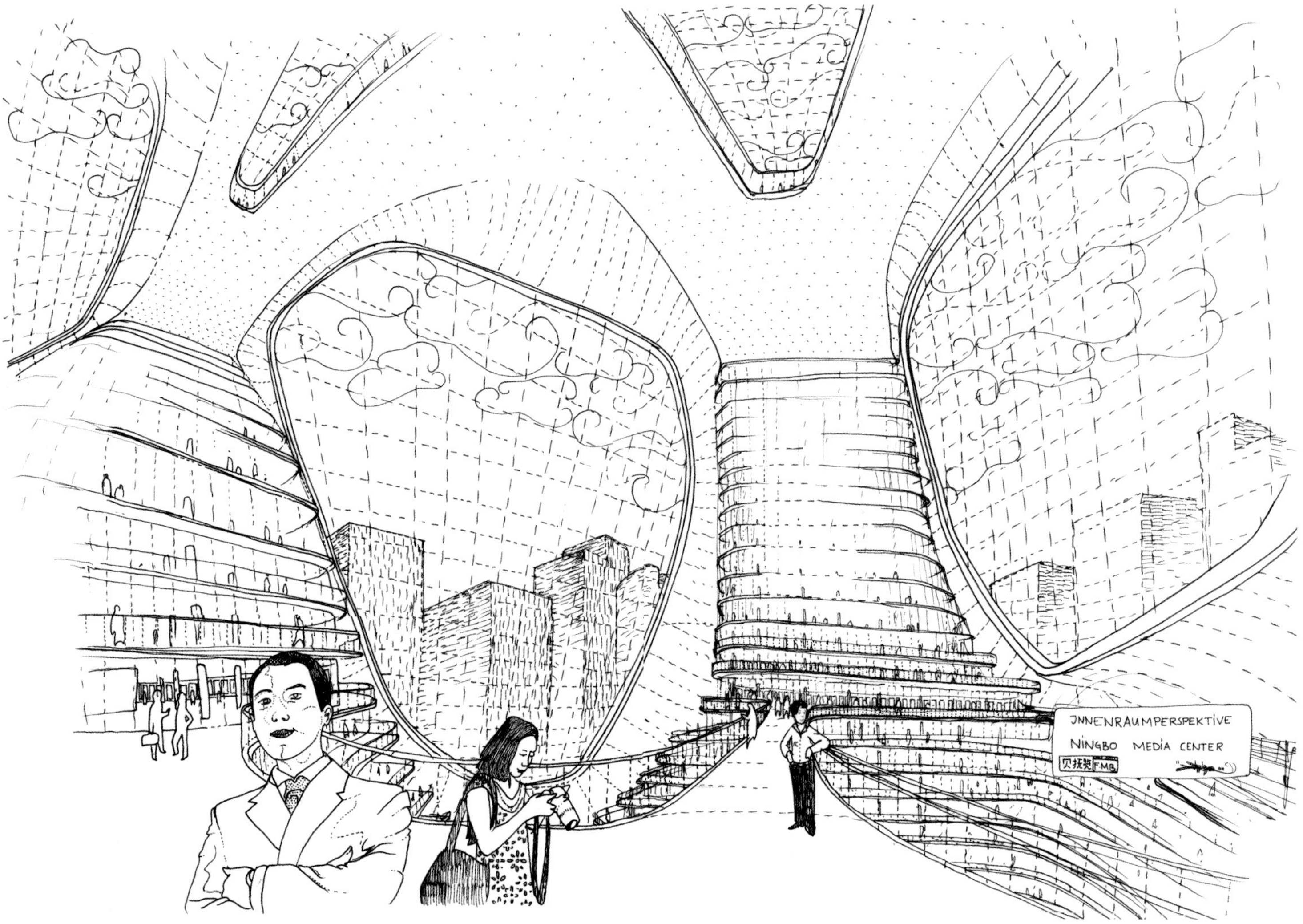

Sketch 草图

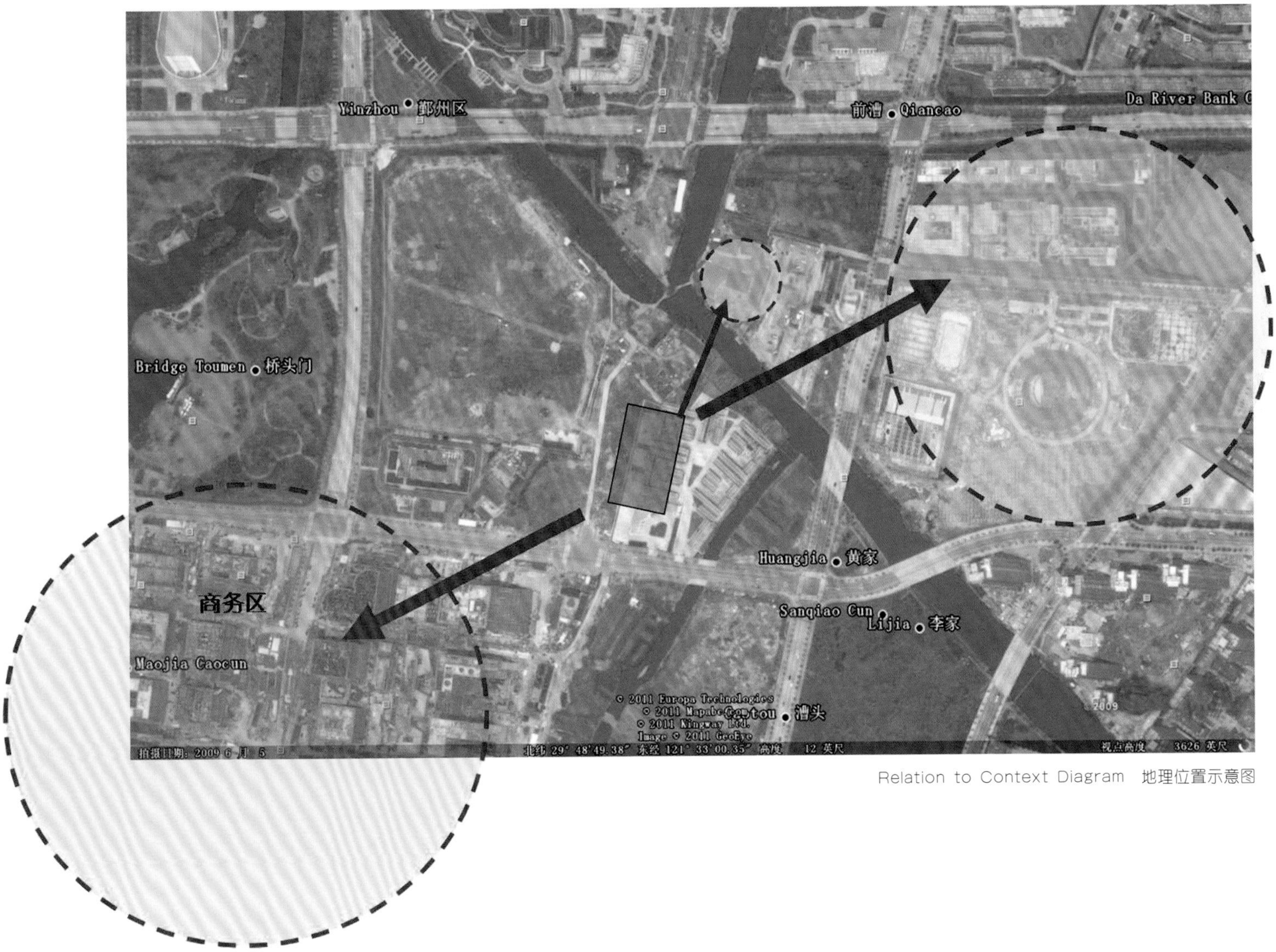

Relation to Context Diagram　地理位置示意图

Genesis Diagram　演生图

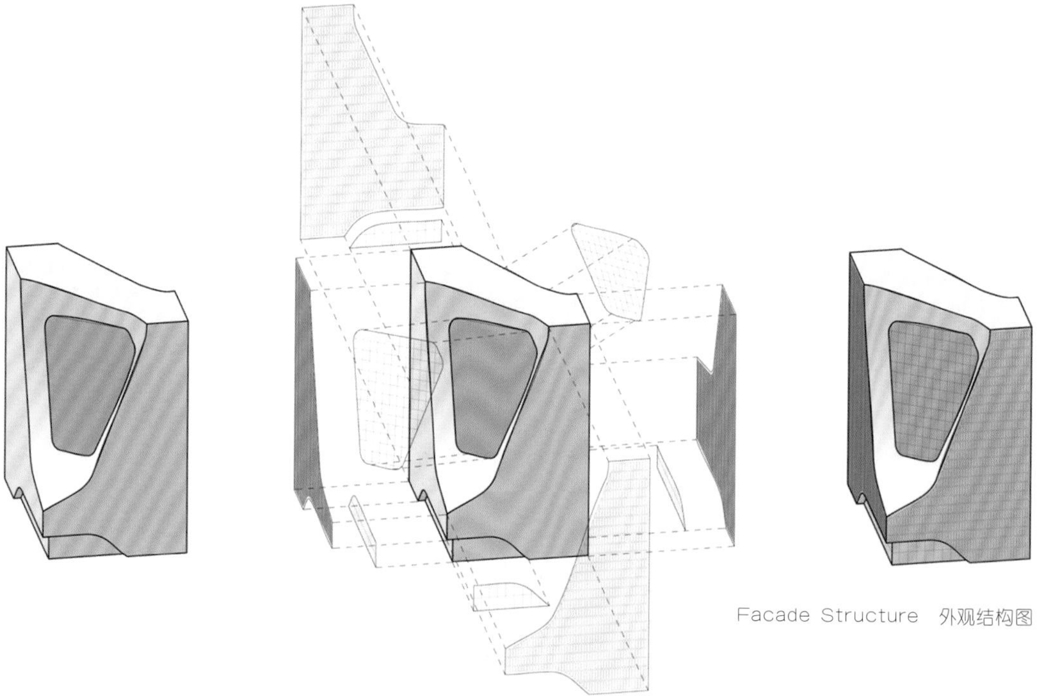

Facade Structure 外观结构图

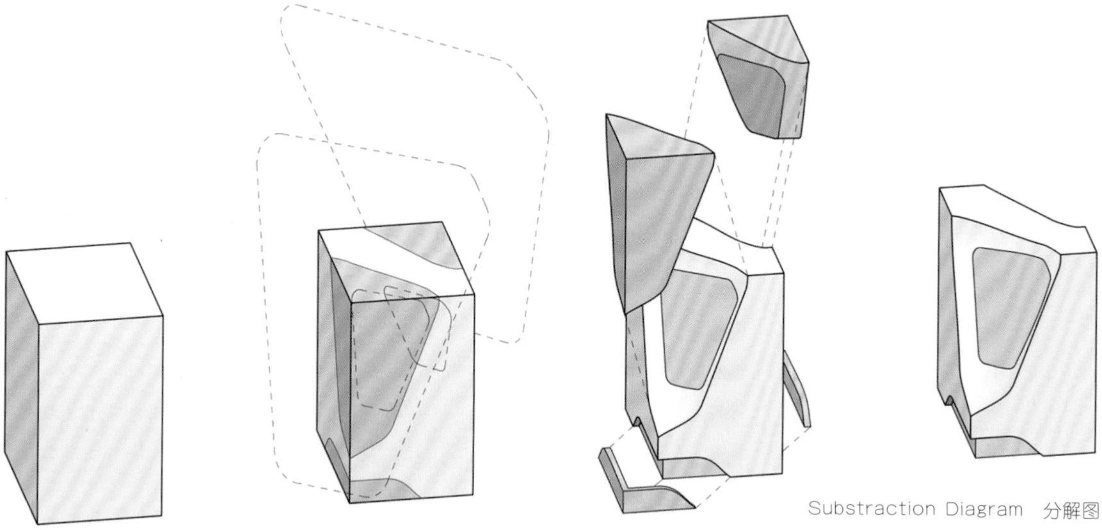

Substraction Diagram 分解图

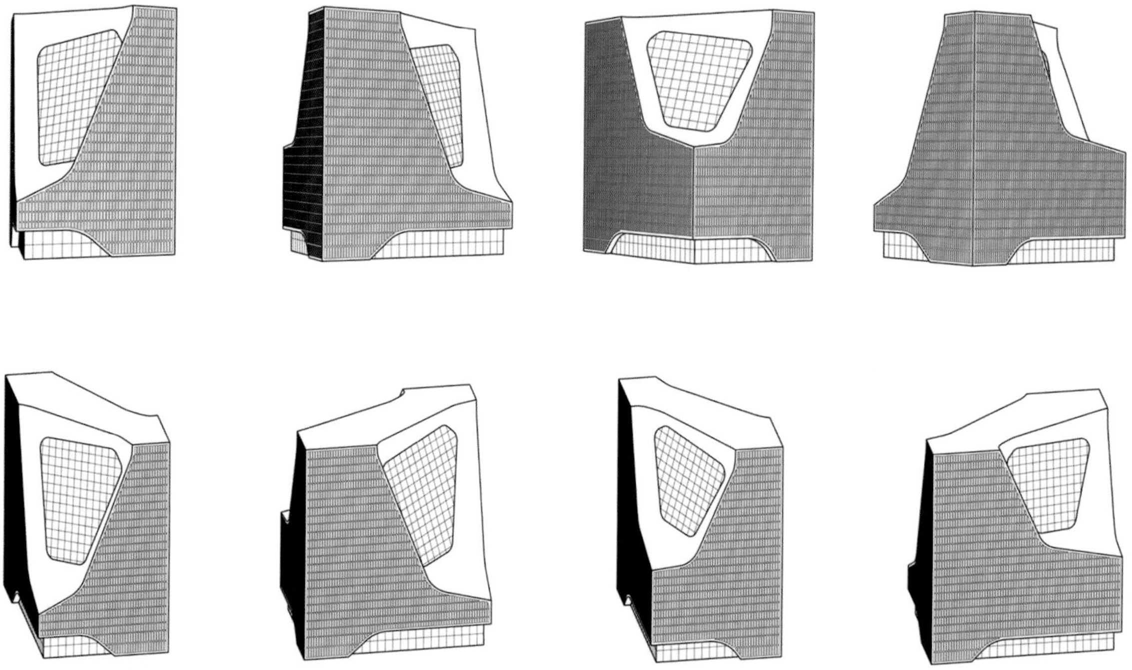

Views 视图

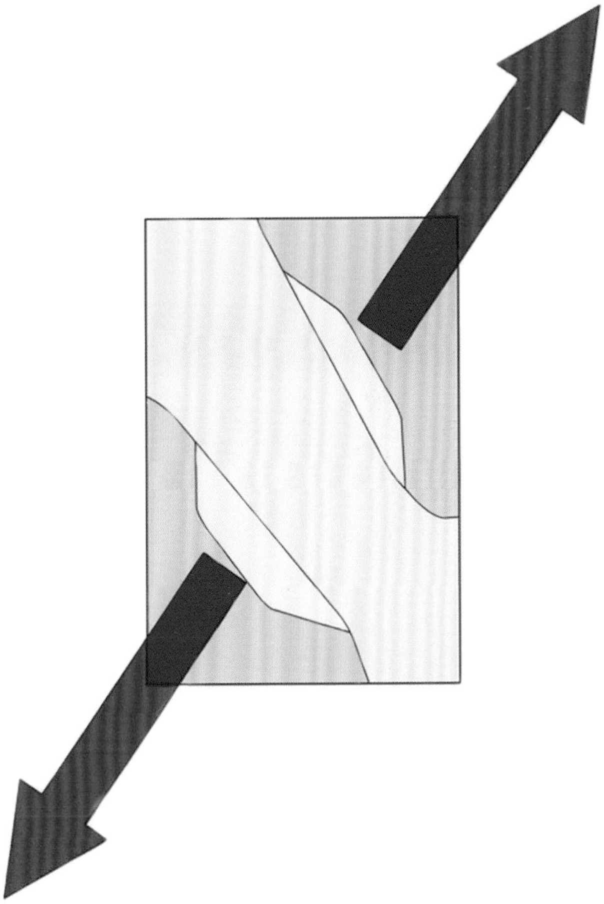

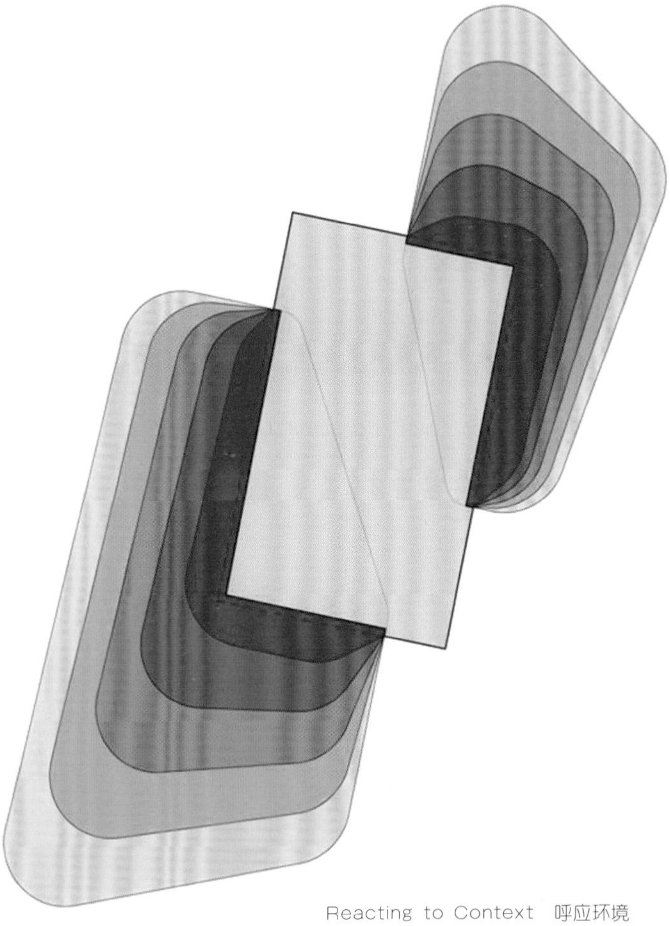

Reacting to Context　呼应环境

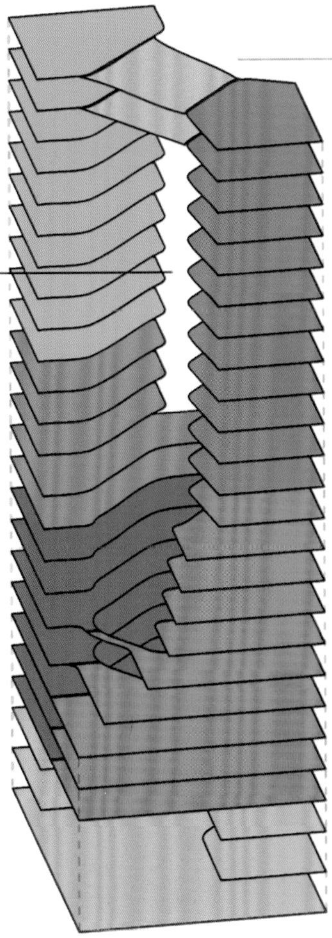

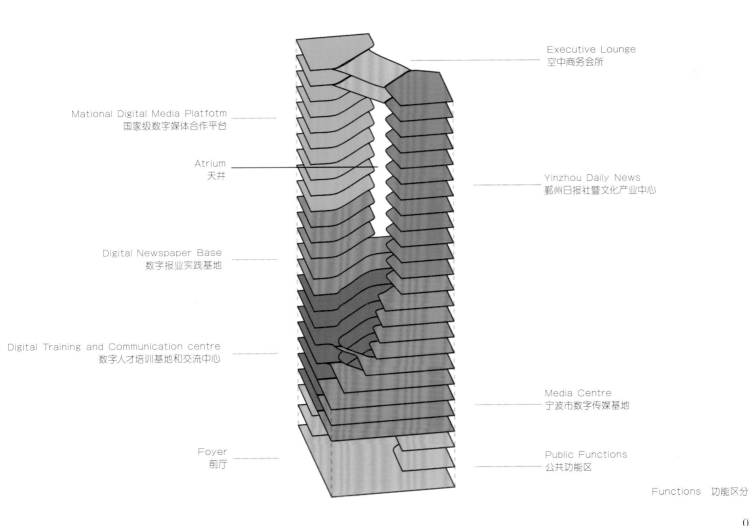

Functions　功能区分

Erdos 20+10 Plot 13
鄂尔多斯 20+10 P13 地块

Keywords 关键词

Integrated Design
整体设计

Terrace Landscape
台地景观

Facade Design
立面设计

Location: Erdos, Inner Mongolia, China
Architectural Design: in+of architecture
Project Leaders: Wang Lu, Li Jian
Design Team: Xu Jie, Lei Yuansheng, Bao Wei, Bodgen, Shi Xiaohui
Land Area: 4,782 m^2
Floor Area: 9,745 m^2
Plot Ratio: 1.32

项目地点：中国内蒙古自治区鄂尔多斯市
建筑设计：壹方建筑
项目主持：王路、李坚
设计团队：徐杰、雷沅胜、鲍薇、Bodgen、石晓辉
用地面积：4 782 m^2
建筑面积：9 745 m^2
容 积 率：1.32

Features 项目亮点

Designers tried to repair the base, intensified the concept of contour lines and integrated it into landscape and even interior & exterior design.

在基地修复的基础上，强化等高线的概念，并将其融入地段景观和建筑室内外的整体设计中。

■ Aechitectural Design

Designers used a 32A module for P13 and the total floor area is 9,745 m^2 (about 6,300 m^2 is calculated for plot ratio). The base is the steep slope, on which there are pits with large height differences. So they tried to repair the base, intensified the concept of contour lines and integrated it into the landscape and even interior & exterior design. The single building stands on the base like a huge stone and the southward arc facade interplays with the contour lines (terrace landscape), which highlight the main facade.

■ 建筑说明

P13采用一个32A模块，总建筑面积9 745 m^2（其中计容积率面积6 300 m^2）。基地为陡坡地，其中有高差较大的土坑。设计的理念是在修复基地的基础上，强化等高线的概念，并将其融入地段景观和建筑室内外的整体设计中。独栋建筑巨石般耸立于基地，只有南向有弧曲的立面呼应户外的等高线（台地景观），突出了主立面。

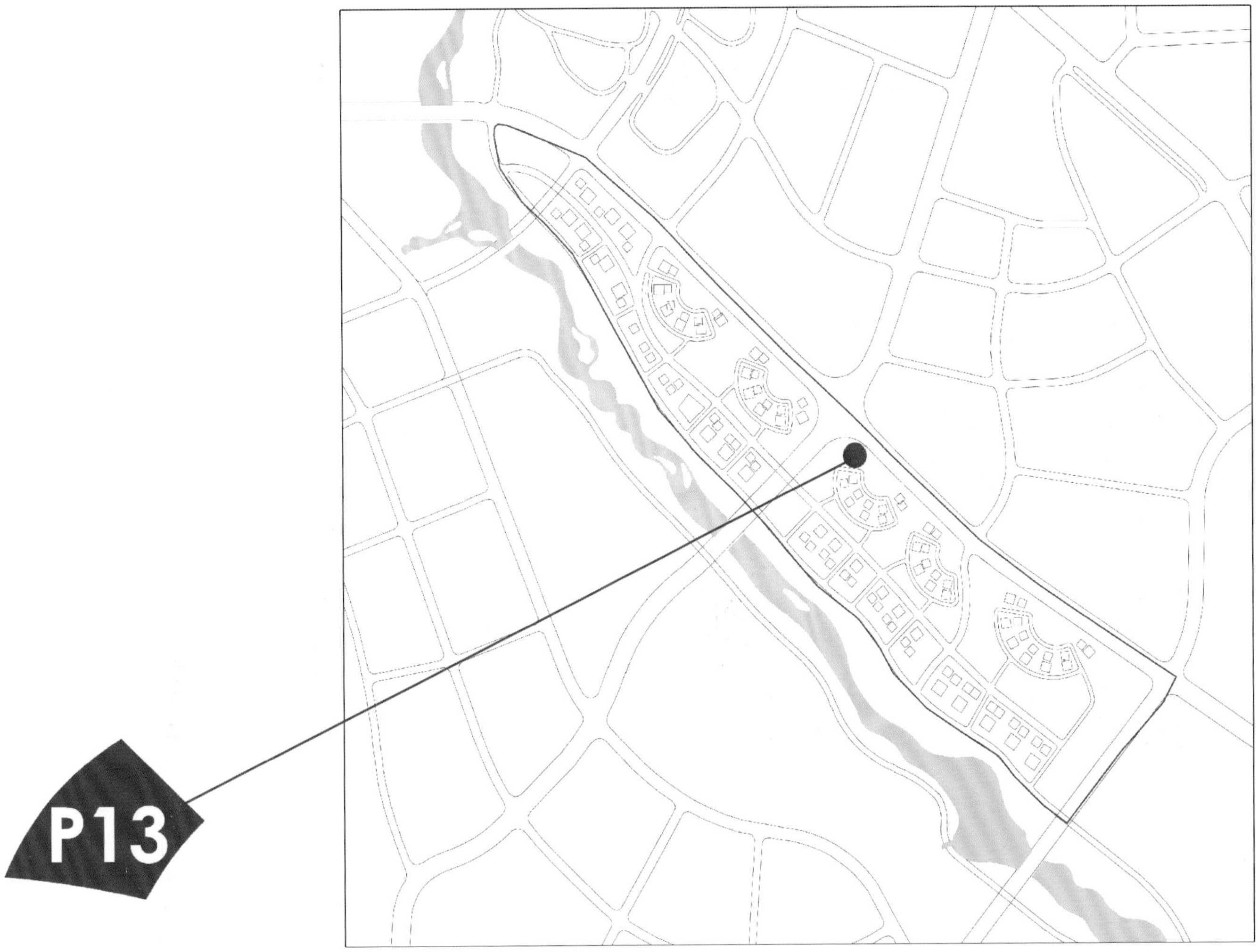

Site Location 基地位置

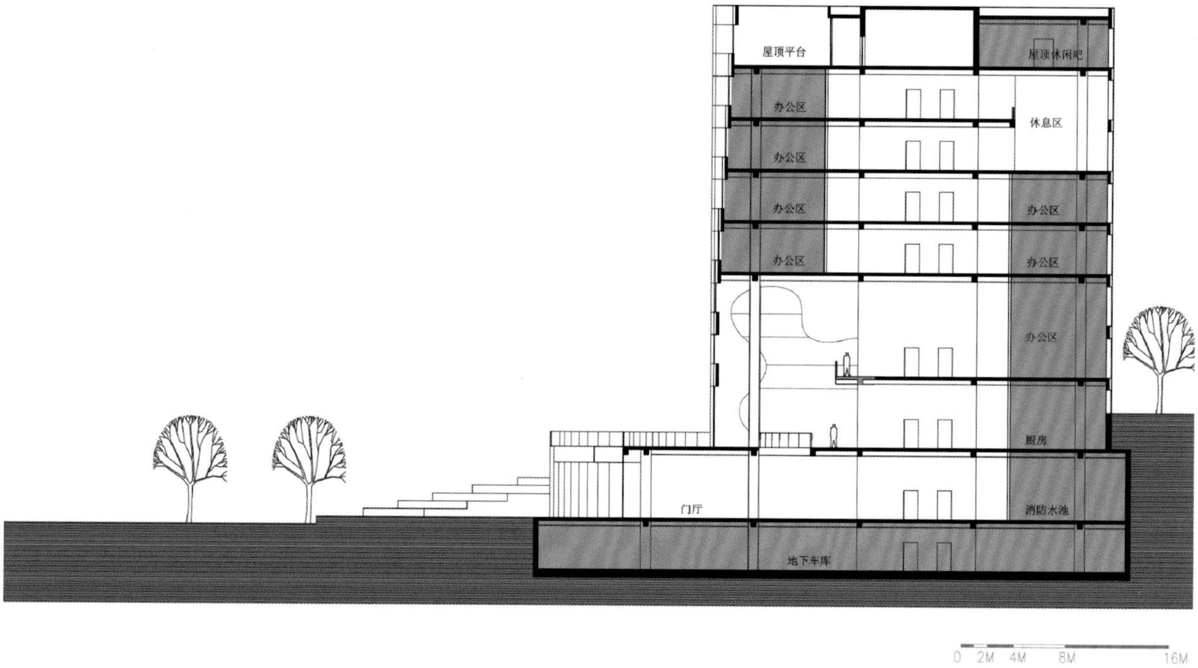
屋顶平台
屋顶休闲吧
办公区
休息区
办公区
办公区
办公区
办公区
办公区
办公区
厨房
门厅
消防水池
地下车库
0 2M 4M 8M 16M

Section 剖面图

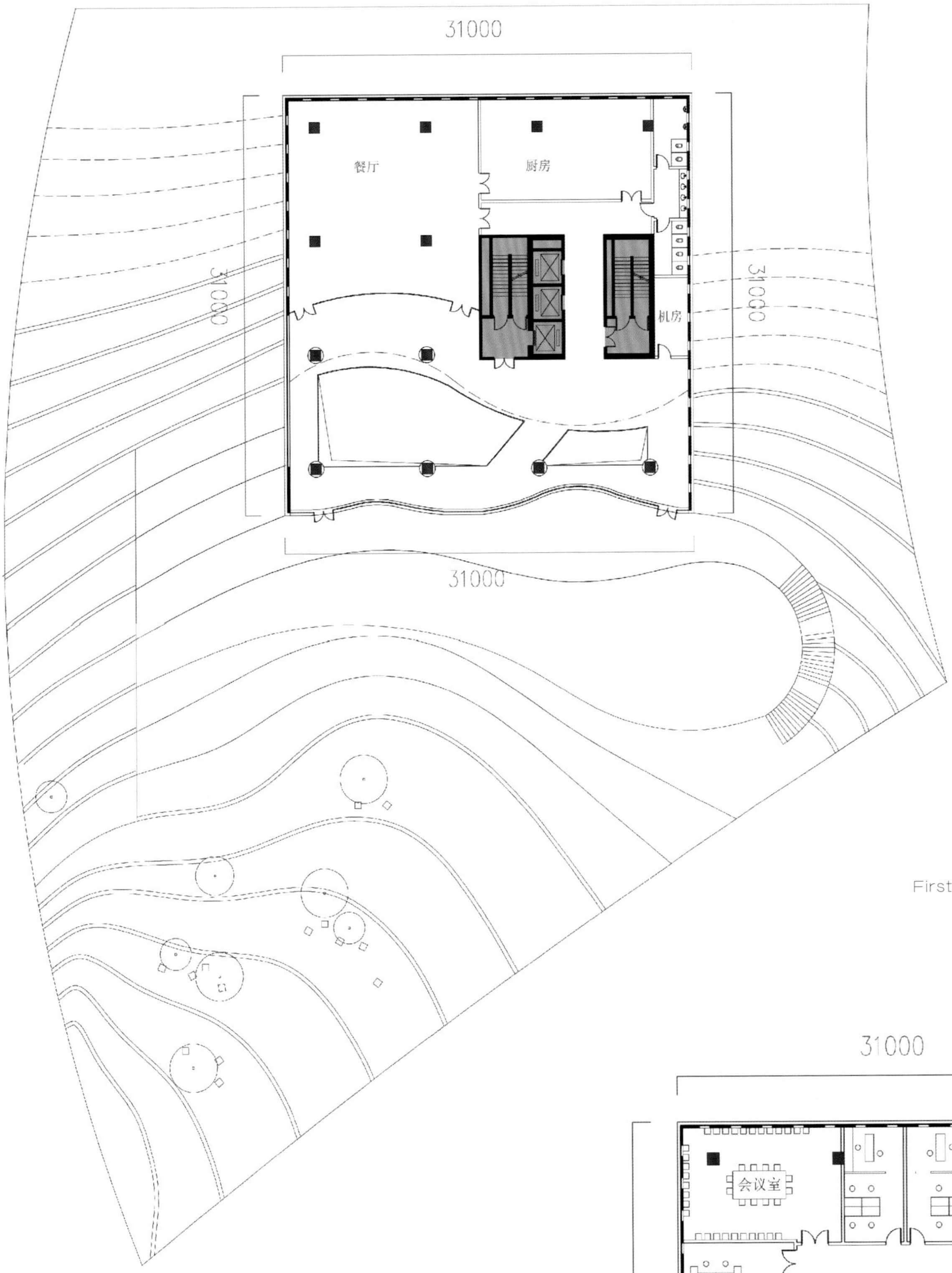

First Floor Plan 一层平面图

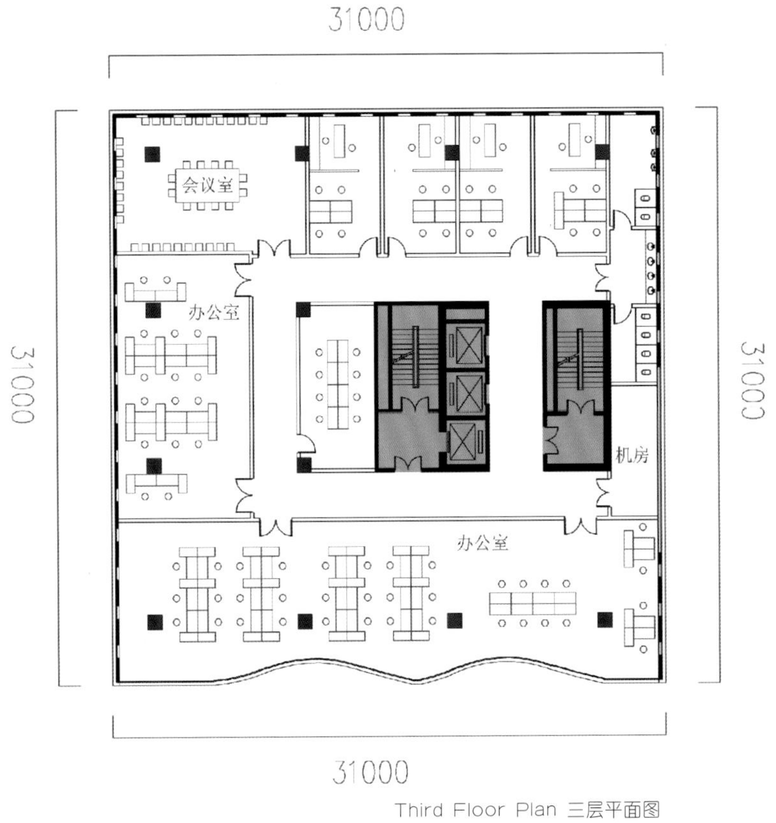

Third Floor Plan 三层平面图

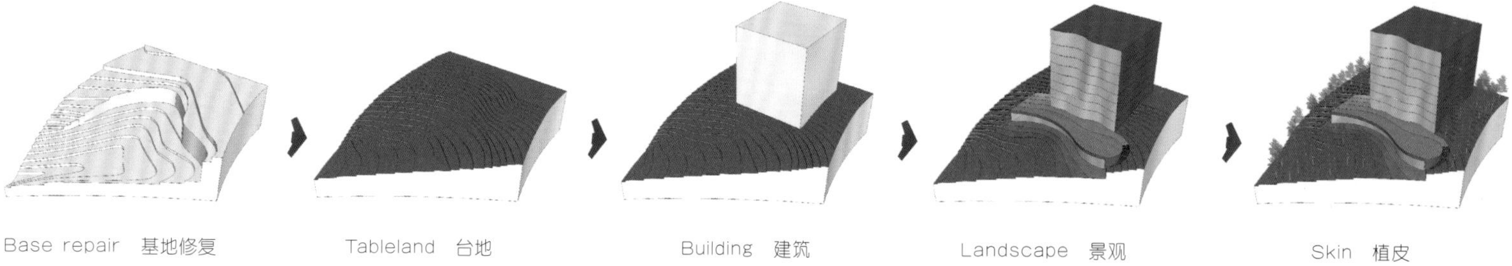

Idea 理念

Northwest Elevation 西北立面图

Southwest Elevation 西南立面图

Southeast Elevation 东南立面图

Northeast Elevation 东北立面图

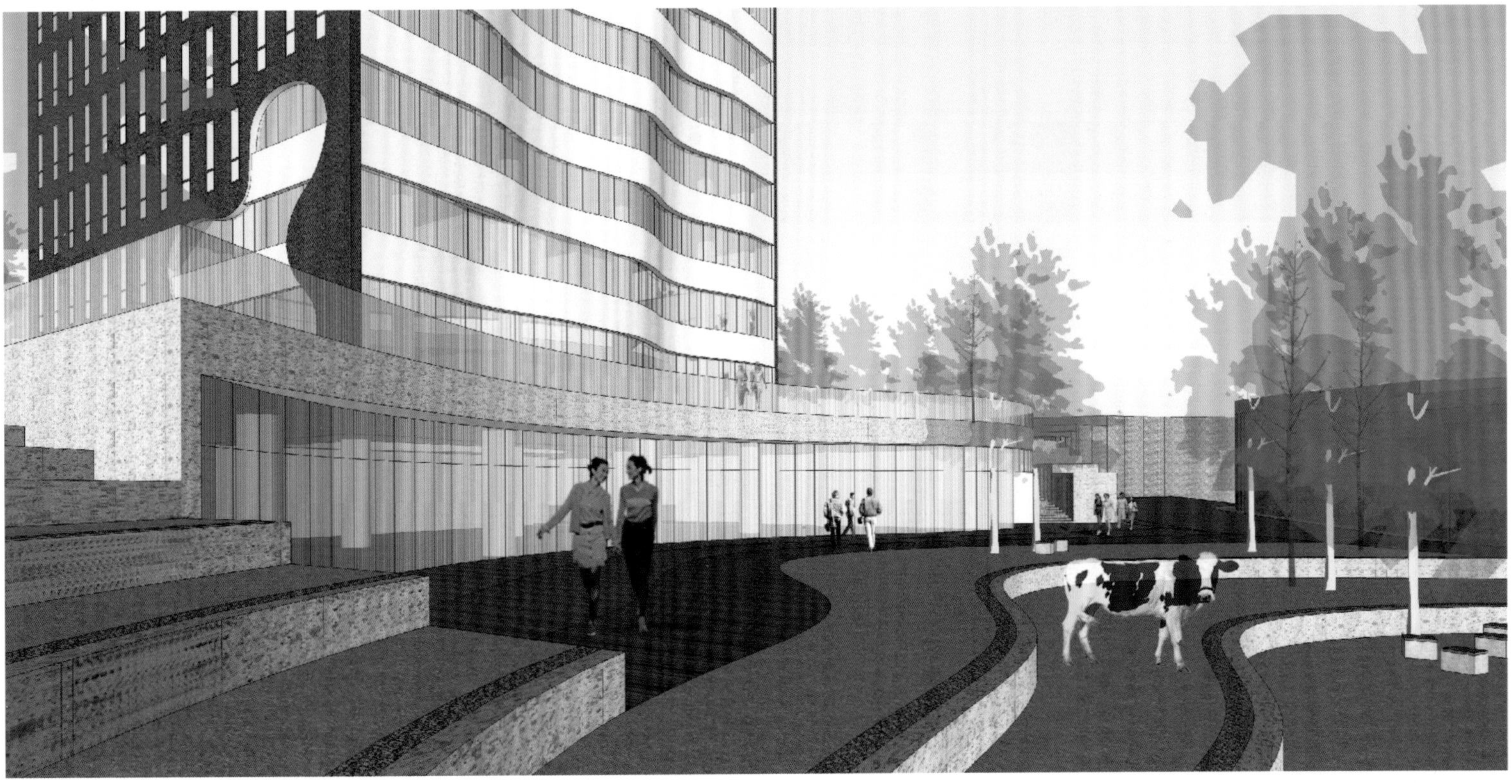

Fujian Haixia Bank
福建海峡银行

Keywords 关键词

Glass Curtain Wall
玻璃幕墙

Daylight-flooded Atrium
采光中庭

Architectural Image
建筑形象

Location: Fuzhou, Fujian, China
Architectural Design: PEDDLE THORP MELBOURNE PTY LTD.
Land Scale: 8,487 m^2
Floor Area: 65,974 m^2
Plot Ratio: 6.24
Result: won the bidding
Status: preliminary design

项目地点：中国福建省福州市
建筑设计：澳大利亚柏涛（墨尔本）建筑设计有限公司
用地面积：8 487 m^2
建筑面积：65 974 m^2
容 积 率：6.24
设计结果：投标中标
状　　态：扩初设计

Features 项目亮点

The space shifts and remoulds itself, extending upwards into a vertical space enclosed by the towers and transparent glass curtain walls, which endows the substantial building a solid and stable image.

建筑变换、上升，结合玻璃幕墙构造出垂直空间，突出银行建筑所需要的坚实稳定的形象。

■ Overview

This project is located on prominent site adjacent to the Minjiang River and on the CBD edge of the financial district of North Fuzhou. The original Fuzhou Commercial Bank was established decades ago but more recently formed a conglomerate that cooperatively incorporated many smaller banks and took its name from its context, as Haixia refers to the strait between mainland China and Taiwan China.

■ 项目概况

项目所在地块毗邻闽江，位于福州北部金融区商务中心边界处。项目前身是福州市商业银行，成立于数十年之前，如今合并了一些小型银行，成为联合企业，正式更名为福建海峡银行。

■ Design Concept

In accordance with the location and land condition, designers fully respected the axis relationship to create a work that interprets the congruent relationship the most basic urban space needed. As an emerging power in banking it is important that strong growth and transparency are represented, and also that the bank maintains an open alliance with its customers. Later analysis lead to the understanding that the proposed form also represents the way the bank functions. In addition, capturing the physical representation of the headquarters required contemplation of its origins, its strengths and its relationship with customers.

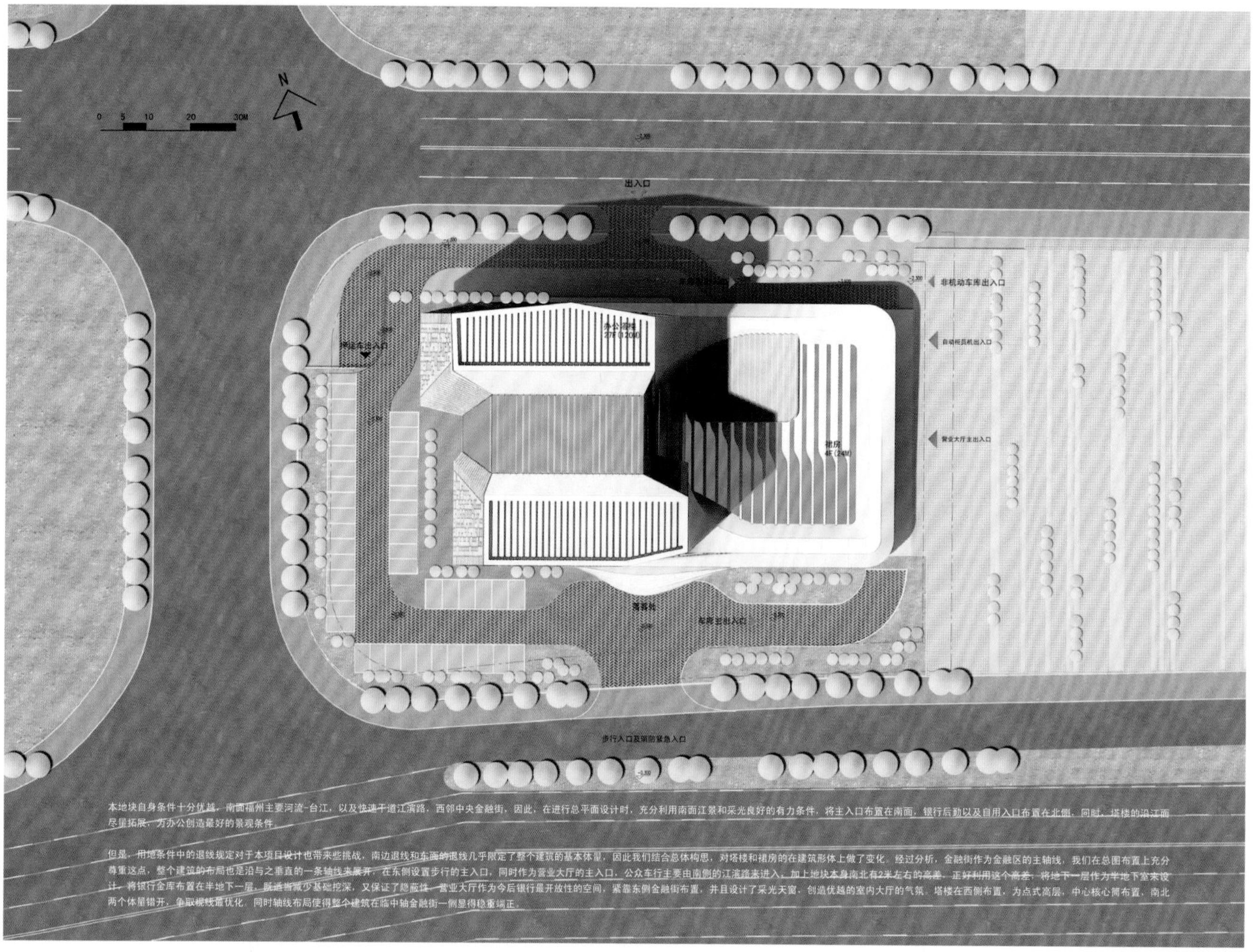

本地块自身条件十分优越，南面福州主要河流-台江，以及快速干道江滨路，西邻中央金融街，因此，在进行总平面设计时，充分利用南面江景和采光良好的有力条件，将主入口布置在南面，银行后勤以及自用入口布置在北侧，同时，塔楼的沿江面尽量拓展，为办公创造最好的景观条件。

但是，用地条件中的退线规定对于本项目设计也带来些挑战，南边退线和东面的退线几乎限定了整个建筑的基本体量，因此我们结合总体构思，对塔楼和裙房的在建筑形体上做了变化。经过分析，金融街作为金融区的主轴线，我们在总图布置上充分尊重这点，整个建筑的布局也是沿与之垂直的一条轴线来展开，在东侧设置步行的主入口，同时作为营业大厅的主入口，公众车行主要由南侧的江滨路来进入，加上地块本身南北有2米左右的高差，正好利用这个高差，将地下一层作为半地下室来设计，将银行金库布置在半地下一层，既适当减少基础挖深，又保证了隐蔽性。营业大厅作为今后银行最开放性的空间，紧靠东侧金融街布置，并且设计了采光天窗，创造优越的室内大厅的气氛。塔楼在西侧布置，为点式高层，中心核心筒布置，南北两个体量错开，争取视线最优化，同时轴线布局使得整个建筑在临中轴金融街一侧显得稳重端正。

Site Plan 总平面图

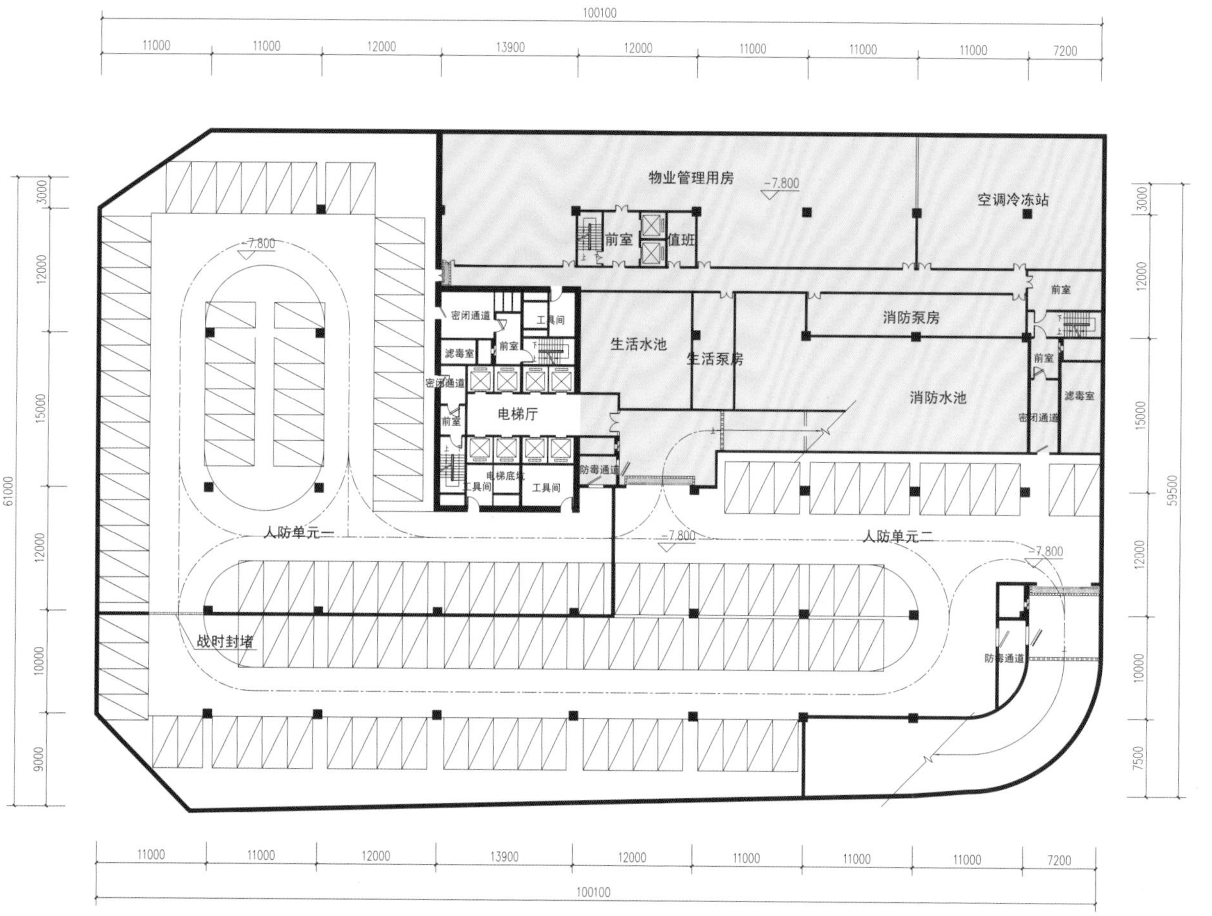

Basement Two Plan 地下二层平面图

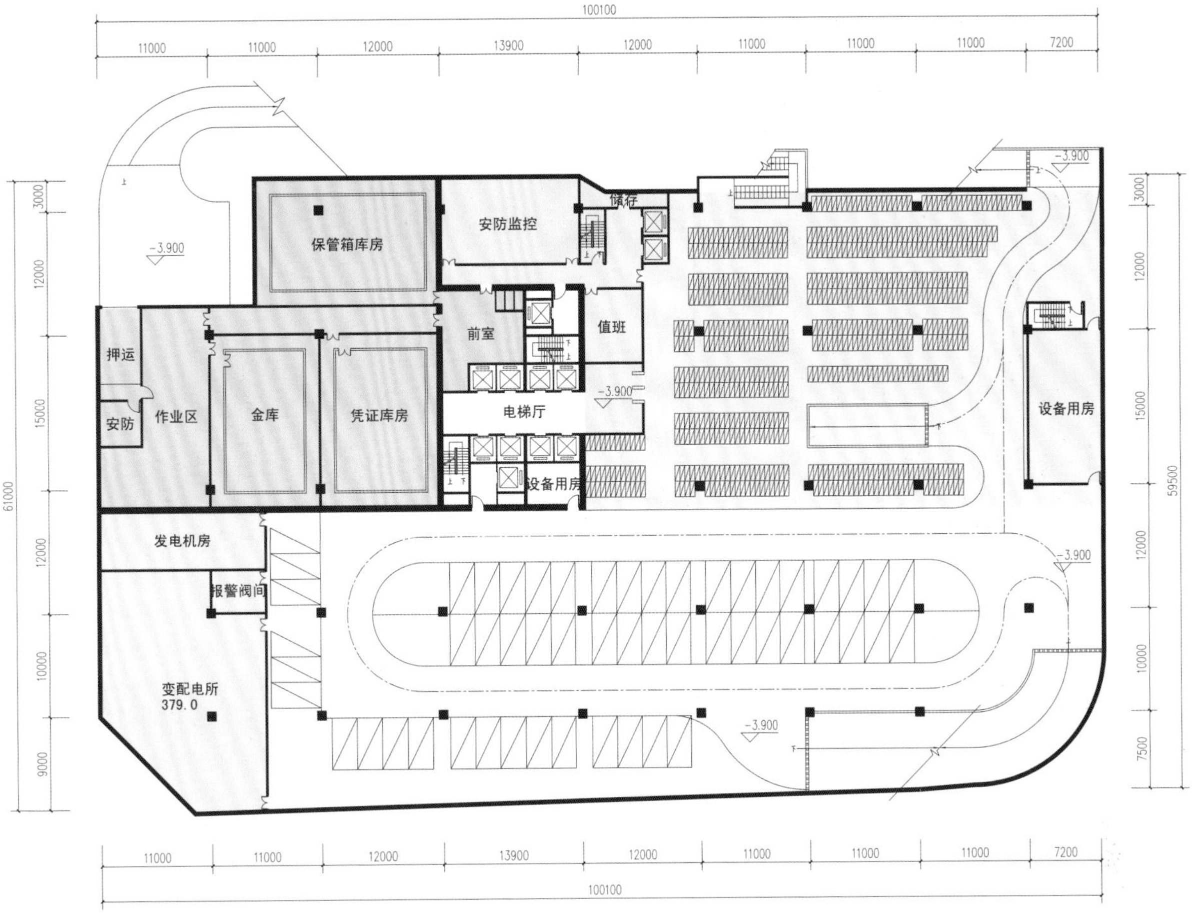

Plan of the First Basement Floor 地下一层平面图

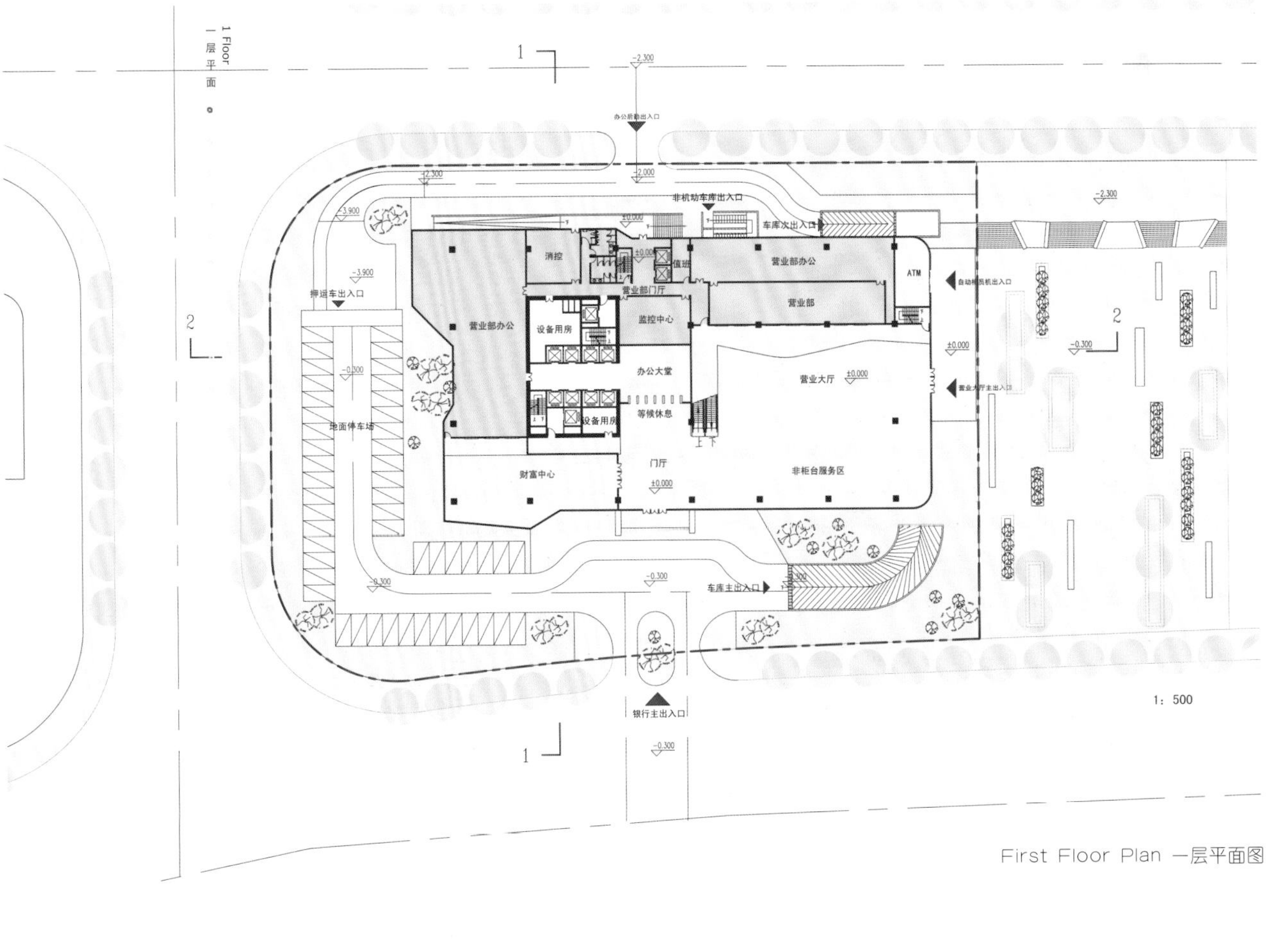

First Floor Plan 一层平面图

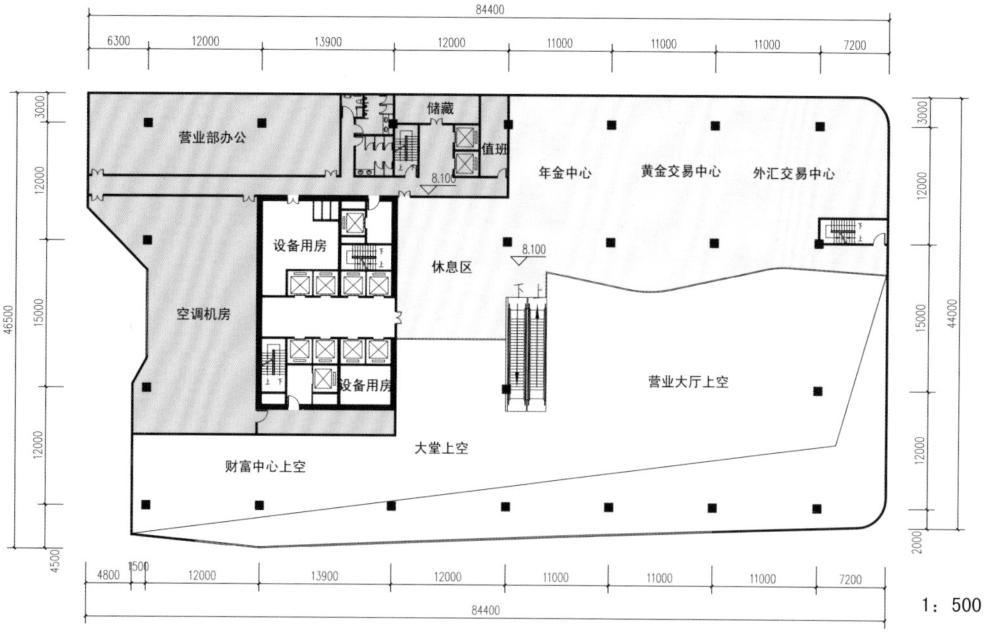

Second Floor Plan 二层平面图

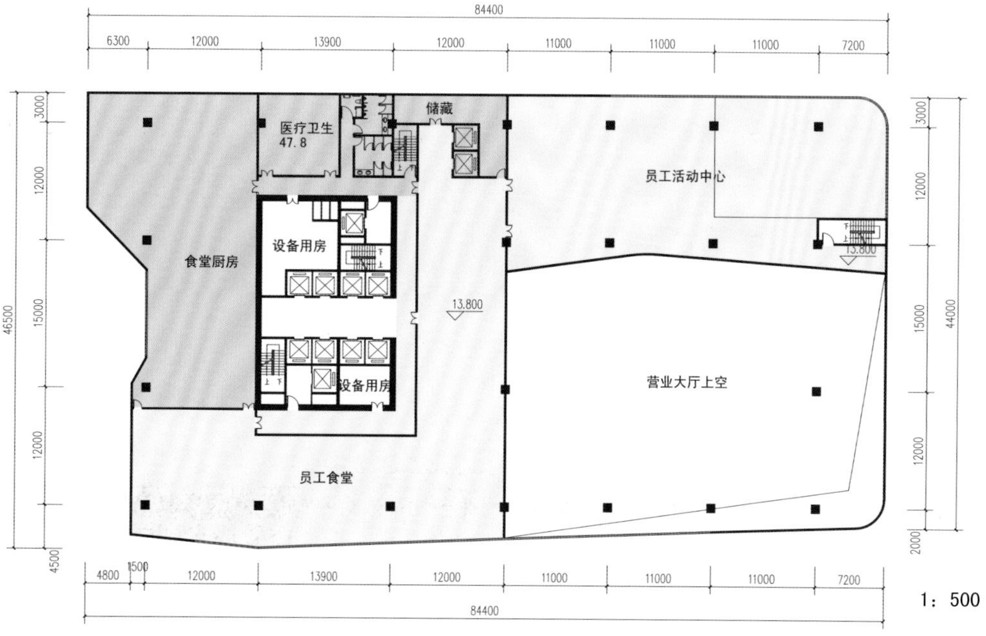

Third Floor Plan 三层平面图

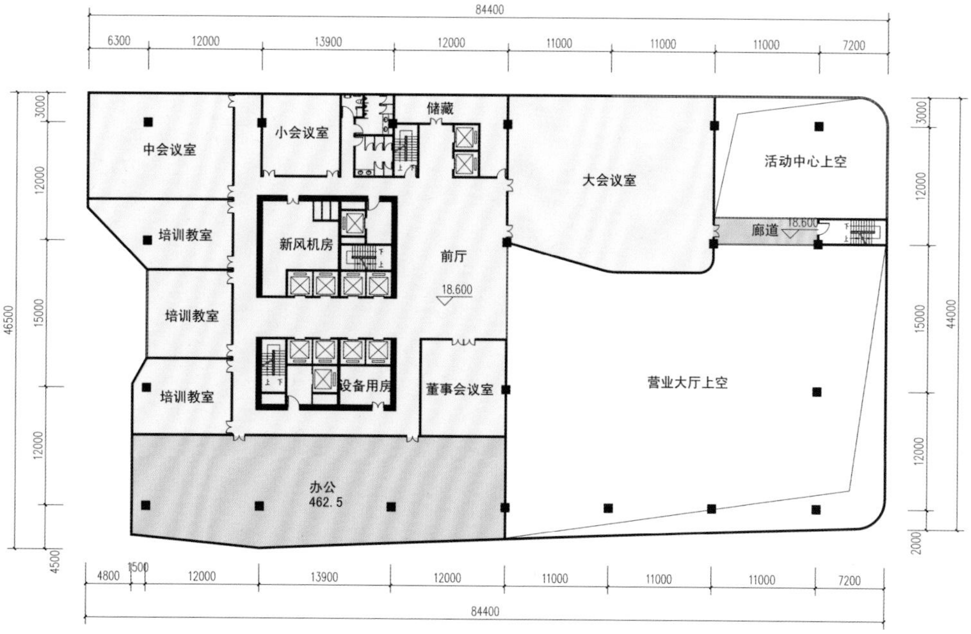

Fourth Floor Plan 四层平面图

■ 设计理念

作为银行办公建筑，本次设计力求遵循大气稳重的基本原则，实现银行建筑对公众展示形象的需求。基于地块所处的位置，设计师利用原有规划和用地条件，充分尊重轴线关系，让设计能够反映和周边环境的关系，建立最基本的城市空间对应关系。在总体构思时，通过对海峡银行自身历史和发展情况的分析，设计团队希望能够从设计上折射出其历史和现状，将多重因素整合概括为两种积极向上的态势，在建筑塔楼上展现象征海峡两岸、银行自身和投资公众、投资需求和收益回报等互相对应的关系。同时，这些互相倚靠、共同成长的关系又植根于银行本身的健康发展中，特别是和公众建立的良好关系，因此，需要将这两种元素在合适的位置融合交汇。

■ Architectural Design

Fusion of these aspects of the bank informed the design of both a tower and a podium encircling "skirt" that becomes integrated into the tower and an undiversified facade treatment unifies the major horizontal and vertical elements. Part of the design strategy was to create a pleasing atmosphere for customers, so the banking floor has ample open spaces, gardens and a glazed atrium and customers will also have access to a sky lobby which has a landscaped garden. The podium area contains the major functions of the bank and is a light filled space. The space shifts and remoulds itself, extending upwards into a vertical space enclosed by the towers and transparent glass curtain walls, which endows the substantial building a solid and stable image.

■ 建筑设计

办公塔楼应由两个形体互相倚靠形成，这两个形体在裙房，也就是本建筑最具开放性的位置围合。四层通高的营业大厅作为裙房的主要功能空间，有着采光天窗，裙房的建筑元素在顶部围合，并向上延伸构成塔楼，结合玻璃幕墙构造出垂直空间，形成相互倚靠、挺拔向上的视觉形象，突出银行建筑所需要的坚实稳定的形象。

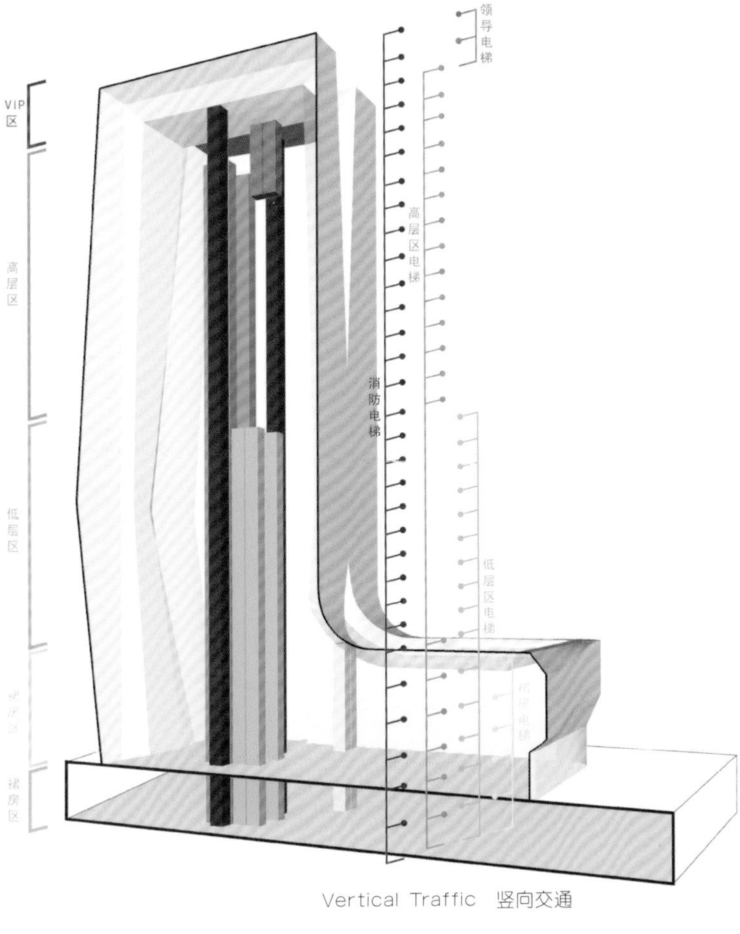

Vertical Traffic 竖向交通

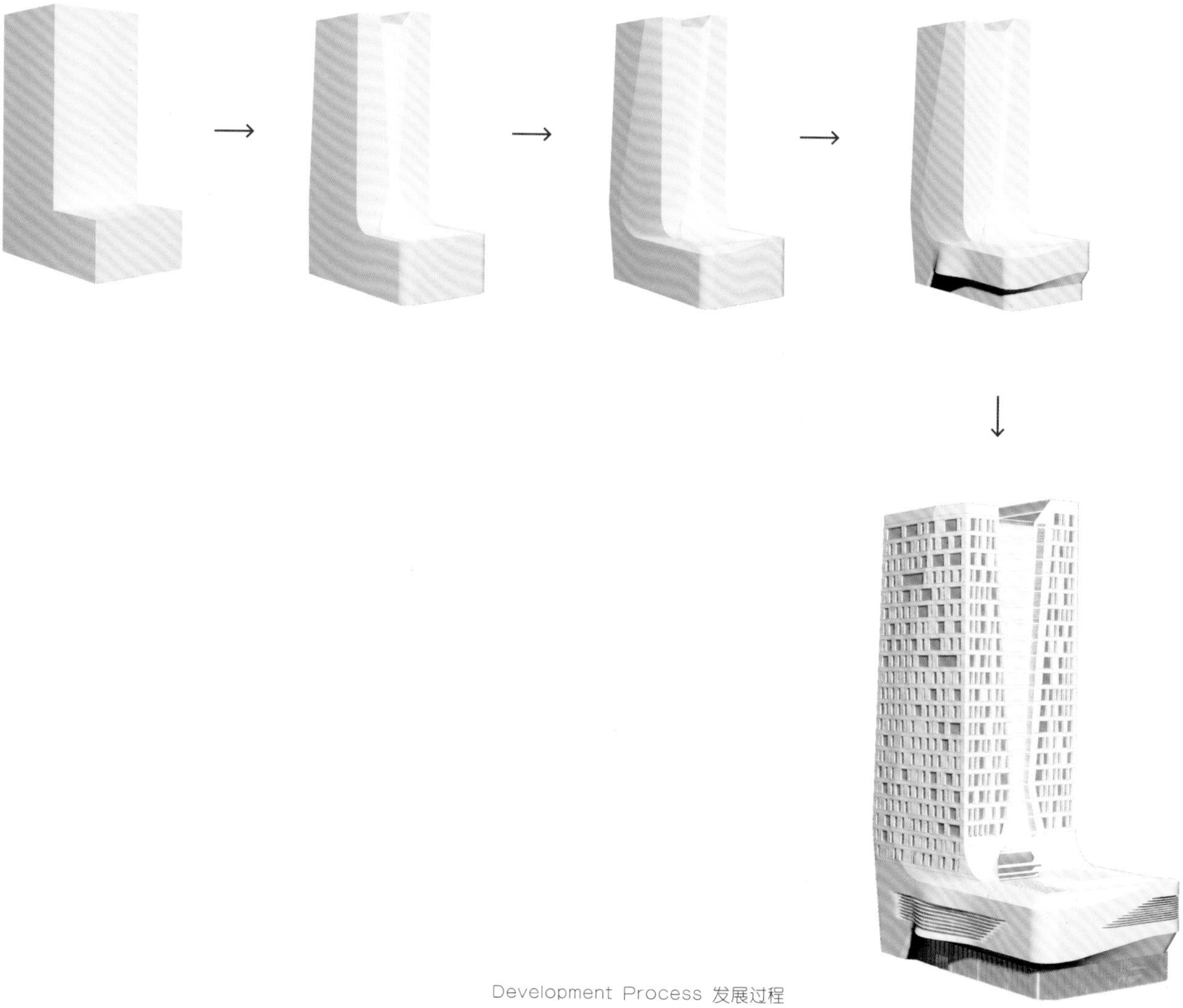

Development Process 发展过程

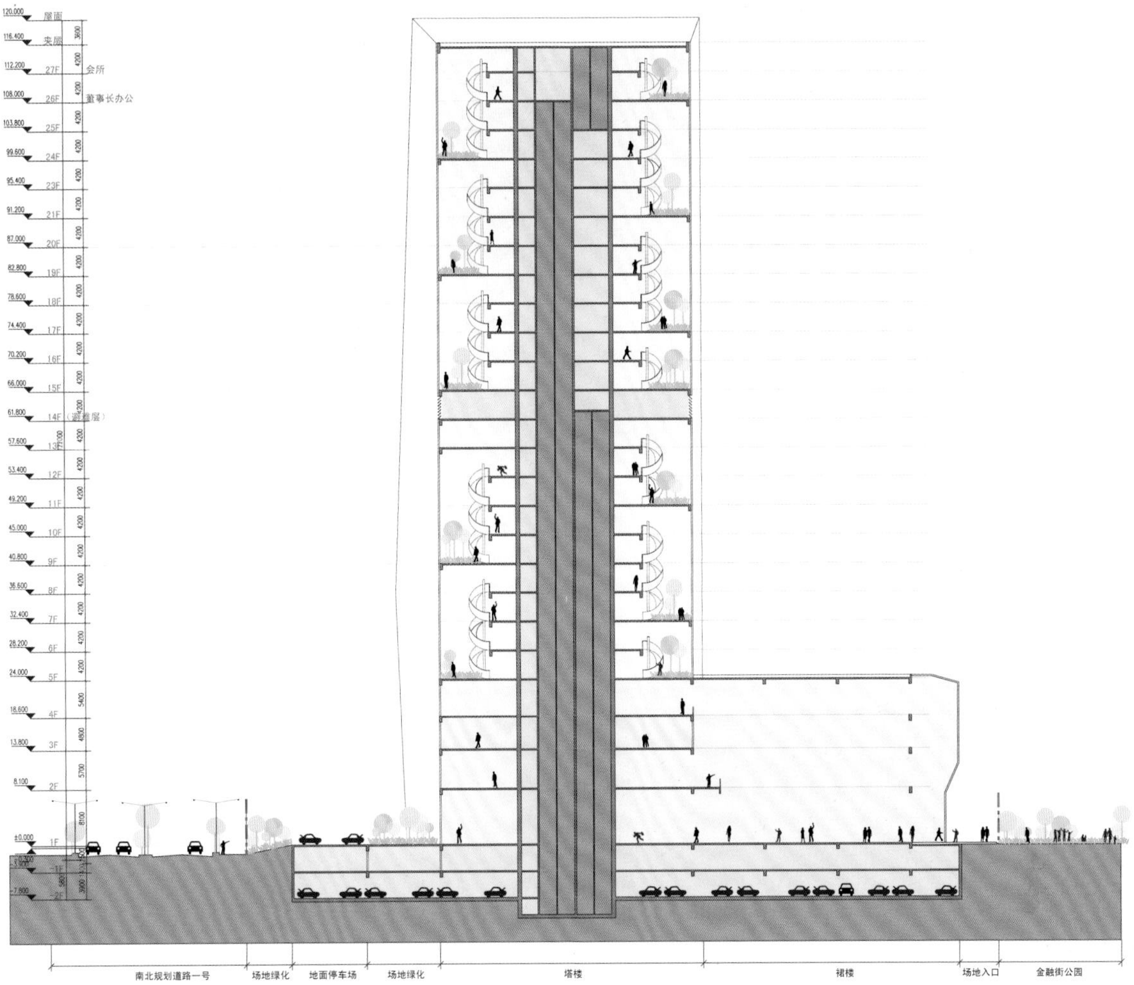

Section 1 剖面图 1

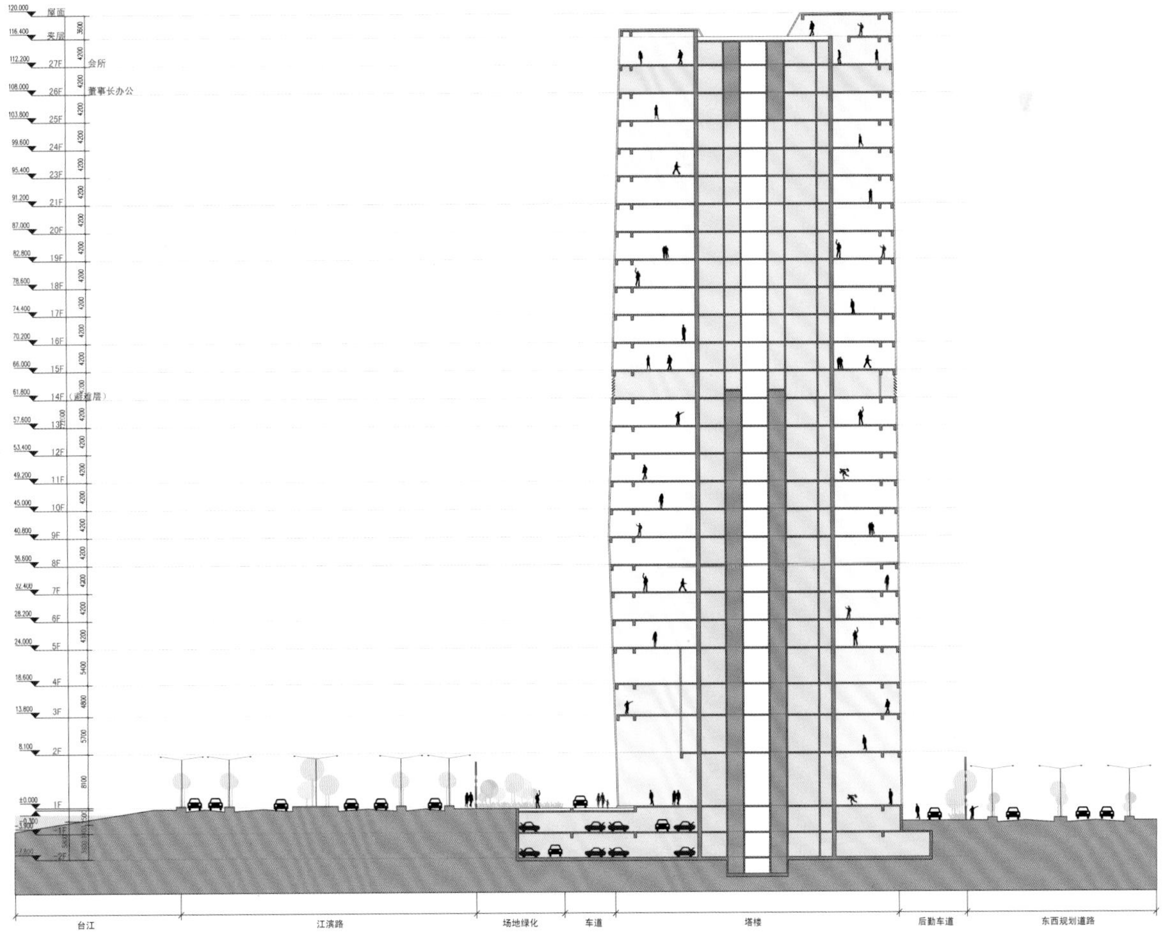

Section 2 剖面图 2

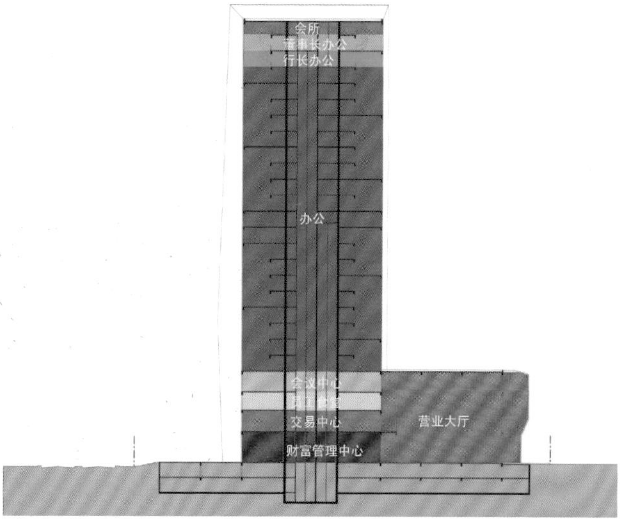

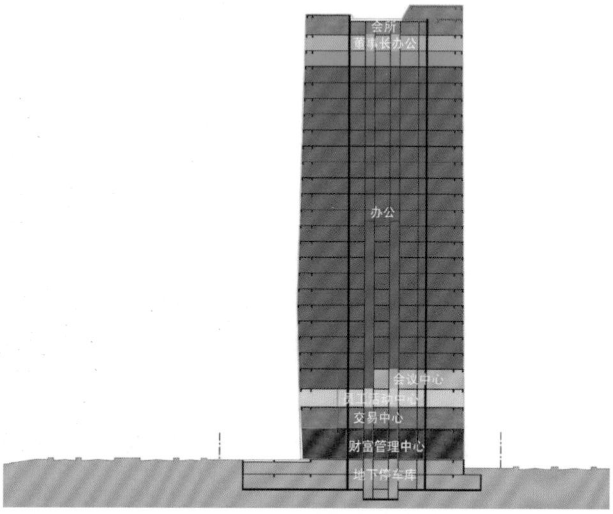

Functional Layout 功能布局

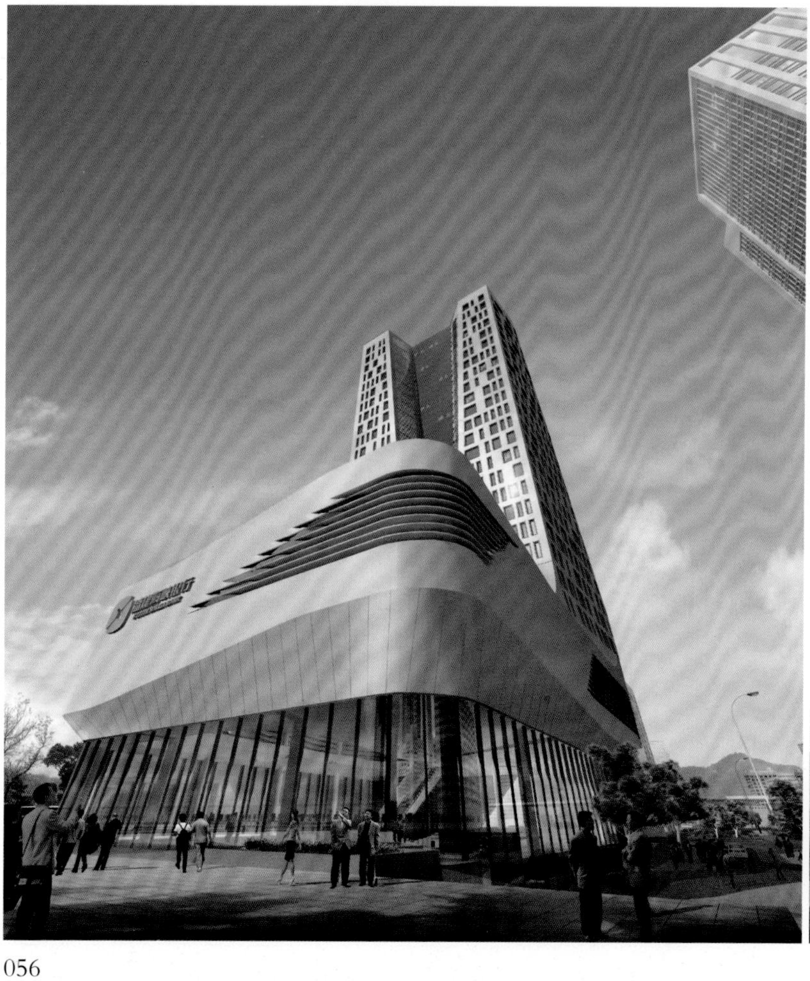

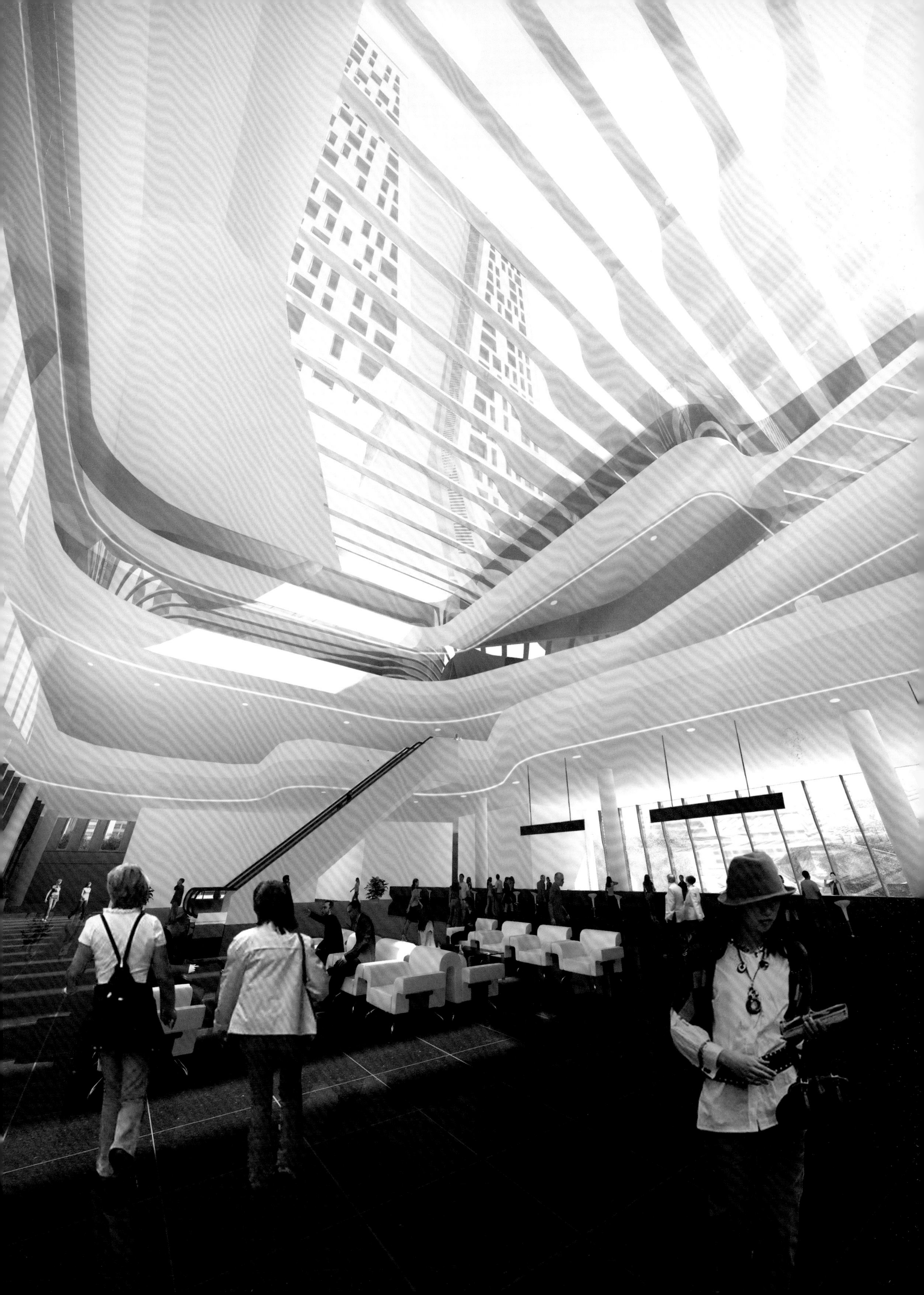

Nanping Wuyi Civic Center
南平市武夷新区市民中心

Keywords 关键词

Cambered Roof
曲面屋顶

Harmonious Environment
环境协调

Local Tradition
地域传统

Features 项目亮点

The building looks elegant and well-proportioned with strong modern sense and local characteristics, showing the high efficiency of modern administration.

整体建筑形态简洁大方、虚实结合、比例尺度和谐，体现出现代行政办公大楼高效精练的办事作风，具有鲜明的地域性和时代感。

Location: Nanping, Fujian, China
Commissioned by Wuyi New District Administrative Committee
Architectural Design: KLP KELLER HU PARTNERSHIP LIMITED
Main Designers: Larry J. Keller, Sun Zhen
Total Land Area: 152,116.56 m^2
Total Floor Area: 145,488.23 m^2 overground, 45,102.23 m^2 underground
Date of Design: January, 2013

项目地点：中国福建省南平市
委托单位：武夷新区管委会
设计公司：美国KLP建筑设计有限公司
主要设计人员：拉里·凯勒（Larry J. Keller）、孙铮
用地面积：152 116.56 m^2
建筑面积：地上145 488.23 m^2、地下45 102.23 m^2
设计时间：2013年1月

■ Aechitectural Design

This project is located on plot A-01 and B-01 of Wuyi District, Nanping. With the design idea derived from the shape of Wuyi Mountain, it abstractly interprets the relations between building base, body and roof in modern ways. The terrace structure shrinks generally as it rises up, achieving the harmony and coexistence between architecture, nature and human beings. Cambered roof, curved windows and arch axis dialogue with each other, echoing the beautiful and magnificent Wuyi Mountains and softening the rigidness of the administration building. At the same time, it makes reference to the form and details of local traditional architecture like the ancient Yuqing Bridge, to create an elegant and well-proportioned building that embodies the high efficiency of modern administration.

■ 设计说明

项目位于南平市武夷新区南林片区A-01、B-01地块，设计理念取自"武夷茶山"的自然形态，运用现代手法抽象化演绎基底、主体、屋顶三者的关系，层层退台式的建筑向上收拢，将自然融入建筑，追求建筑、自然、人三者和谐共生，体现"天人合一"的朱熹理学思想。弧形线条打造的曲面屋顶和弧形墙窗与弧形主拱轴相呼应，映衬着闽北连绵秀美的武夷山系，也柔化了行政办公建筑刚硬刻板的形象。同时对古余庆桥等武夷山固有传统建筑的形态、细节的研究及对其精髓的提炼再加工，使得整体建筑形态简洁大方、虚实结合、比例尺度和谐，体现现代行政办公大楼高效精练的办事作风，具有鲜明的地域性和时代感。

总综合经济技术指标

名称			数量	单位
一、总规划用地面积：			152116.56	M^2
二、总建筑面积：			190590.46	M^2
其中	地上建筑面积（计入容积率）：		145488.23	M^2
	其中	市民中心面积：	131219.73	M^2
		会议中心面积：	14268.5	M^2
	地下建筑面积(不计入容积率)：		45102.23	M^2
	其中	市民中心地下室面积：	31285.70	M^2
		会议中心地下室面积：	13816.53	M^2
三、容积率			0.96	
四、基底面积			39424.1	M^2
五、建筑密度			25.92	%
六、绿地率			43.87	%
七、机动车位：			1271	辆
其中	地上机动车位：		158	辆
	地下机动车位：		1113	辆
	其中	市民中心地下室车位：	821	辆
		会议中心地下室车位：	292	辆

Site Plan 总平面图

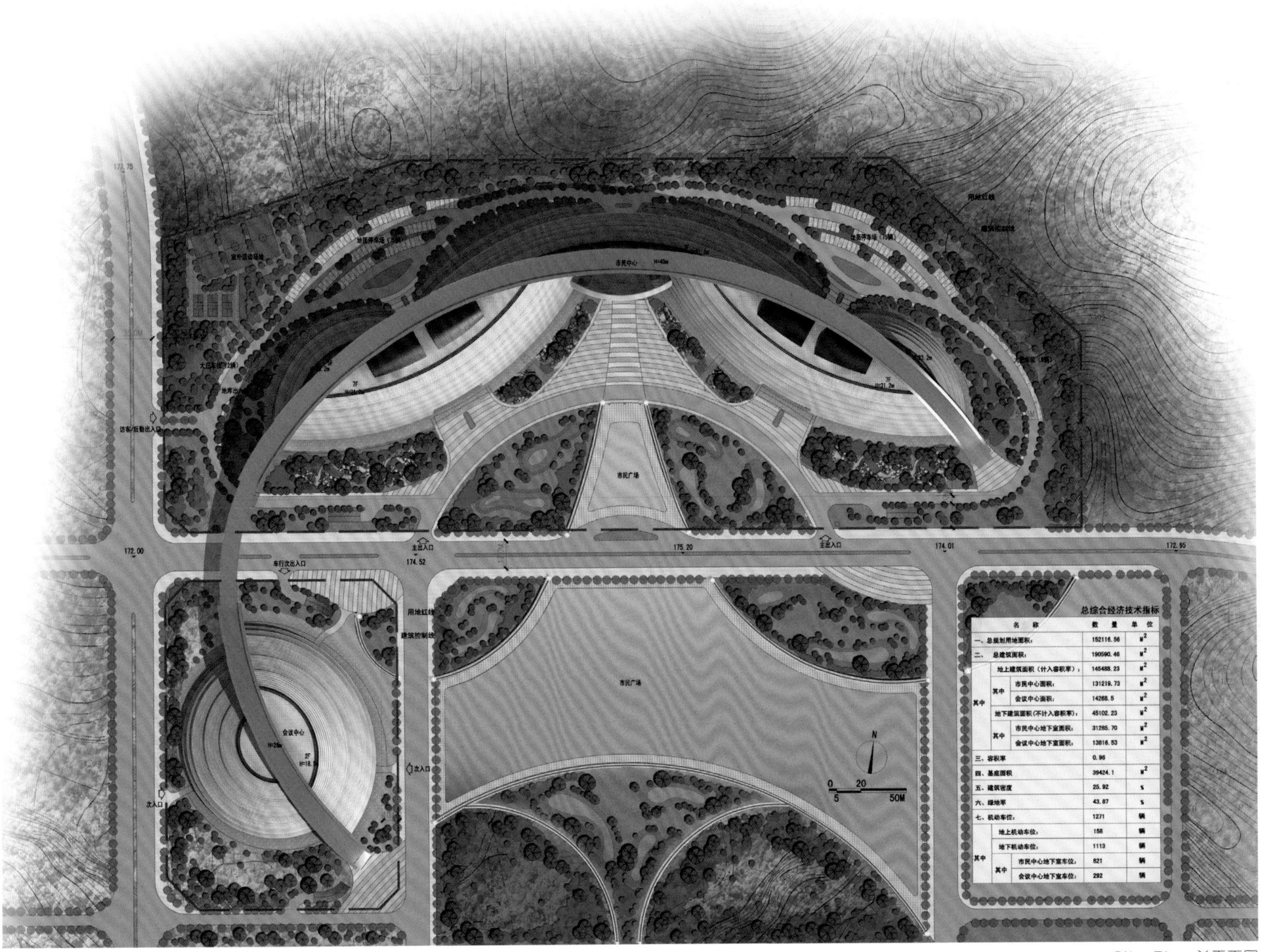

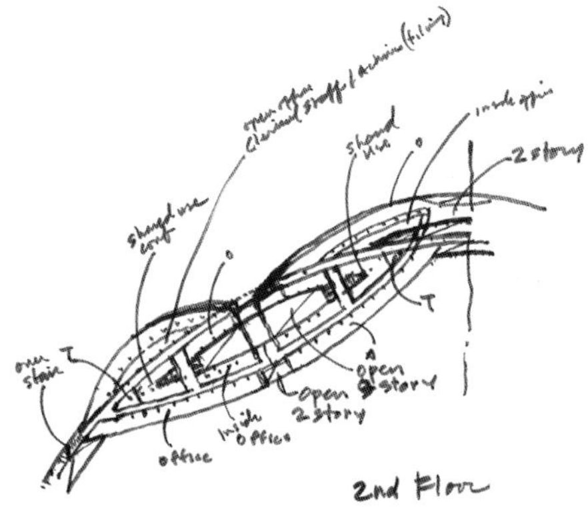

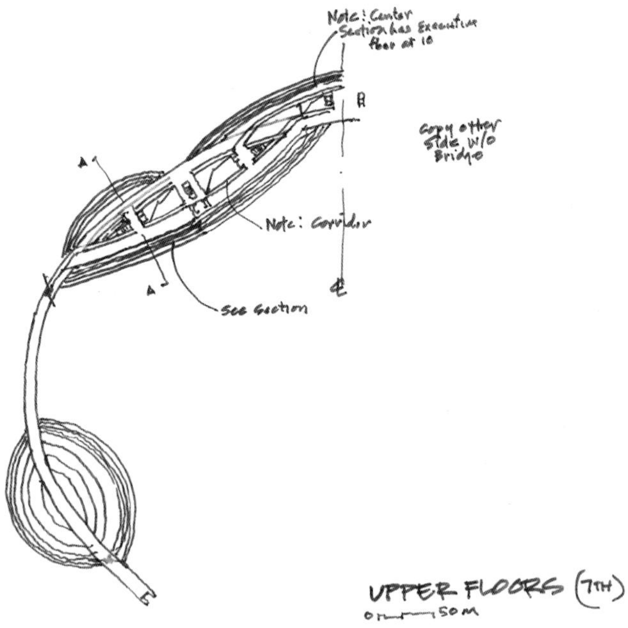

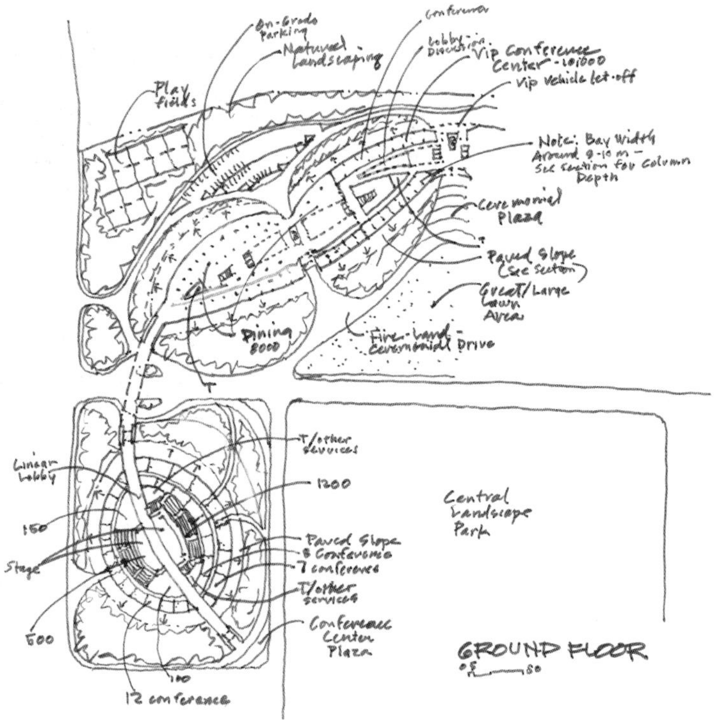

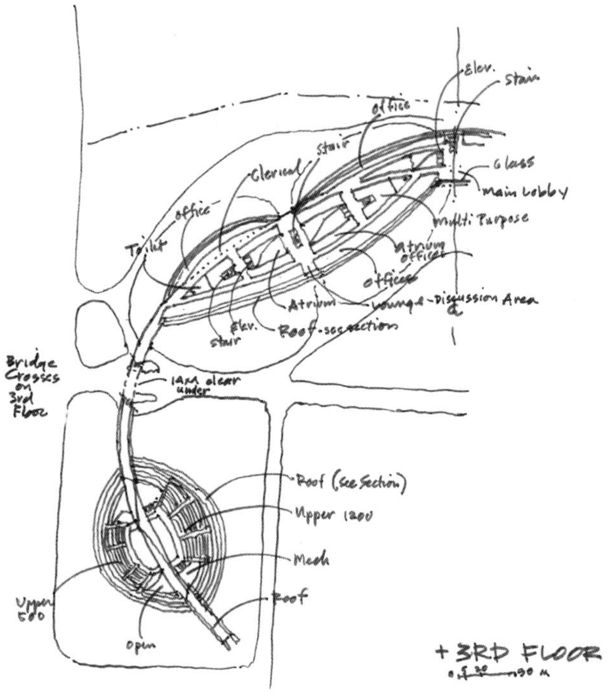

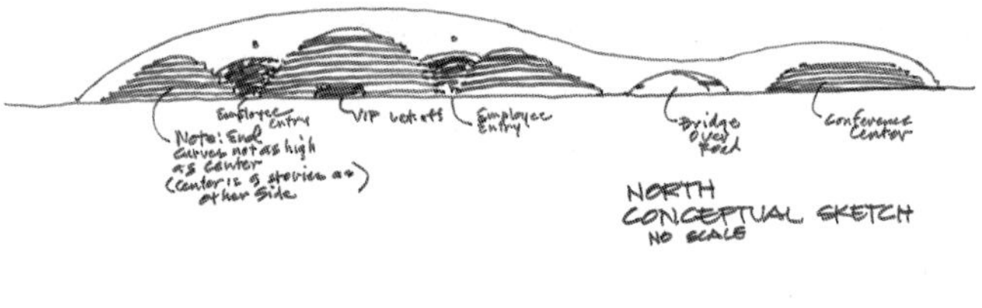

North Elevation 北立面

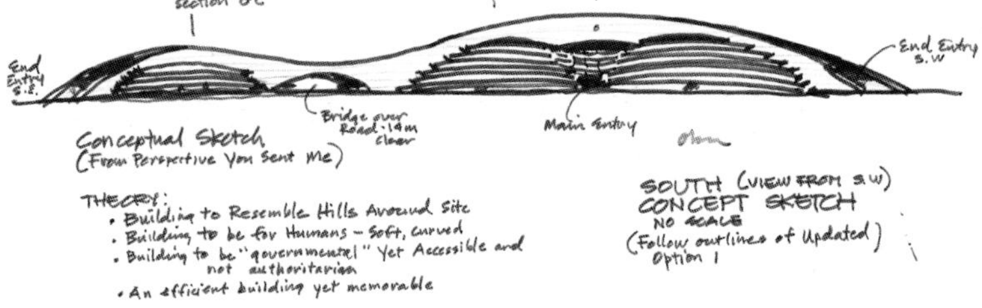

South Elevation 南立面

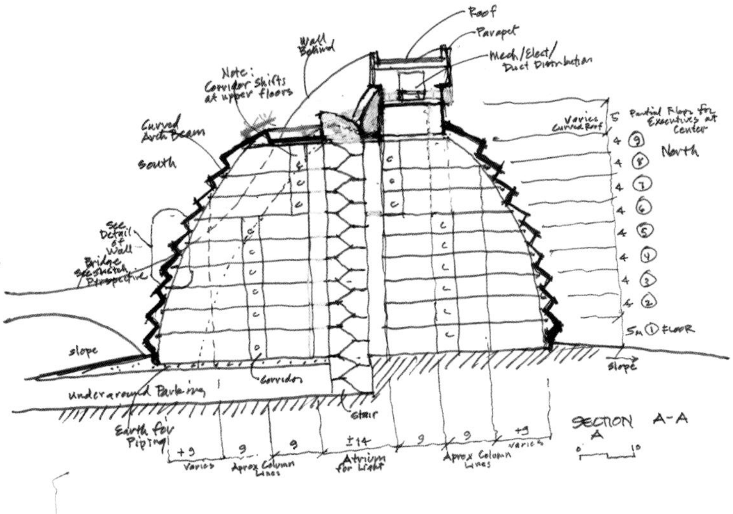

Section A-A 剖面图 A-A

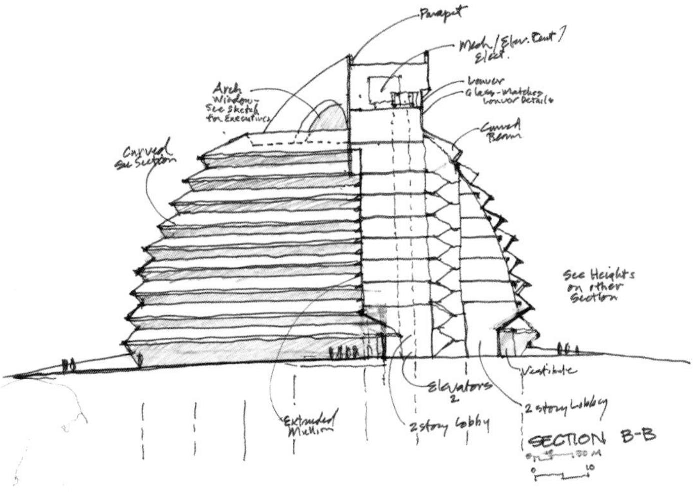

Section B-B 剖面图 B-B

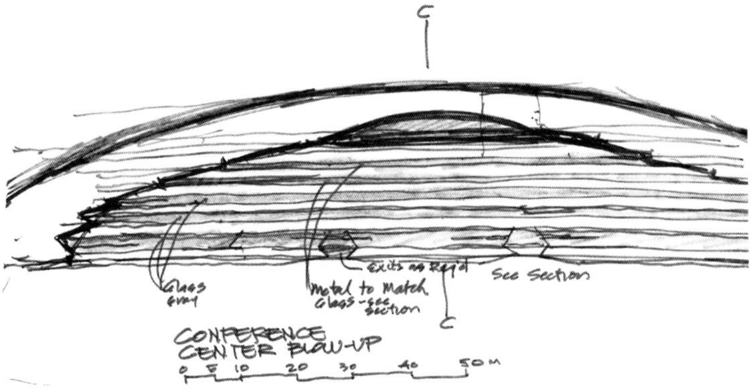

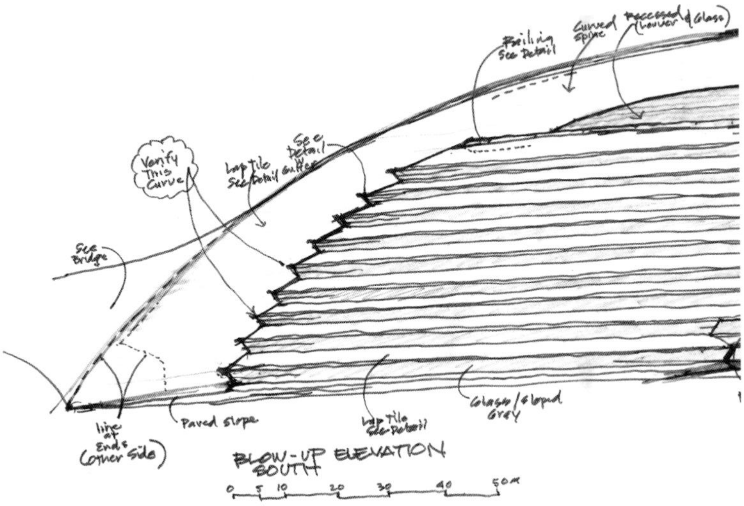

Elevation 立面图

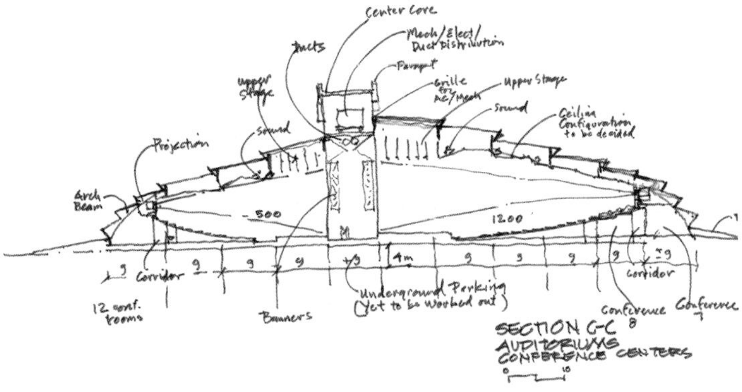

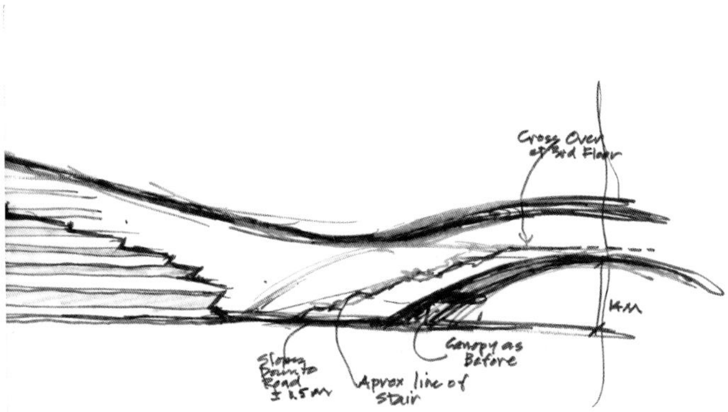

Section C-C 剖面图 C-C

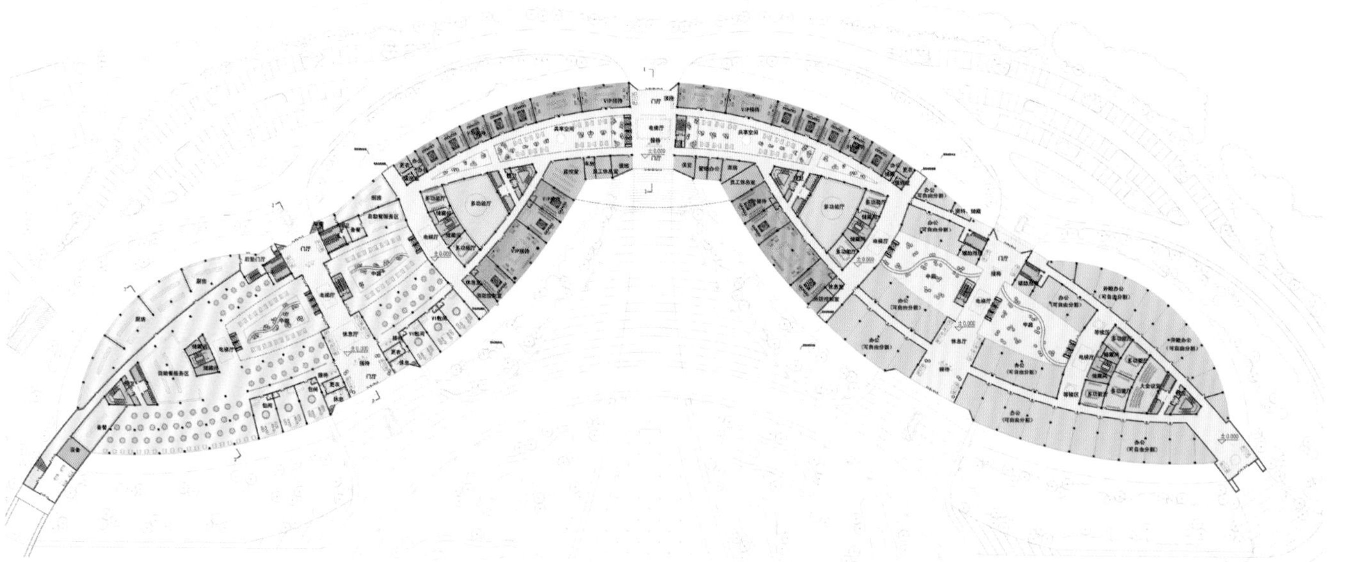

First Floor Plan 一层平面图

- 接待中心
- 餐厅
- 办公空间
- 交通
- 公共空间
- 多功能室/会议室
- 辅助空间

餐厅

接待中心

中庭

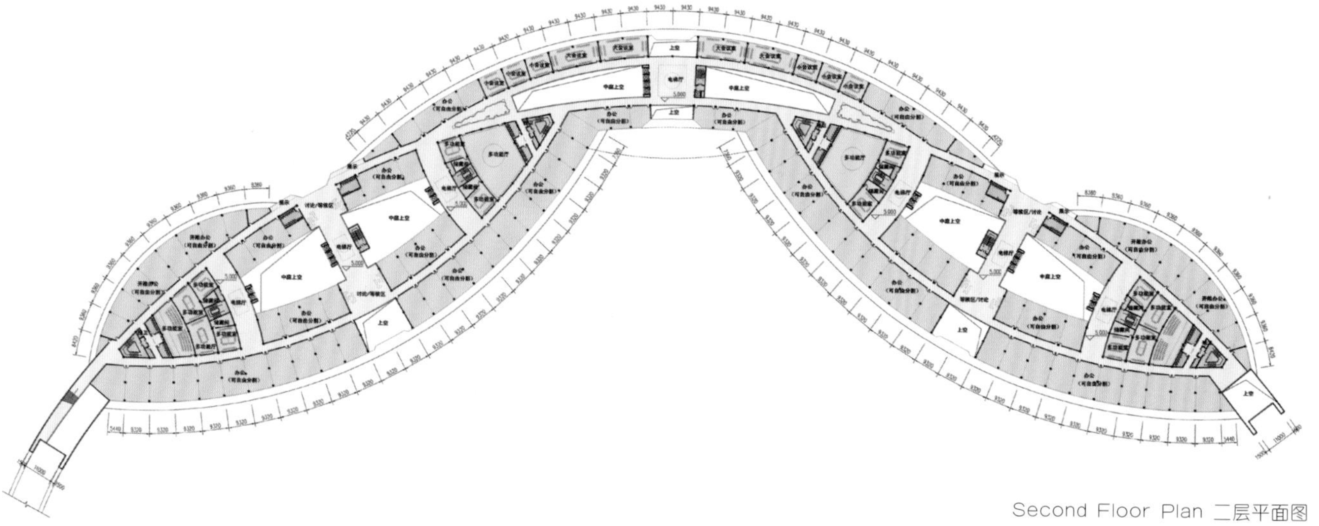

Second Floor Plan 二层平面图

- 办公空间
- 交通
- 公共空间
- 多功能室/会议室
- 辅助空间

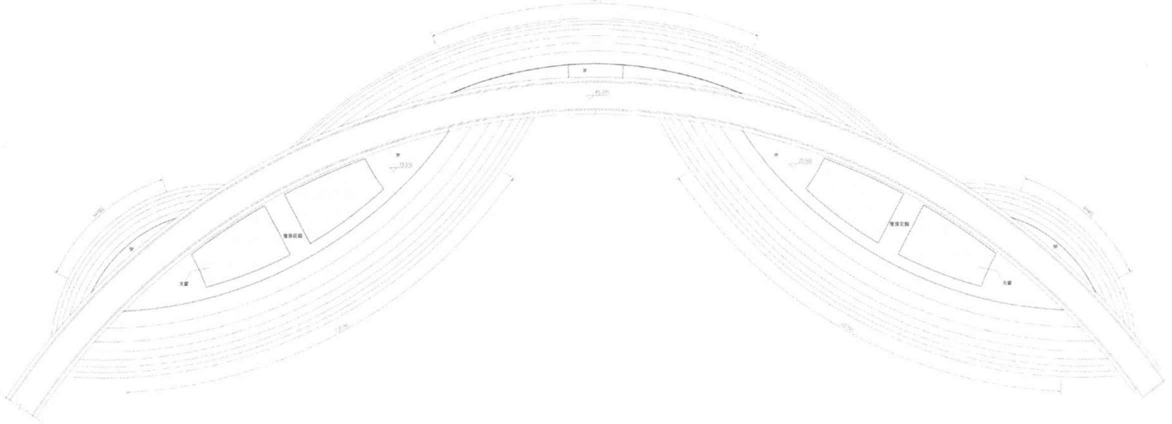

Roof Floor Plan 屋顶层平面图

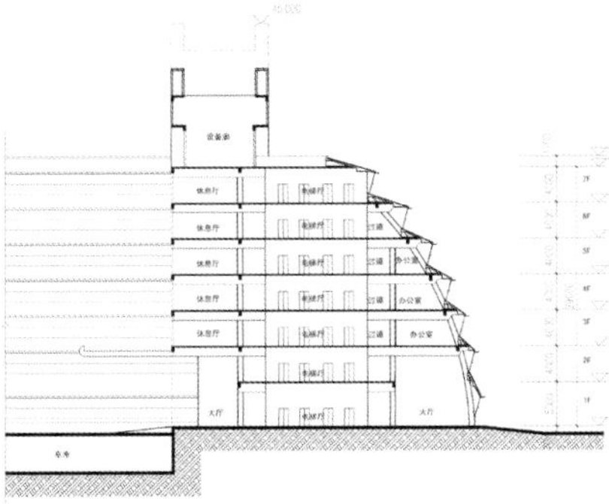

1-1 Section 1-1 剖面图

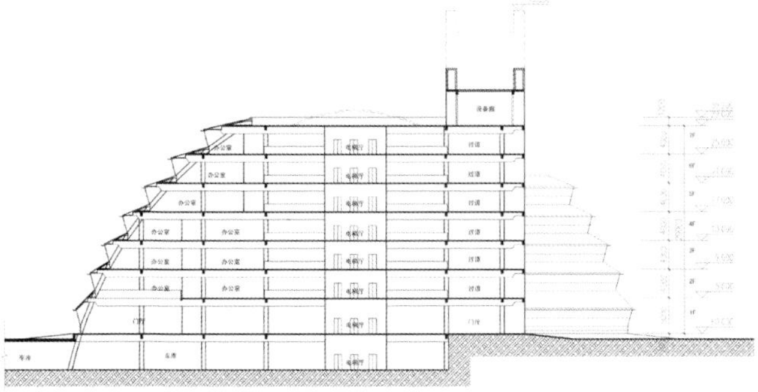

2-2 Section 2-2 剖面图

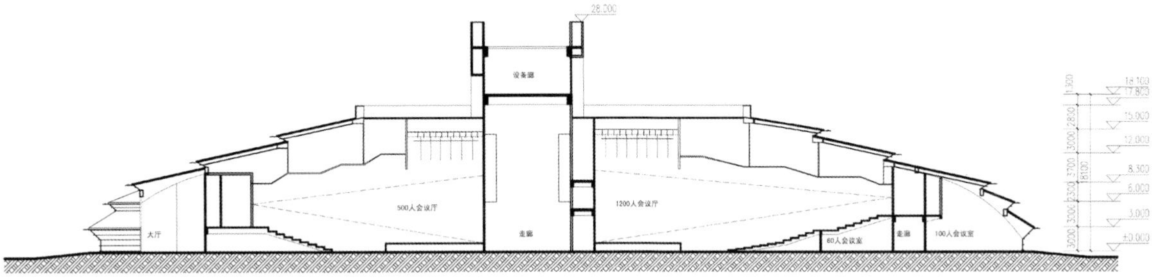

Conference Center for Section 会议中心剖面图

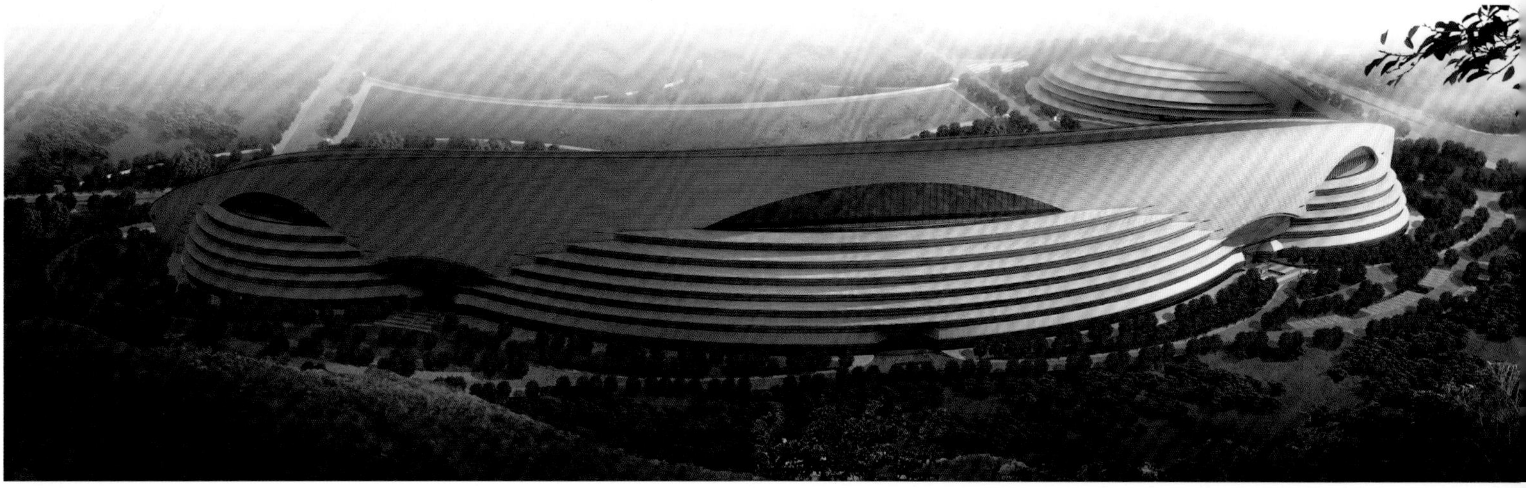

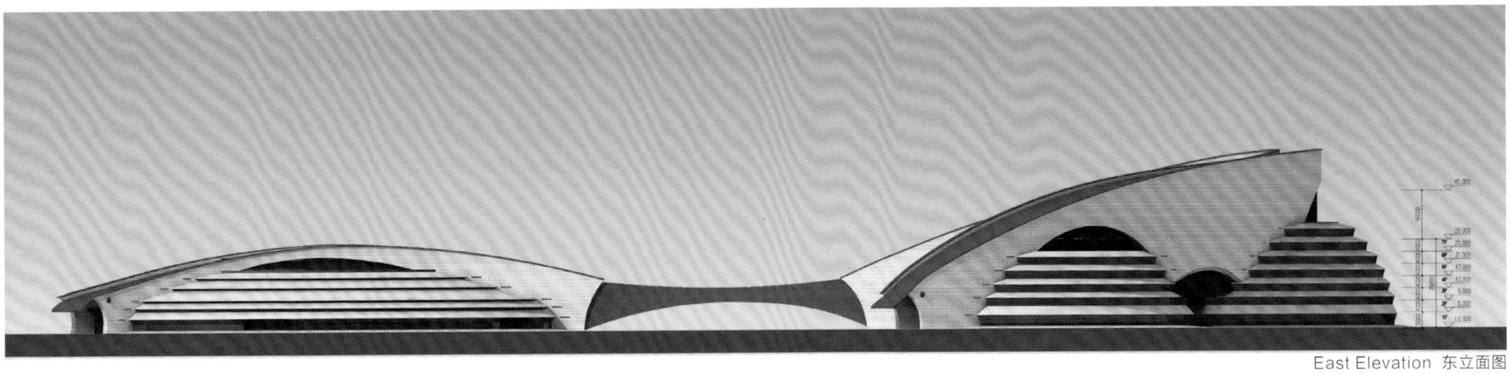
East Elevation 东立面图

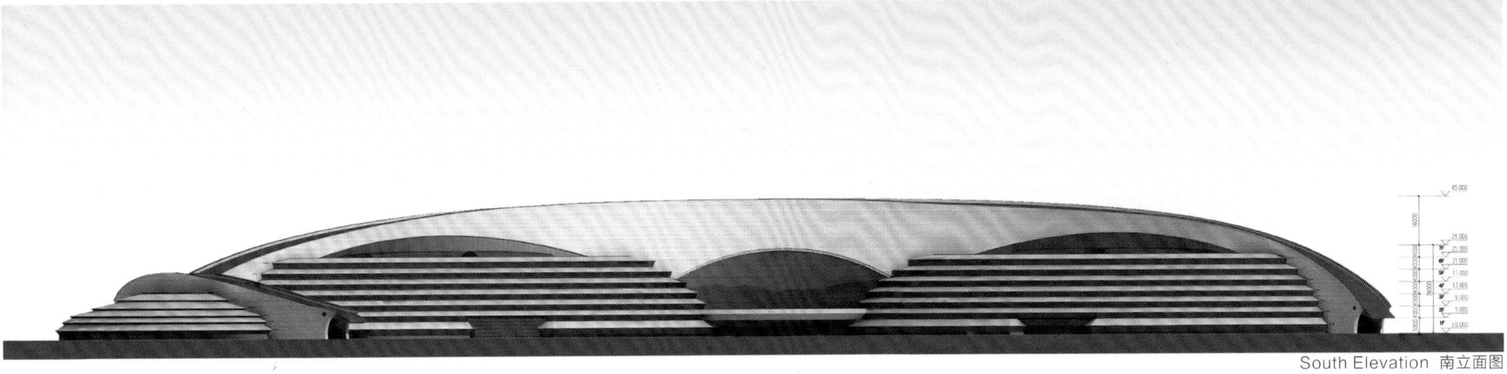
South Elevation 南立面图

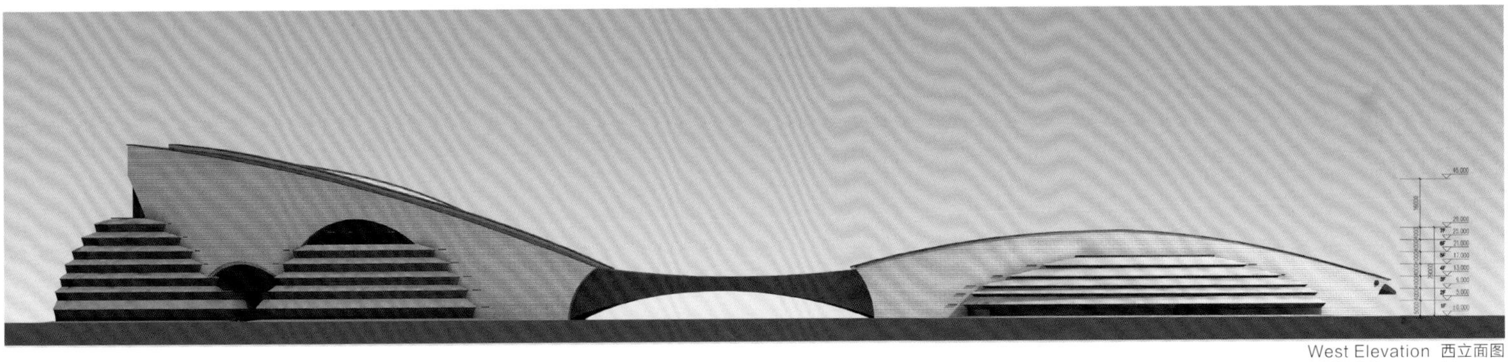
West Elevation 西立面图

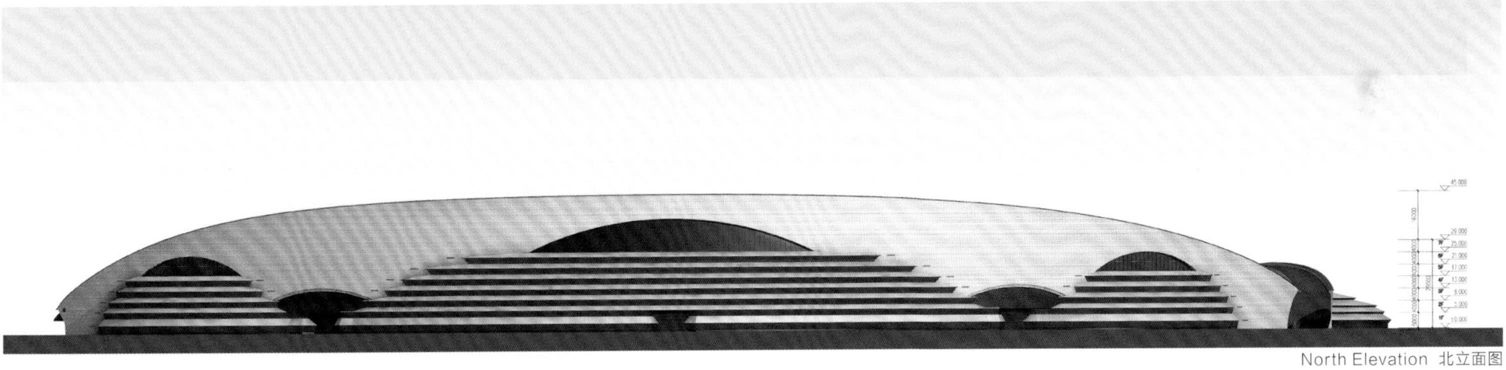
North Elevation 北立面图

CDB Tower and Minsheng Financial Tower
国银金融中心大厦 - 民生金融大厦

Keywords 关键词

Green & Low-carbon
绿色低碳

Dome
穹顶

Space Joint
空间节点

Location: Shenzhen, Guangdong, China
Architectural Design: ZHUBO DESIGN CO., LTD.
Floor Area: 100,000 m^2
Design: 2011

项目地点：中国广东省深圳市
设计单位：筑博设计股份有限公司
建筑面积：100 000 m^2
设计时间：2011年

Features 项目亮点

Boasting height advantage, CDB Tower looks upright. While Minsheng Financial Tower looks stable and boasts wider landscape view. They reach a balance on visual impact.

国银金融中心大厦利用高度优势凸显挺拔感，民生板式塔楼则因面宽的加大获得稳重感和更广的景观视野，两座塔楼在视觉上形成一种均衡。

■ Overview

This project is located close to the intersection of Haitian Road and Fuzhong 3rd Road in the east area of Northern Futian CBD in Shenzhen. The twin towers belong respectively to China Development Bank and Mingsheng Bank, boasting overground floor area of 60,000 m^2 and 40,000 m^2. Each tower contains two parts, office and commercial, and they share the basement.

■ 项目概况

本项目位于深圳福田区CBD北区东片区，海田路、福中三路路口。双塔分别属于国家开发银行和民生银行，其地上建筑面积分别为60 000 m^{2}和40 000 m^2。每座塔楼均包含办公和商业两部分，地下室合建，共享使用。

■ Design Concept

The project is in the central region, which on one hand leads to a narrow display surface for the towers along city main road, and on the other hand requires them to do something for the public space and lifestyle in the surrounding area. Therefore, the focus of this project is not to highlight the avant-garde building image, but to provide two convenient and efficient office towers and create an active urban node.

In the center of the site, a passageway opening to the public is designed to connect the low-storey retails of the Phenix Mansion with the commercial area that will be built in the north. Along the passageway, commercial facilities are arranged to activate the business spaces. At the south end of the development, the entrance to the passageway is expanded to connect with the urban square. And a public hall is designed nearby the hallway of the twin towers. This ceremonial hall embodies the building' s urban identity and highlights the social responsibility of China Development Bank (CDB) and China Minsheng Bank.

CUC

Aerial View 鸟瞰图

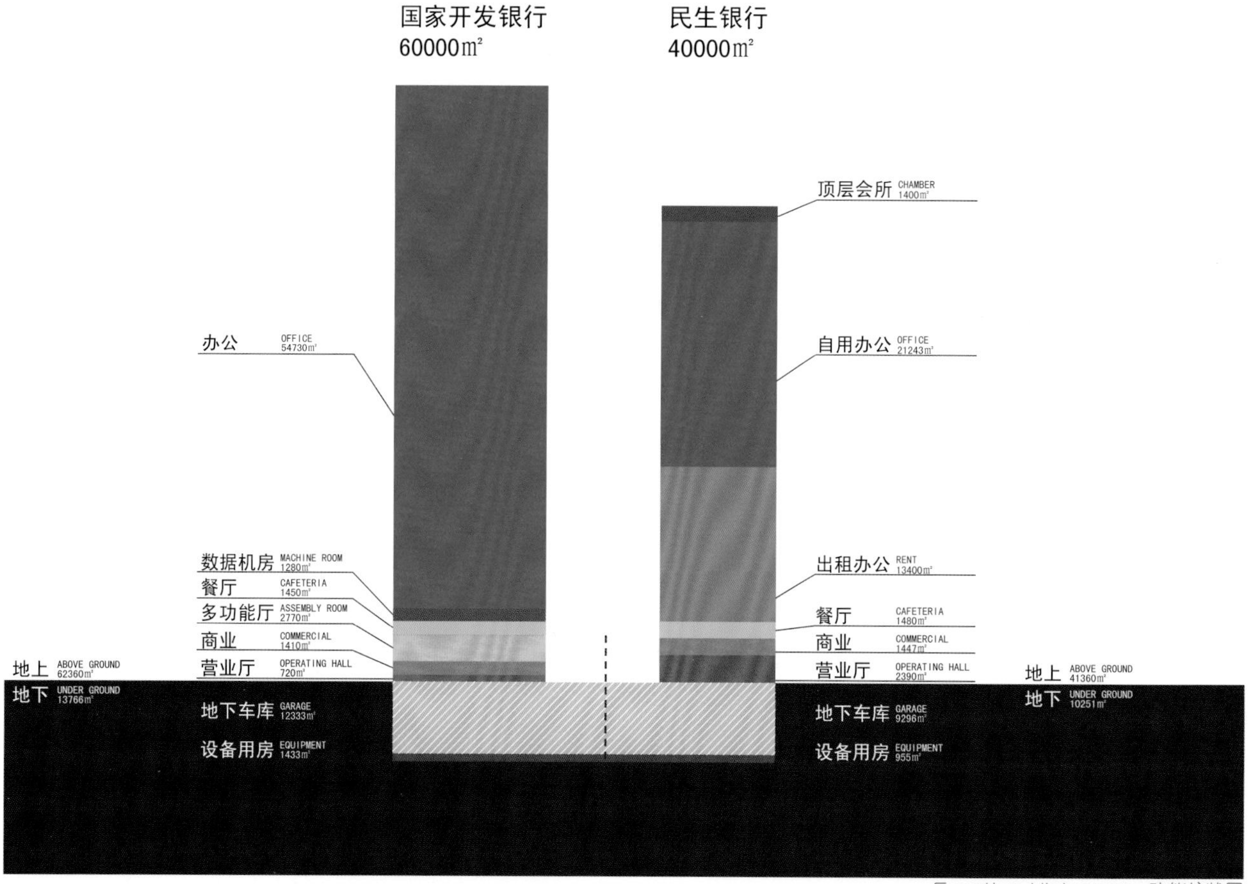
国家开发银行
60000㎡
民生银行
40000㎡
办公 OFFICE 54730㎡
数据机房 MACHINE ROOM 1280㎡
餐厅 CAFETERIA 1450㎡
多功能厅 ASSEMBLY ROOM 2770㎡
商业 COMMERCIAL 1410㎡
营业厅 OPERATING HALL 720㎡
地上 ABOVE GROUND 62360㎡
地下 UNDER GROUND 13766㎡
地下车库 GARAGE 12333㎡
设备用房 EQUIPMENT 1433㎡
顶层会所 CHAMBER 1400㎡
自用办公 OFFICE 21243㎡
出租办公 RENT 13400㎡
餐厅 CAFETERIA 1480㎡
商业 COMMERCIAL 1447㎡
营业厅 OPERATING HALL 2390㎡
地上 ABOVE GROUND 41360㎡
地下 UNDER GROUND 10251㎡
地下车库 GARAGE 9296㎡
设备用房 EQUIPMENT 955㎡

Function Histogram 功能柱状图

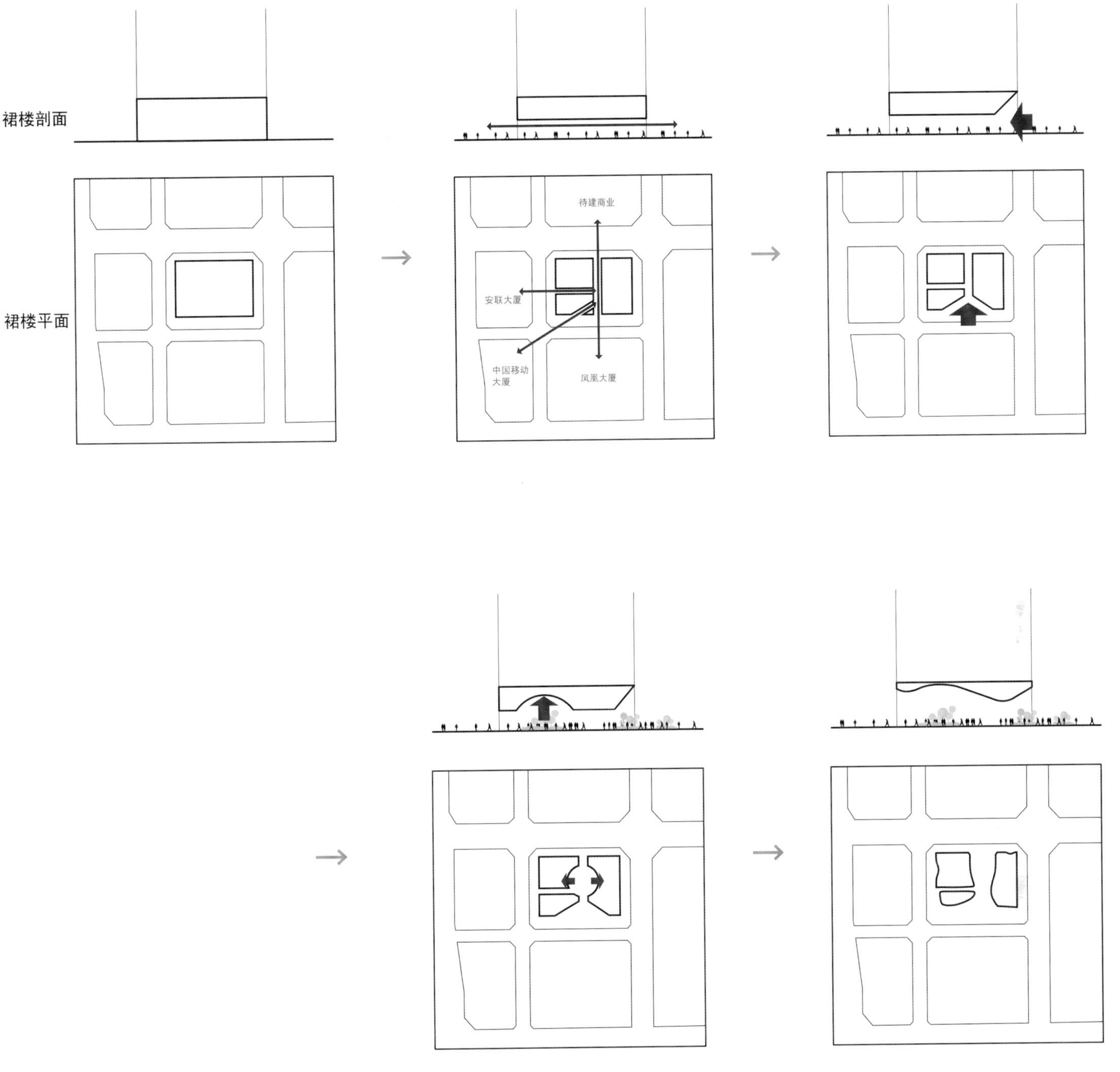

Podium Building Concept-The Hall Shared With the City (Concept Generation)
裙楼概念——与城市共享的大厅（概念生成）

■ 设计理念

本项目位处片区腹地，这个特点一方面导致塔楼沿城市主干道的展示面狭小，形象不突出；另一方面又使得它对片区公共空间和生活模式的辐射作用尤为重要。因此，本项目的重点不在于凸显标新立异的塔楼，而在于提供两栋高效实用的办公建筑的同时为城市营造便利的生活节点。

本案在项目用地的中央切出对城市开放的通道，把凤凰大厦的底层商业和北面地块的待建商业区联系起来。沿通道配置办公配套商业，激活空间。建筑在南端后退，扩大通道开口，与城市设计要求的广场融为一体；在两栋塔楼的门厅附近则形成塔楼与城市共享的大厅。这个大厅有很强的建筑主入口仪式感，同时又具有城市的属性，体现了国银、民生两家银行强烈的社会责任感。

■ **Architectural Design**

Complying with the requirement of urban design, the towers are situated in the north of the lot. To prevent visual interference from the Phoenix Mansion gable, Minsheng Financial Tower is designed in a slab type. Boasting height advantage, CDB Tower looks upright. While Minsheng Financial Tower looks stable and boasts wider landscape view. They reach a balance on visual impact that avoids the impression of primary and secondary.

In accordance with the summer solar radiation, designers use Low-E glass with different transmission rates on the facade to meet the green and low-carbon requirement. Vertical shutters are rotated in different angles for sunshade and view, and to stagger the sights of two towers. Inside, the office staff could feel the comfortable environment; outside, the public could appreciate the simple and subtle facade.

The hall is streamlined, which leaves an impression of sculpture when seen from the southern city square. The solemn office annex building is transformed into an intimate and lively space. Central node is presented as a unique dome to show the imposing image of the bank building and forms a clear spatial sequence with the interior hallway. Peripheral annex adjacent to the road is extended along the regular interface to keep the solemn of the central business district.

■ 建筑设计

塔楼布局遵从城市设计的要求，贴近基地北部。为避免凤凰大厦山墙对民生塔楼南面的视觉干扰，将民生塔设计为板楼。国银金融中心大厦利用高度优势凸显挺拔感；民生板式塔楼则因面宽的加大获得稳重感和更广的景观视野。两座塔楼在视觉上形成一种均衡，避免了一主一次的印象。

立面设计以夏季太阳辐射在建筑立面的分布为依据，采用透射率不同的低辐射玻璃，满足绿色低碳的诉求。竖向百叶为达到遮阳和视野的要求旋转为不同的角度，同时错开两楼视线，使大楼内部给办公人群创造了舒适的物理环境和心理感受，外部形成简洁而微妙的立面效果。

城市大厅呈流线型，从南面城市广场的角度看具有强烈的雕塑感，把严肃的办公裙楼转化为一个亲切而活跃的地方；中心节点处以独特的穹顶姿态展现银行建筑的气势，并与建筑内部的门厅组成清晰的空间序列；毗邻道路的外围裙楼则延续方整的界面，避免破坏中心商务区的庄重气质。

Shenzhen Nanhai Scenic Houhai
深圳南海御景后海

Keywords 关键词

Tensile Structure
形体张力

Facade Design
立面设计

Sky Garden
空中花园

Features 项目亮点

In designing the rectangle tower, designers adopted the techniques of twisting and beveling that create flexible hanging gardens and extend the dynamic of the central vortex, thus to unite the changing space and the harmonious shape as one.

矩形塔楼运用斜切、扭转、衍生等设计手法，创造出灵活多变的空中花园，延续了中心旋涡动感汇聚的趋势，流转变化的空间与和谐统一的造型融为一体。

Location: Shenzhen, Guangdong, China
Architectural Design: CAPA Design
Total Land Area: 24,860 m^2
Total Floor Area: 401,410 m^2
Plot Ratio: 18.41

项目地点：中国广东省深圳市
建筑设计：美国开朴建筑设计顾问（深圳）有限公司
总用地面积：24 860 m^2
总建筑面积：401 410 m^2
容 积 率：18.41

■ Overview

This project aims to create such a space where people can exchange freely and unbounded, relax physically and mentally, and life of different dimensions can be vitalized. There are a variety of vortexes in nature, which unite the surroundings with their natural tension and then present themselves perfectly. Inspired by vortex, designers not only exhibit the unique characteristic of the building group, but also create a powerful control for the entire plot through the rotating form.

■ 项目概况

项目旨在创造这样一种空间，人们可以自由无界地交流，身心情绪可以得到充分的休闲释放，不同维度的生活在这里充满活力。自然界中存在各种各样的漩，它有一种天生的张力，将周围紧密联系，将自身完美呈现。项目将漩涡引申为方案灵感的来源，不仅表现出建筑组群自身独特的个性，同时通过旋转的形态使其对整个片区产生强大的控制力。

■ Logic of Form

Central roundabout integrated the four split lands, which shaped a brand new internal relation inside the plot. Annex buildings around the roundabout are splitting to form recessive volumes that decline level by level. Three groups of overpasses are constructed centripetally that strengthen the interaction between the land and the city further. In order to prevent stiff connection between the annexes and the tower, linked symbiotic tension is generated while cutting and twisting the tower.

■ 形体逻辑

通过中心环岛的设定，整合了四块分置的用地，全新地塑造了用地的内在联系。围绕中心环岛，裙房分割体量形成了逐层跌落的退台，从而向中心发散衍生出三组过街天桥，进一步强化了用地与城市的联系和互动。为避免裙房与塔楼的生硬连接，在塔楼的体量切割及扭转交接上，上下和谐统一形成了联动共生的形体张力。

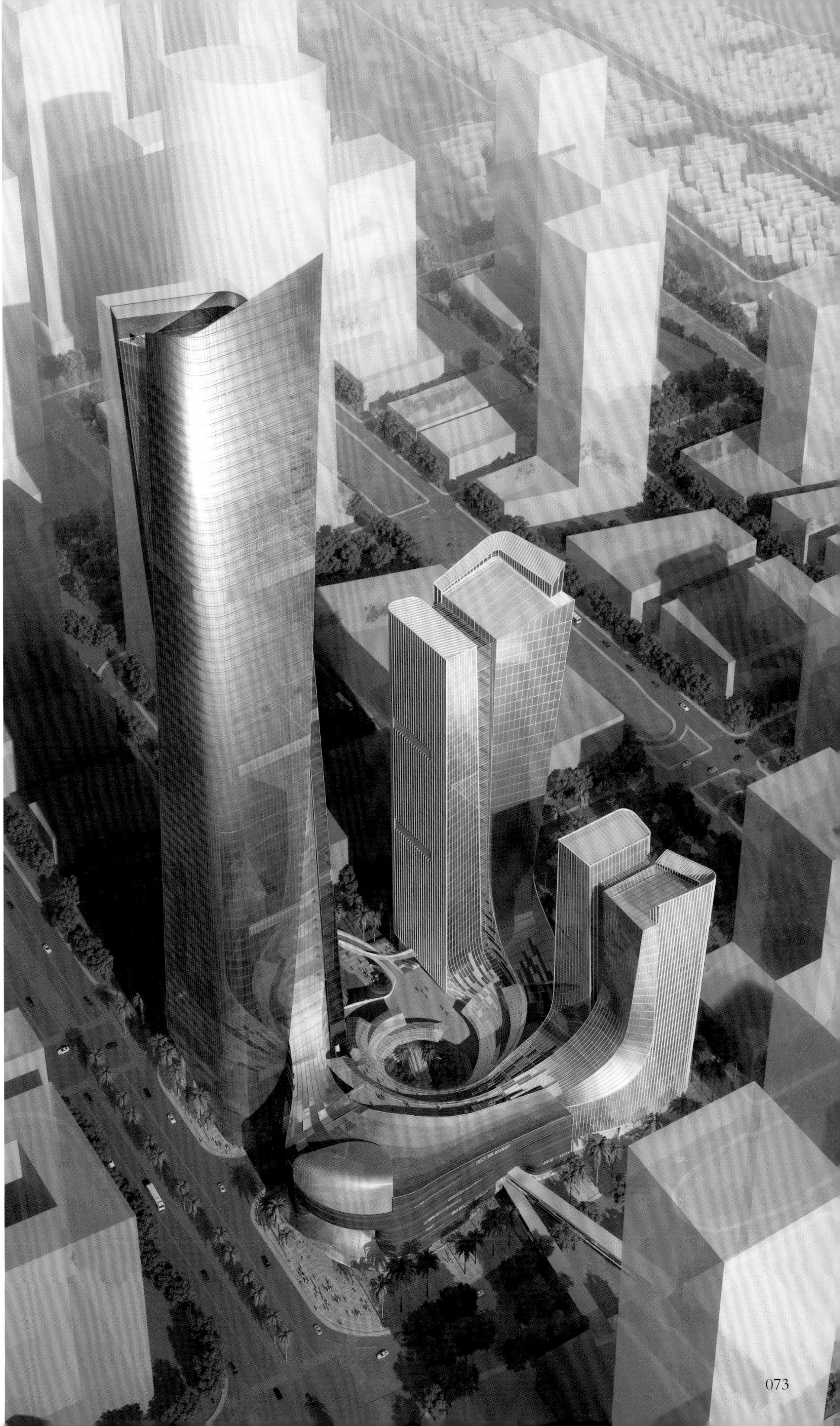

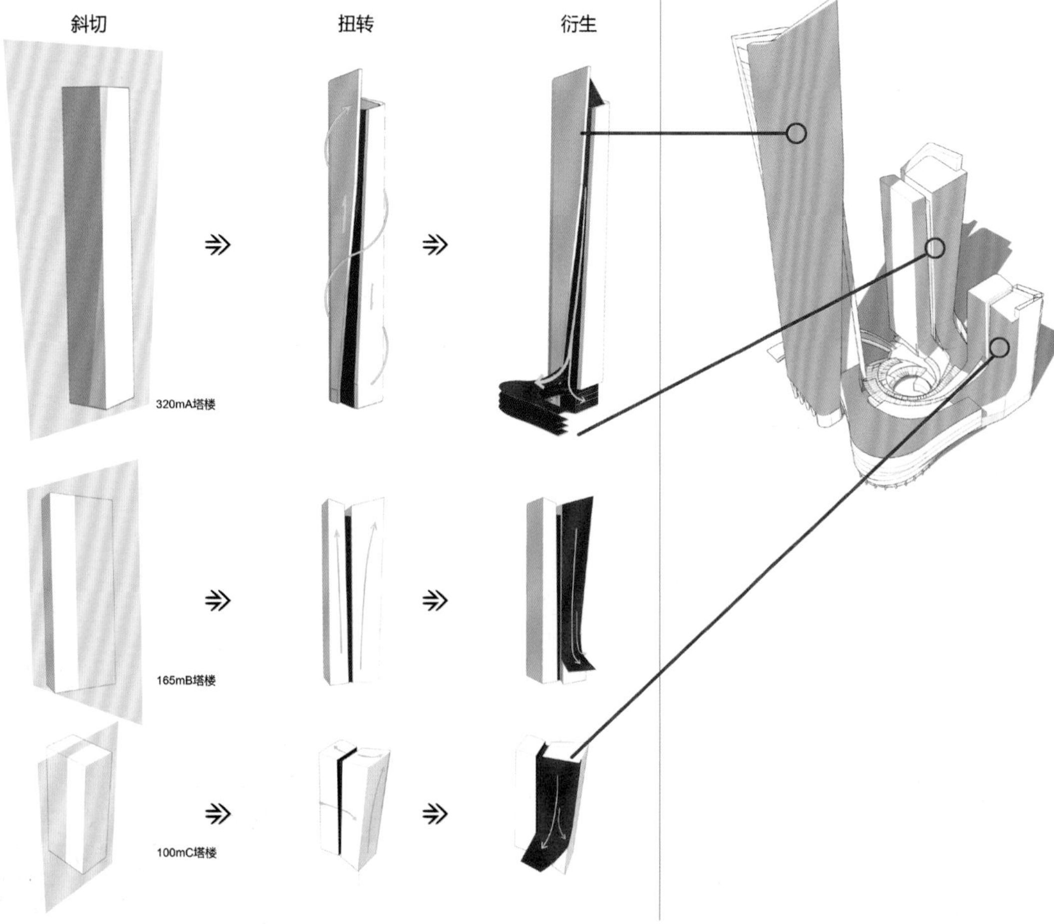

Office　办公

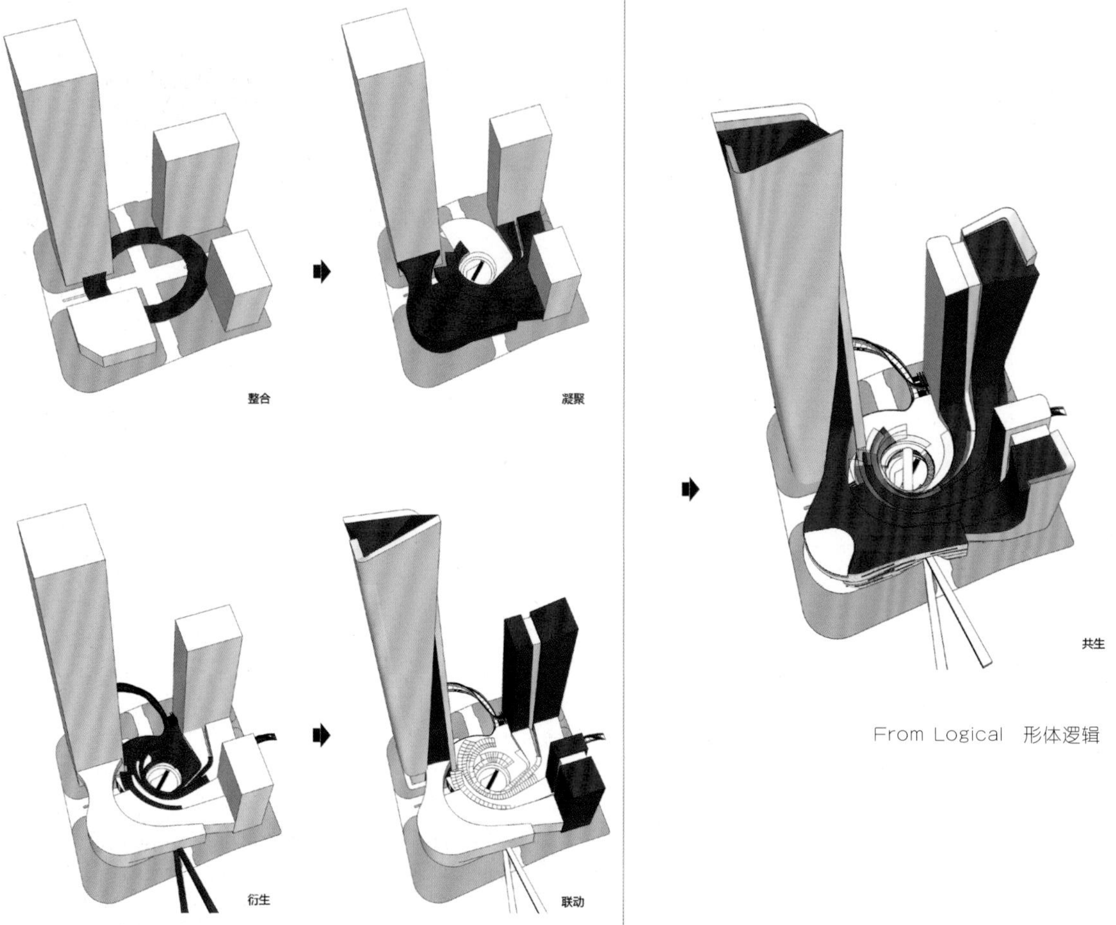

From Logical　形体逻辑

Architectural Design

n designing the office tower, designers adopted the :echniques of twisting and beveling which not only create flexible hanging gardens but also show a stylized :ower shape. The tower shares a unified facade with the commercial annex, so that people can feel the changing spatial layout and the harmonious shape characteristics.

建筑设计

在设计办公塔楼的过程中，设计师运用扭转、斜切等手法，不仅创造出灵活多变的空中花园，同时也呈现出独具一格的塔楼造型。塔楼表皮与商业裙房相接，形成完整统一的立面造型，使人们在建筑中感受到流转变化的空间布局及和谐统一的造型特征。

VERSACE
AMONNI
Dior
Be Jeweled
SEIKO
COACH
FENDI
BOUCHERON
ESTEE LAUDER
IVECO
Chanel
ANNASUI
ARMANI

LOUIS VUITTON

Sichuan Media Group Industry Center
四川传媒集团·西部文化产业中心

Keywords 关键词

Glass Curtain Wall
玻璃幕墙

Stone
石材

Green Roof
绿化屋面

Features 项目亮点

The facade of the tower is well designed to meet different sun-shading requirements, presenting different facade effects and get the maximum use of the daylight.

塔楼外观结合不同的遮阳需求设计出不同的立面肌理，达到不同的生动的立面效果，同时将日光利用最大化并减小潜在的热量增幅。

Location: Chengdu, Sichuan, China
Architectural Design: KAZIA.LI Design Collaborative
Total Floor Area: 93,193.40 m^2

项目地点：中国四川省成都市
建筑设计：天津凯佳李建筑设计事务所
总建筑面积：93 193.40 m^2

■ **Overview**

Phase two of the Sichuan Media Group Complex is located at the northwest corner of the site, and consists of a 173 m tower with retails, SOHO, and 5A office functions as well as the adaptive reuse of an existing building to the south. The site circulation must be considered with respect to public and private entries, and underground parking should link the buildings.

■ **项目概况**

本案为四川传媒集团项目的第二阶段，位于地块的西北角，主体为一栋173 m的大楼，整体包括零售、SOHO、5A写字楼功能，以及地块南部的现有建筑物再利用。该地块内部流线必须考虑到公共和私人区域，并且需要和地下停车场联系起来。

■ **Design concept**

In keeping with the design concept of phase one, phase two is also designed like a newspaper rolled up and held in hand by an avid reader. It expresses our persistence on paper media. Unlike phase one to be easily accessible, this 173 m high landmark building will remind us of the signal tower, which symbolizes the wide and rapid transmission of information in modern society.

■ **设计理念**

方案的灵感来自对一期方案的延续。同样是一份报纸，卷起它，拿在手里，舍不得放开，表达对纸质信息媒介的坚持。一期方案体现一种读者近距离阅读的感觉，而本项目是一栋高达173 m的地标大楼，将之与信号塔联系起来，形成信息传播范围更大更广的视觉感受。

SMG
西部文化产业中心

SMG 四川传媒集团
西部文化产业中心

East Elevation 东立面

North Elevation 北立面

South Elevation 南立面

West Elevation 西立面

■ Office Environment

The goal for this building is to create a comfortable and healthy environment for the end users while reducing energy use and providing ultimate flexibility. With an understanding to the local climate, the architects proposed a mixed ventilation system by mechanical VAV and natural wind, providing comfort and saving energy. An open floor is incorporated to help accommodate the ventilation system and also to provide great flexibility for office arrangement. The use of hollow slabs can reduce depth of the interior space and get maximum day lighting.

■ 办公环境

设计的初衷是为最终的使用者创造舒适健康的办公环境，同时又能减少建筑使用过程中的能耗并保证室内空间的灵活性。根据对当地气候的理解，在设计中采用了人工与自然混合的通风系统，在节约能耗的同时提供了舒适的环境。设计同时采用了架空式楼板系统适宜通风系统的同时使室内空间在使用布置中获得更大的灵活性。采用空心楼板可以减少建筑进深以获得最大程度的采光。

Site Plan 总平面图

SMG
Dior
TOD'S

■ **Building Materials**

The structure is mainly built with reinforced concrete and supplemented by steel. The facade materials are well selected to keep harmonious with the surrounding buildings. Warm colored stones, glass curtain walls and metal frames are combined together to match buildings of phase one with the same colors. Moreover, glass and metal pergolas are designed to connect new and old buildings, which presents colorful facade effect.

■ **建筑材料**

结构部分主要采用钢筋混凝土为主，钢结构为辅。立面选材为与周边建筑形成协调与对比，墙面部分采用暖色石材与玻璃幕墙结合，部分使用金属框料，颜色采用与一期建筑相协调的颜色，同时配有部分玻璃金属廊架，与原有建筑形成组合，打造丰富的立面层次与富有变化的材料质感。

■ **Green Roof System**

The podium and part of the tower use green roof system to make full use of the spaces and provide more green landscapes. Green roof plants can increase the oxygen content in the air and reduce radiant heat on the roof. The use of local plants can improve the survival rate and reduce the maintenance cost.

■ **绿化屋面系统**

本方案裙房屋顶及塔楼局部采用绿化屋面系统，不占用城市地面面积进行绿化。屋顶绿化的植物可以增加城市空气中的氧气含量、减少建筑物屋顶的辐射热。采用当地植物，提高成活率，可以减少维护成本。

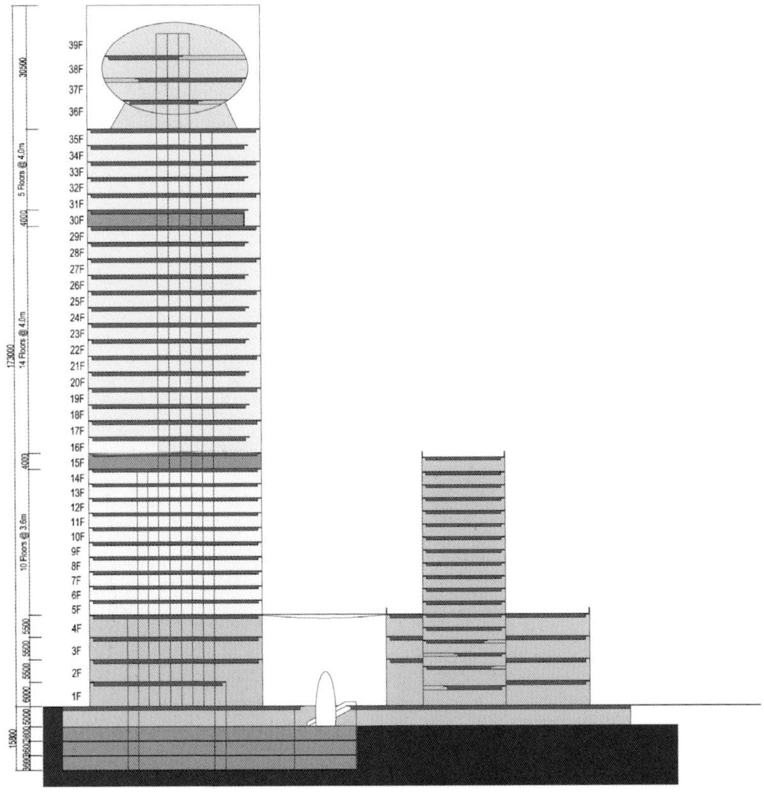

Section 剖面图

Taiyuan Coal Chemistry Center
太原煤化工技术研发大厦

Keywords 关键词

Double Curtain Wall System
双层幕墙系统

Green Roof
屋面绿化

Building Materials
建筑材料

Features 项目亮点

Through the axis control, the design integrates hotel, office, trading floor and other supporting facilities forming a clear framework for context.

设计对酒店、办公、交易大厅及其他配套进行功能整合，通过轴线控制，形成脉络清晰的架构。

Location: Taiyuan, Shanxi, China
Architectural Design: KAZIA.LI Design Collaborative
Land Area: 36,900 m^2
Total Floor Area: 124,238 m^2

项目地点：中国山西省太原市
建筑设计：天津凯佳李建筑设计事务所
用地面积：36 900 m^2
总建筑面积：124 238 m^2

■ **Overview**

This project is located in the business incubator of the Taiyuan National Hi-tech Industrial Park in Xiaodian District. It is nearby Jinyang Street in the north, and 15 m far from planning roads in the other directions, 8.6 km away from the city hall and 7 km to Wusu Airport. The site was built with a complex for convention and exhibition, office, hotel, R&D, trading and so on. Upon completion, it has become a landmark in this area.

■ **项目概况**

项目位于太原市小店区太原国家高新技术产业区的火炬创业园内，北临晋阳街，南、东、西邻创新创业园15 m规划路，距市政府8.6 km，距武宿机场7 km。该地块位于太原国家高新技术产业开发区内，是集会展、办公、酒店、研发、交易等多种功能为一体的综合性大厦。该项目是太原市高新区具有时代特色的标志性建筑。

■ **Overall Planning**

The complex is composed of two towers and one podium which accommodate hotel rooms, offices, trading floors and other supporting facilities. All of these functions are controlled by an axis to form a clear structure.

■ **总体布局**

整个基地由两个塔楼及一个裙房三个部分组成，设计对酒店、办公、交易大厅及其他配套进行功能整合，通过轴线控制，形成脉络清晰的架构。

■ **Facade and Materials**

Sitting on the central axis of the high-tech park, the complex is the core and icon of this area. Just like layers on the stratigraphic section, sunken cracks on the central axis also show history of different periods. Colors of the building materials change from deep and dark to light and bright as the building rising up, indicating the history from the past to the future. Glass symbolizes the future, showing that materials can change from particle to fluid, and then to solid. While stone represents the past, and it is the basic element of the earth and the core of coal-mining industry.

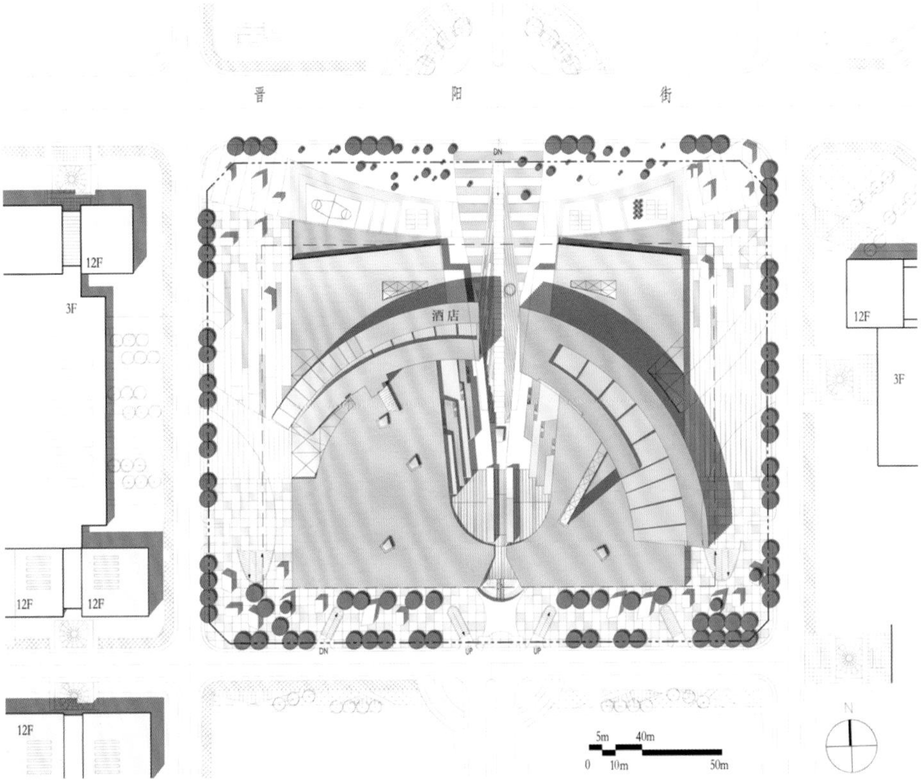

■ **建筑外观与材料**

建筑位于创业园区的中轴线上，是此区域的核心。就像被剖开的地层，中轴上的下沉裂缝所展现的代表不同年代的历史，显得厚重、有力。建筑主体由低逐渐升高，材料由厚重、深暗向轻巧、明亮过渡，预示着历史从过去向未来的演进。玻璃代表着未来，显示着材料从粒子到液体再到固体的转换。石头是过去的标志，是地球构造的基础和煤矿工业基础的核心。

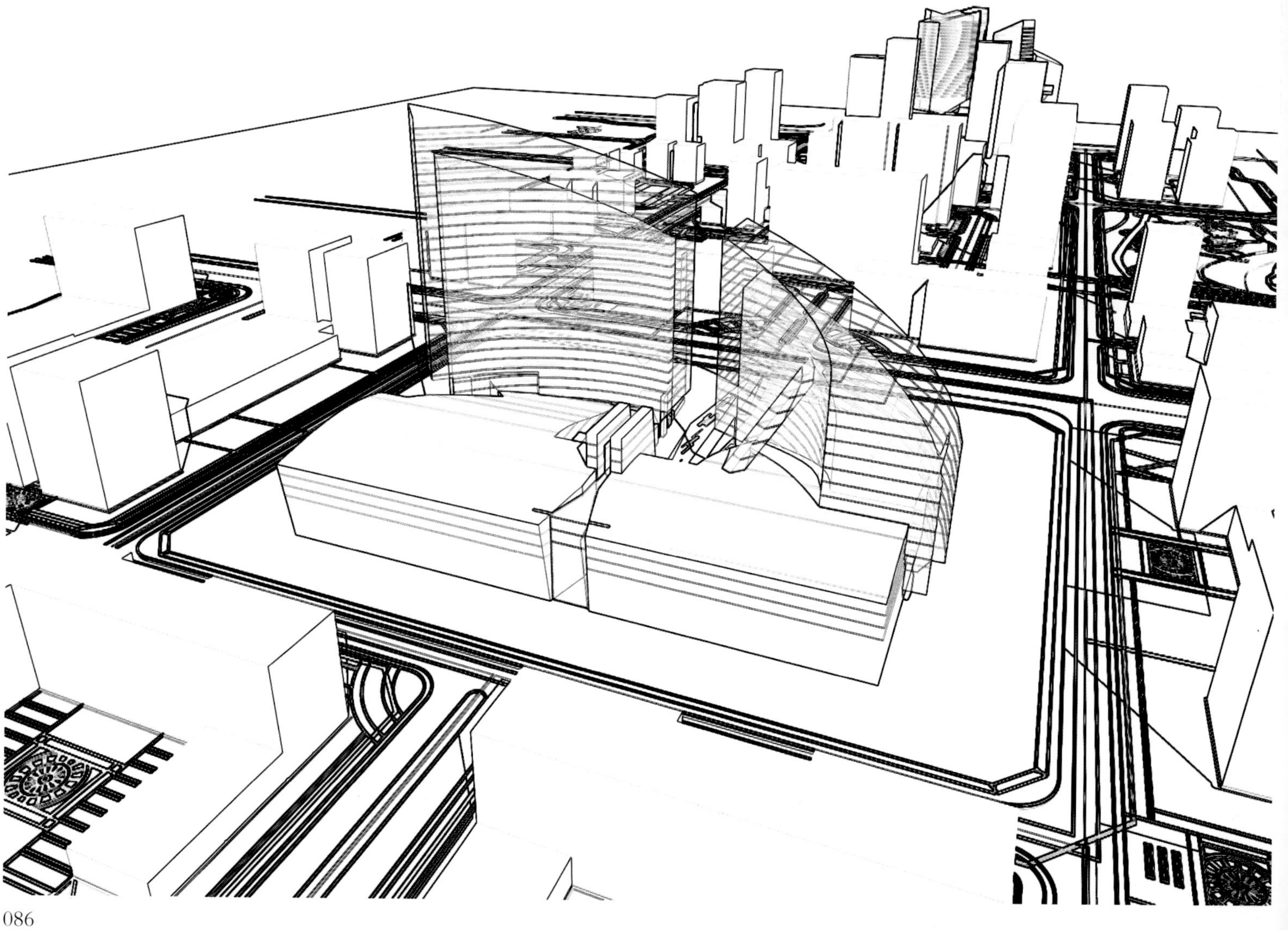

Energy-saving measures

1. Double curtain wall system: the program uses the double curtain wall system which s composed of metal mesh wall outside and glass curtain wall inside. Blinds between the double walls block most of the solar rays and release heat by natural ventilation system, which greatly reduce the energy consumption on cooling system.

2. Solar energy utilization: solar power system formed by solar modules is designed for public power use and lighting system. Thus it creates an environmentally friendly building with clean energy.

3. Green roof system: top of the podium and part of the tower are designed with green roof system which provides more green spaces, increase more oxygen and reduce thermal radiation to the building.

节能措施

1. 双层幕墙系统。此方案采用双层幕墙系统，外侧为金属网幕墙，内侧为玻璃幕墙。遮阳百叶在双层幕墙间，绝大部分太阳光线可被挡住，并通过自然通风系统排到建筑以外，大大减少室内空调制冷负荷。

2. 太阳能利用。在方案中设置公众用电到灯光系统用电的太阳能发电系统，使大厦成为清洁的环保建筑。太阳能发电系统由太阳能发电模块构成。

3. 绿化屋面系统。此方案裙房屋顶及塔楼局部采用绿化屋面系统，不占用城市地面面积进行绿化。屋顶绿化的植物可以增加城市空气中的氧气含量、减少建筑物屋顶的辐射热。

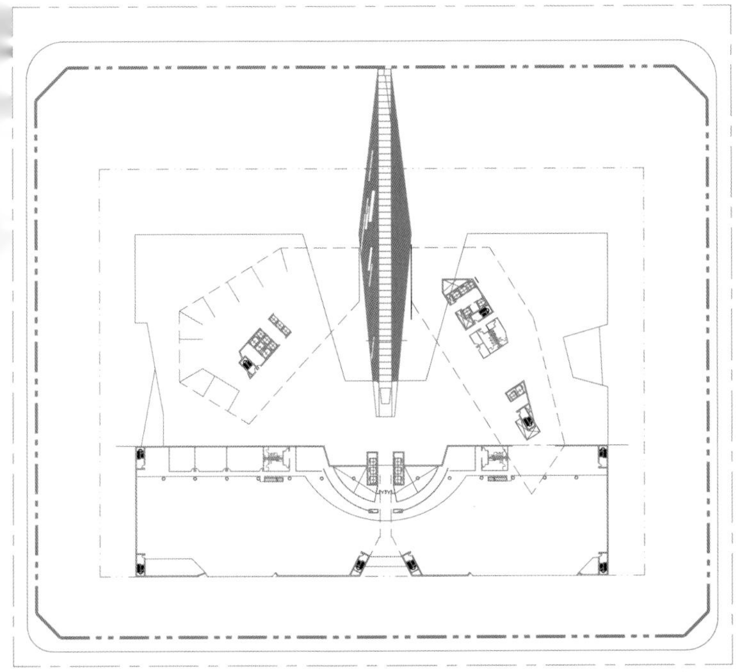

Plan of the First Basement Floor 地下一层平面图

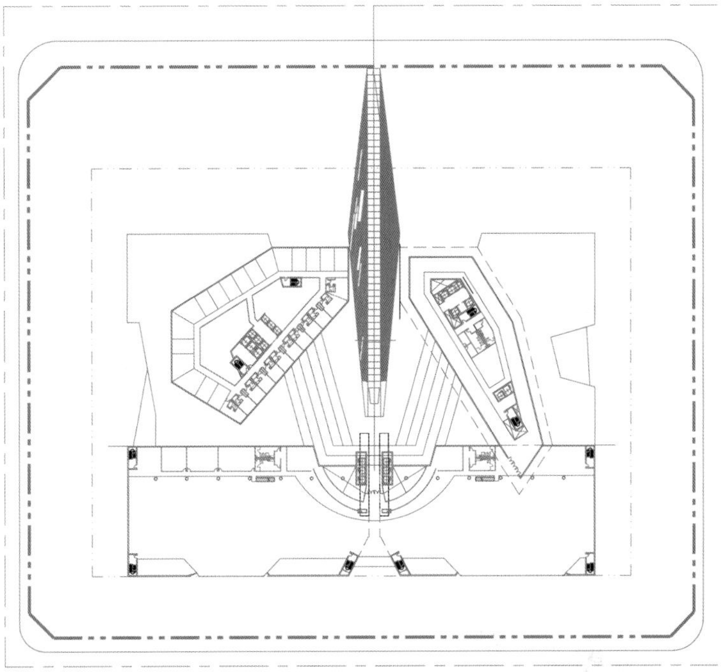

Ground Floor Plan 地面层平面图

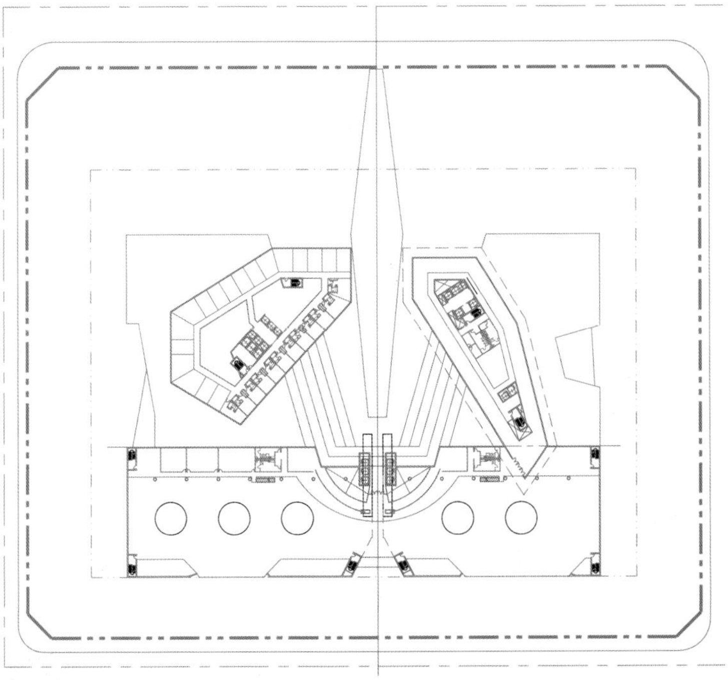

Lower Trading Floor Mezzanine 低层交易大厅夹层

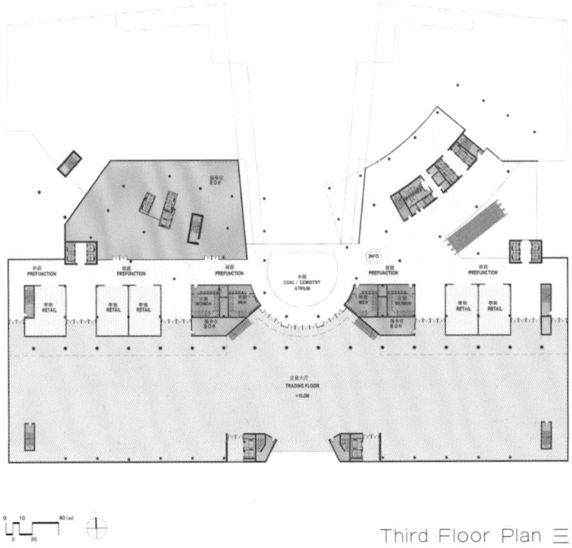

Third Floor Plan 三层平面图

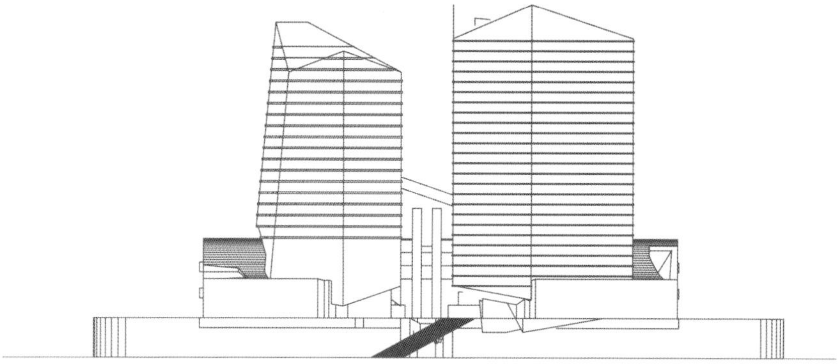

North Elevation 北立面图

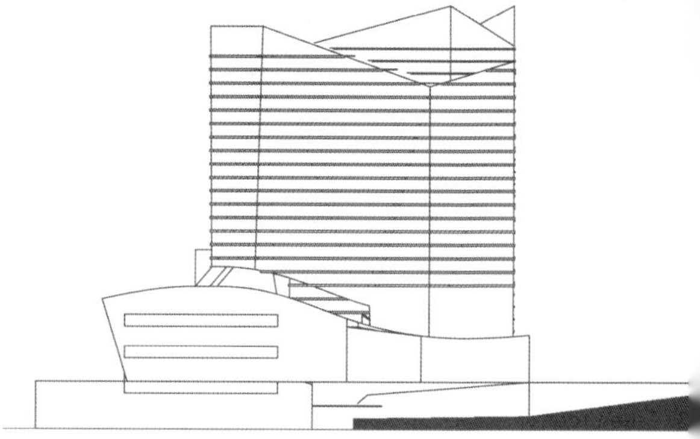

East Elevation 东立面图

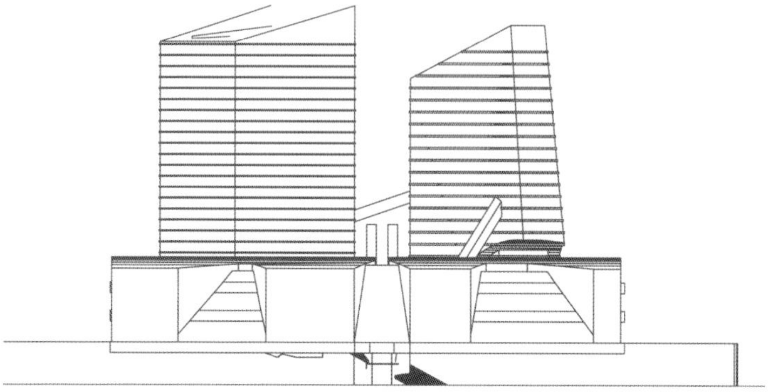

South Elevation 南立面图

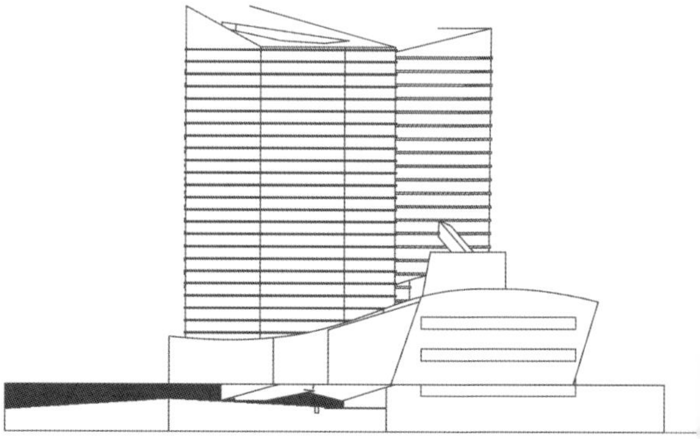

West Elevation 西立面图

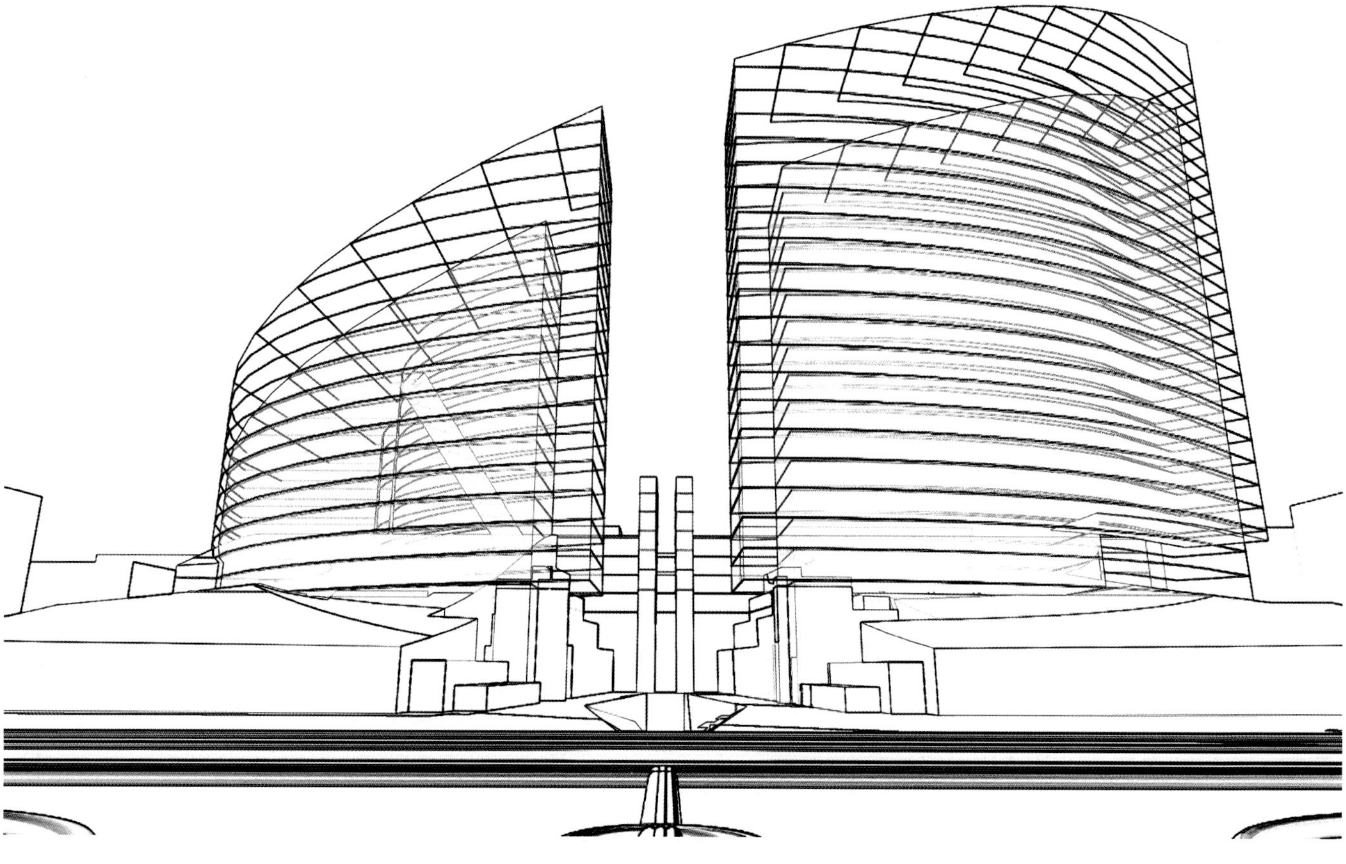

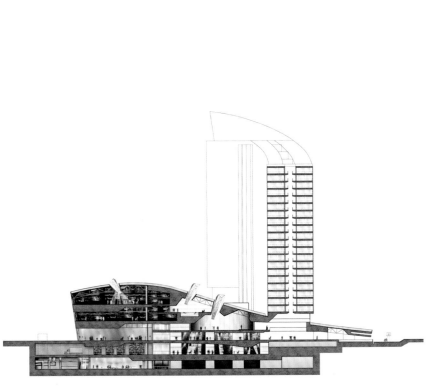

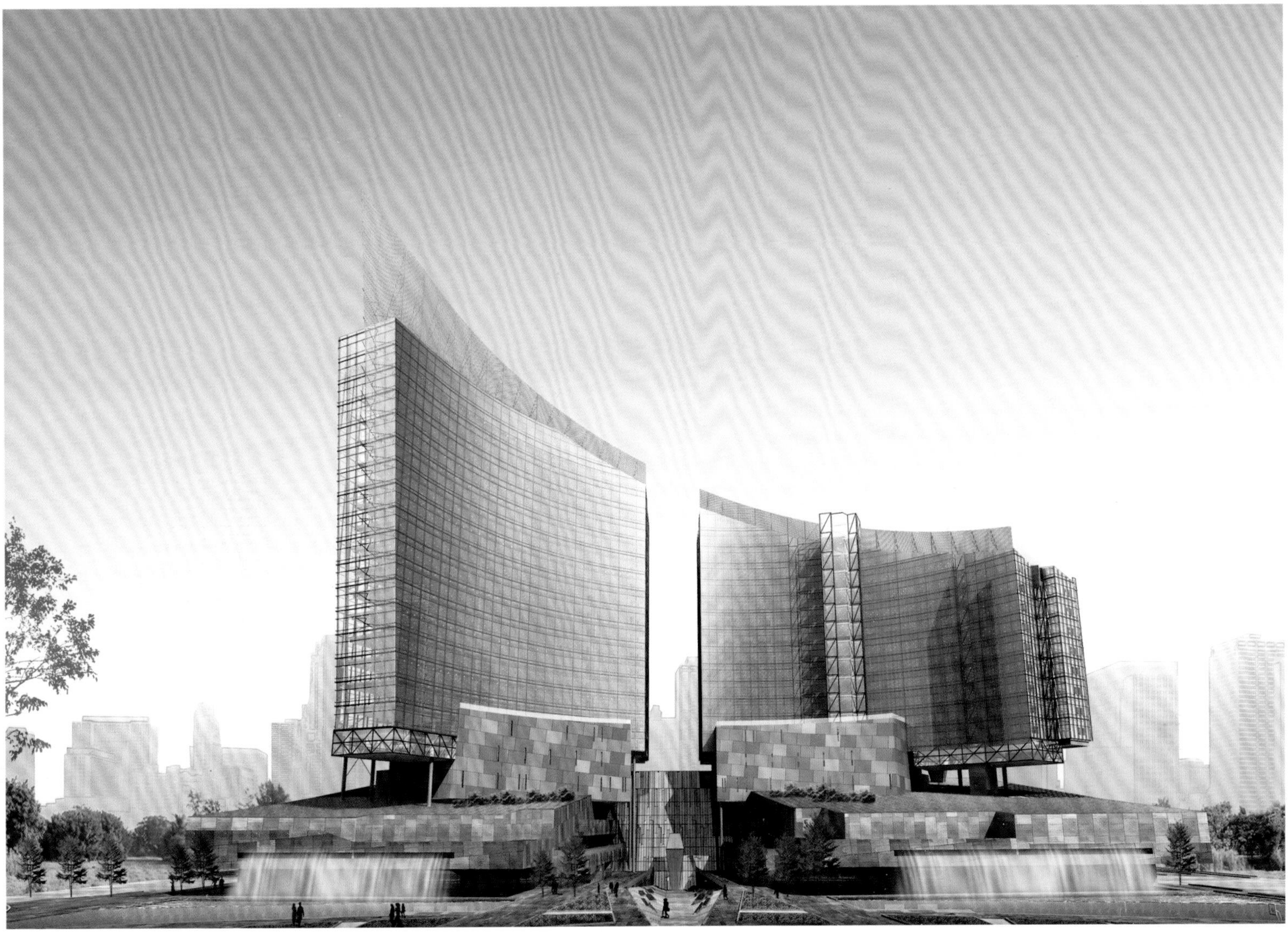

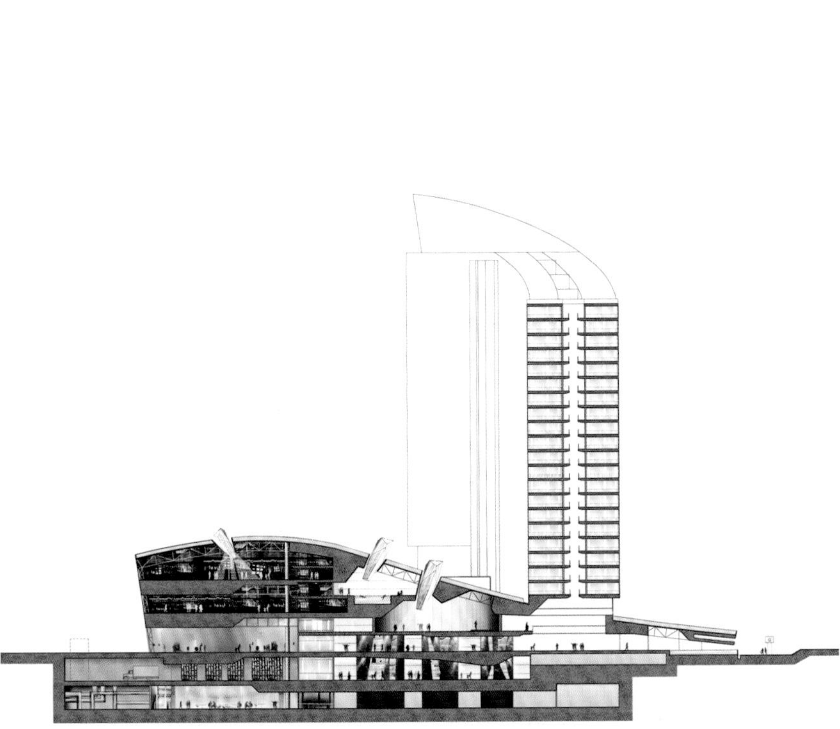

Tangshan Silicon Valley Tower
唐山硅谷大厦

Keywords 关键词

Glass Curtain Wall
玻璃幕墙

Oblique Cutting Technique
斜向切割手法

Visual Effect
视觉效果

Location: Tangshan, Hebei, China
Architectural Design: Tangshan Planning And Architectural Design & Research Institute
Zeidler Partnership Architects (Canada)
Site Area: 5,462 m^2
Total Floor Area: 62,927 m^2
Plot Ratio: 8.0

项目地点：中国河北省唐山市
设计单位：河北省唐山市规划建筑设计研究院
加拿大蔡德勒建筑师事务所
基地面积：5 462 m^2
总建筑面积：62 927 m^2
容 积 率：8.0

Features 项目亮点

The facade looks cool, on which the glass in different reflectivity is used to achieve the effect like diamond and crystal. In addition, there are LED lights emit colorful lights, making it one of the highlights of the city.

建筑立面整体为冷色调，用不同反射率的玻璃来实现钻石和水晶般的效果；建筑立面装有彩色 LED 灯，夜晚放出五彩斑斓的光芒，成为城市中的一大亮点。

■ Overview

Located in Central Tangshan, this project is on the east of Jianshe Road and the north of Tiyuguan Road, facing to the city cultural center in planning.

■ 项目概况

项目地处唐山市中心区域，西邻南北向主干道建设路，南邻体育馆道，与规划中的城市文化中心相望。

■ Design Concept

Throughout the ages, people often express their good wishes through diamond and crystal. The building is cut into several ramps using the similar crafts for cutting diamonds, so that they can refract light in different directions, producing the purity and clarity like diamond and crystal on the building, which is inlaid in the city, exuding a unique temperament, a strong character and timeless values.

■ 设计理念

项目设计灵感来自于钻石、水晶，采用类似于切割钻石的手法将建筑切成数个斜面，让其对光线产生不同方向的折射，令建筑呈现出钻石和水晶般的纯净无瑕、晶莹剔透，镶嵌于城市中，散发独特气质，透露出刚强的品格和永恒的价值。

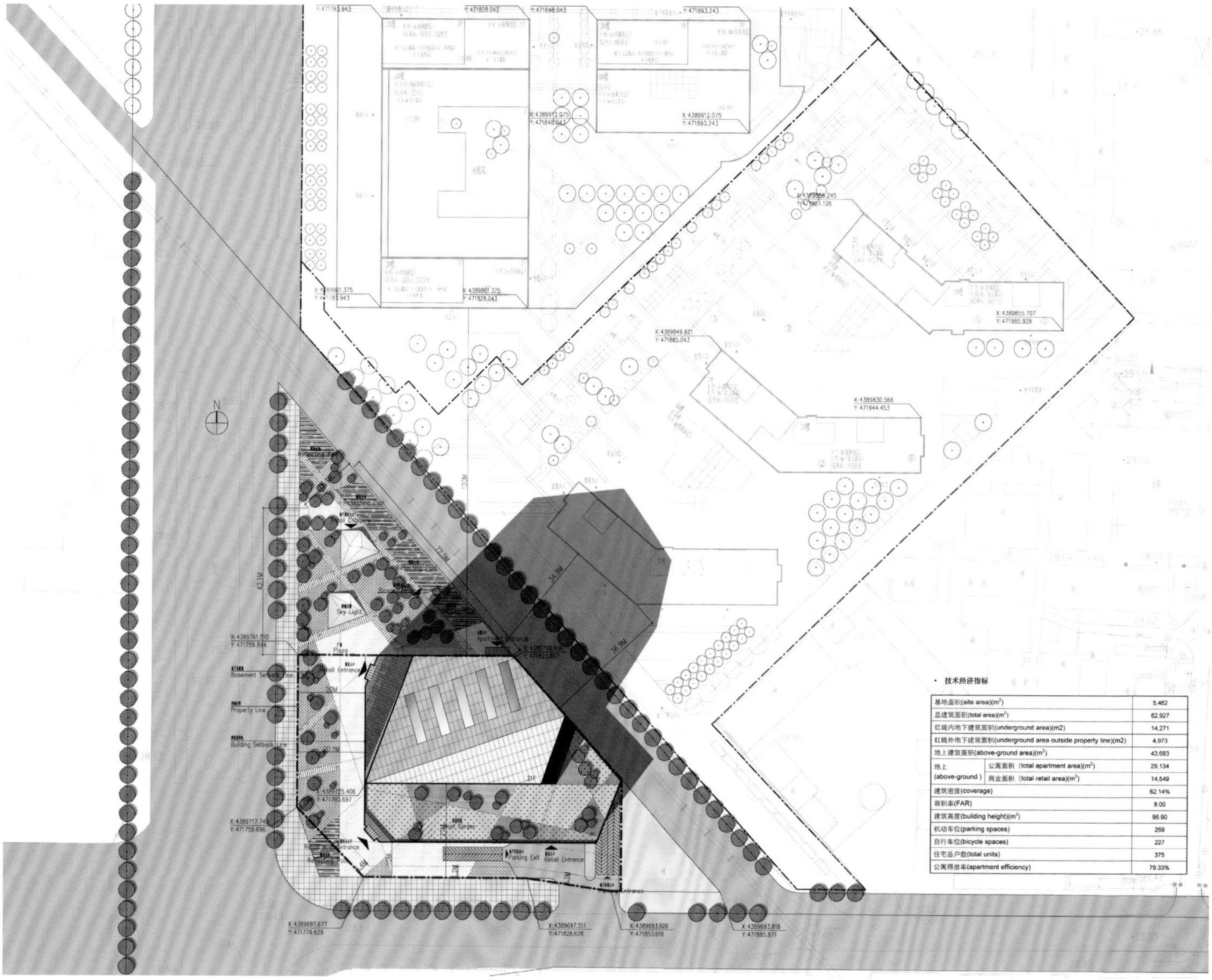

• 技术经济指标

基地面积(site area)(m^2)		5,462
总建筑面积(total area)(m^2)		62,927
红线内地下建筑面积(underground area)(m2)		14,271
红线外地下建筑面积(underground area outside property line)(m2)		4,973
地上建筑面积(above-ground area)(m^2)		43,683
地上 (above-ground)	公寓面积（total apartment area)(m^2)	29,134
	商业面积（total retail area)(m^2)	14,549
建筑密度(coverage)		62.14%
容积率(FAR)		8.00
建筑高度(building height)(m^2)		96.90
机动车位(parking spaces)		259
自行车位(bicycle spaces)		227
住宅总户数(total units)		375
公寓得房率(apartment efficiency)		79.33%

Site Plan 总平面图

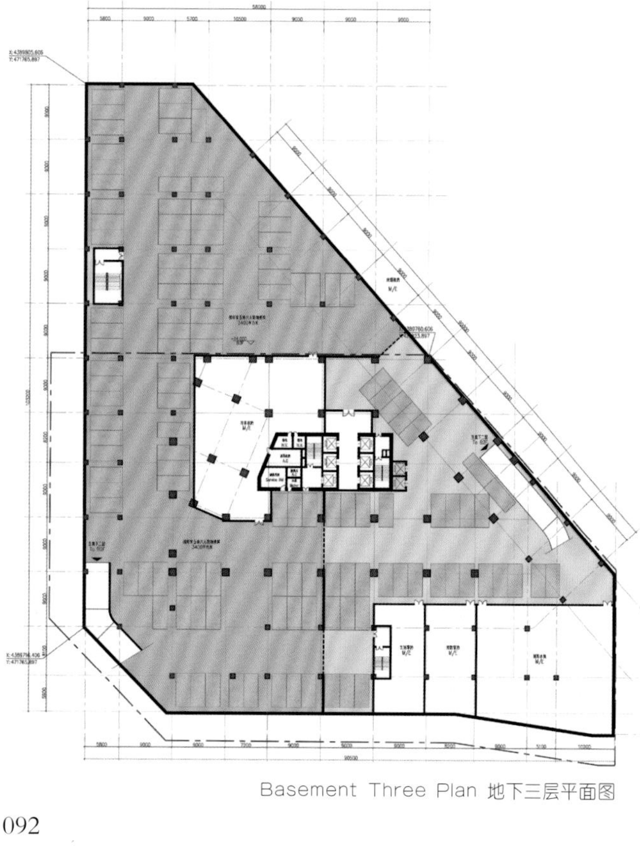

Basement Three Plan 地下三层平面图

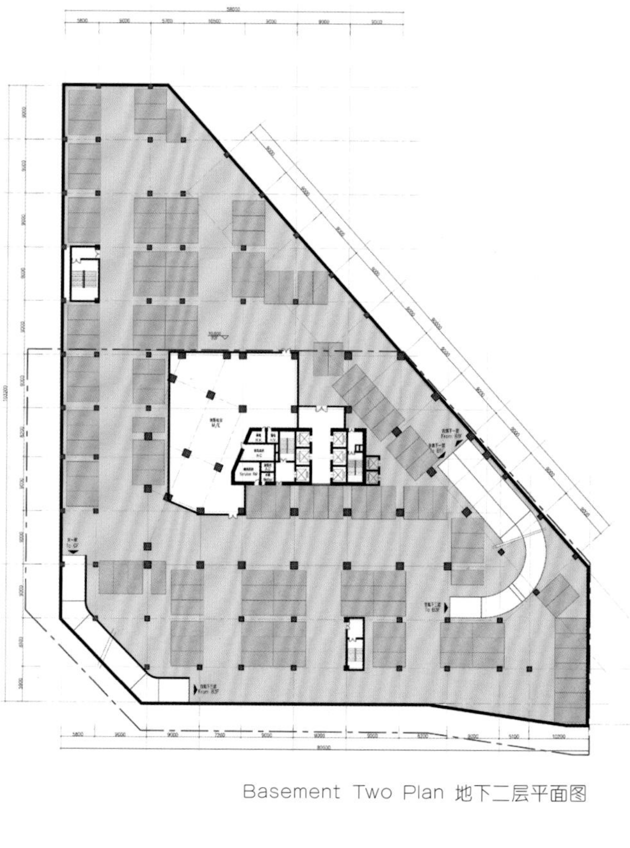

Basement Two Plan 地下二层平面图

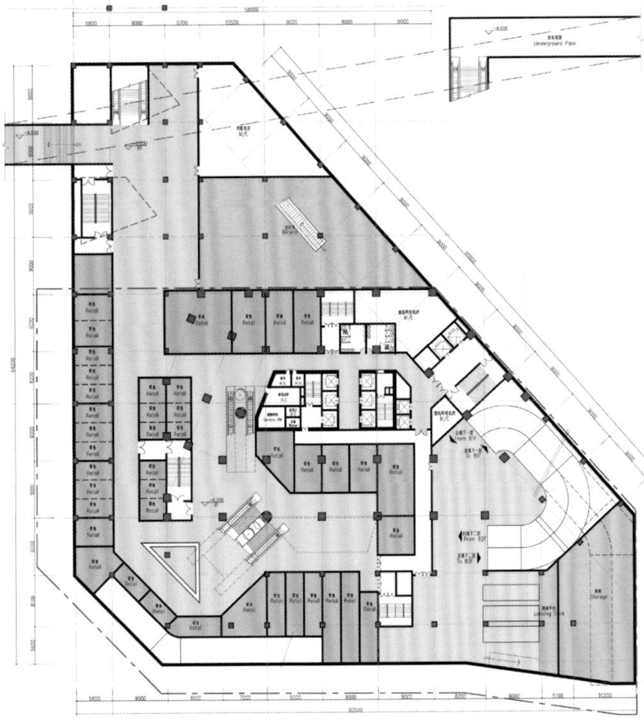

Basement Plan 地下一层平面图

First Floor Plan 一层平面图

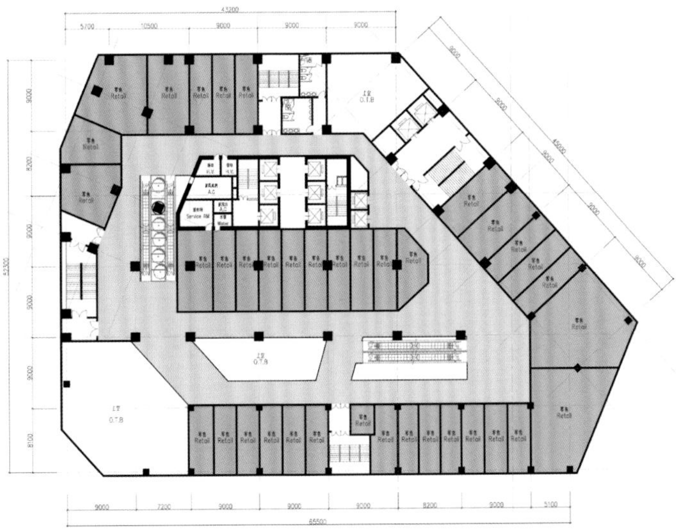

Second Floor Plan 二层平面图

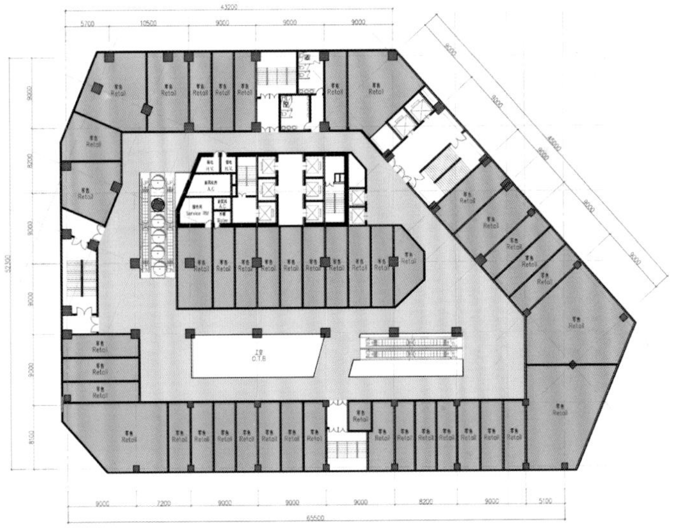

Third Floor Plan 三层平面图

■ Architectural Design

The facade looks cool, on which the glass in different reflectivity is used to achieve the effect like diamond and crystal. In addition, there are LED lights emit colorful lights, making it one of the highlights of the city. Being composed of mirror glass and common glass, the glass curtain wall looks like a mirror on the exterior that reflects the sky and the surrounding view, creating colorful images and endless variation along with the changing lights.

■ 建筑设计

建筑立面整体为冷色调，用不同反射率的玻璃来实现钻石和水晶般的效果。建筑立面装有彩色LED 灯，夜晚放出五彩斑斓的光芒，成为城市中的一大亮点。玻璃幕墙采用镜面玻璃与普通玻璃的组合，外观上看整片外墙犹如一面镜子，将天空和周围环境的景色映入其中，光线变化时，影像色彩斑斓、变化无穷。

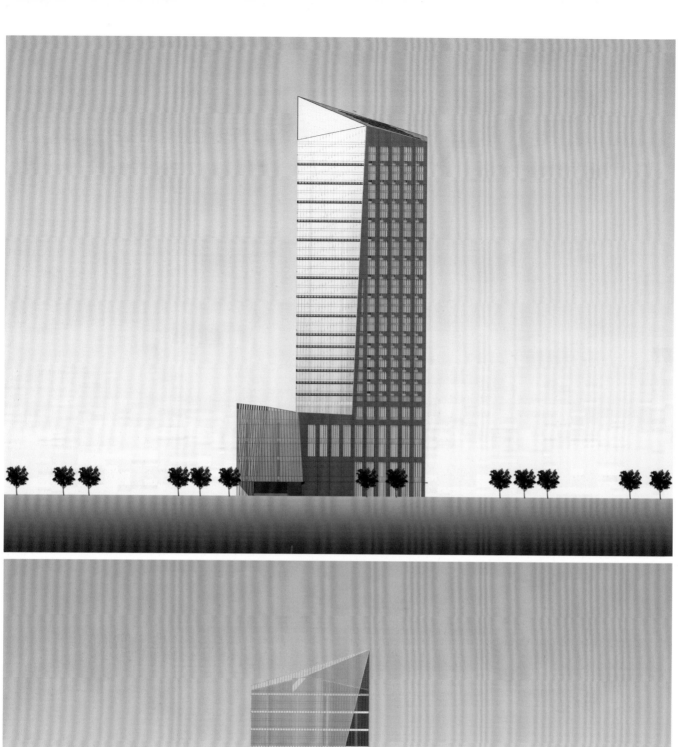

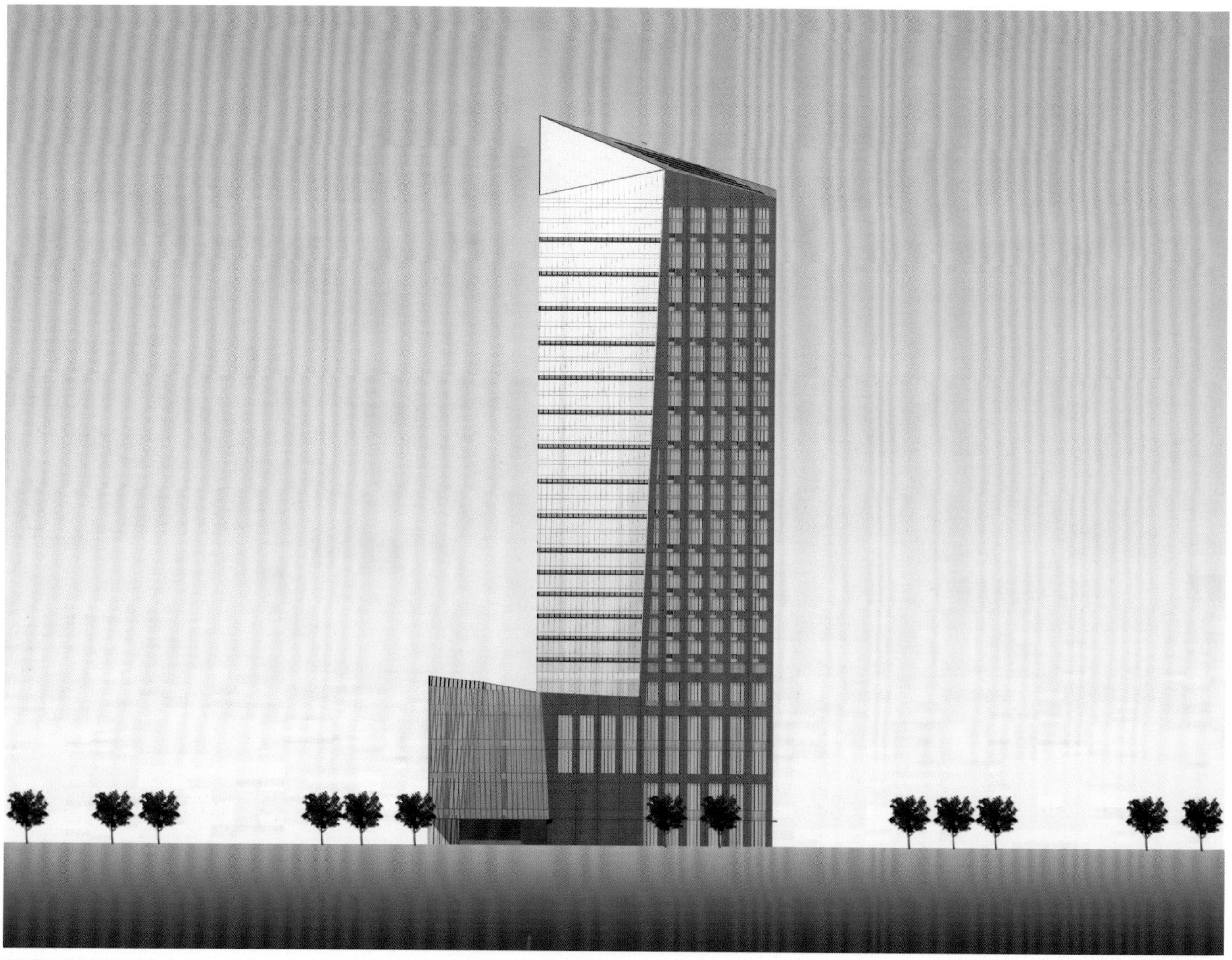

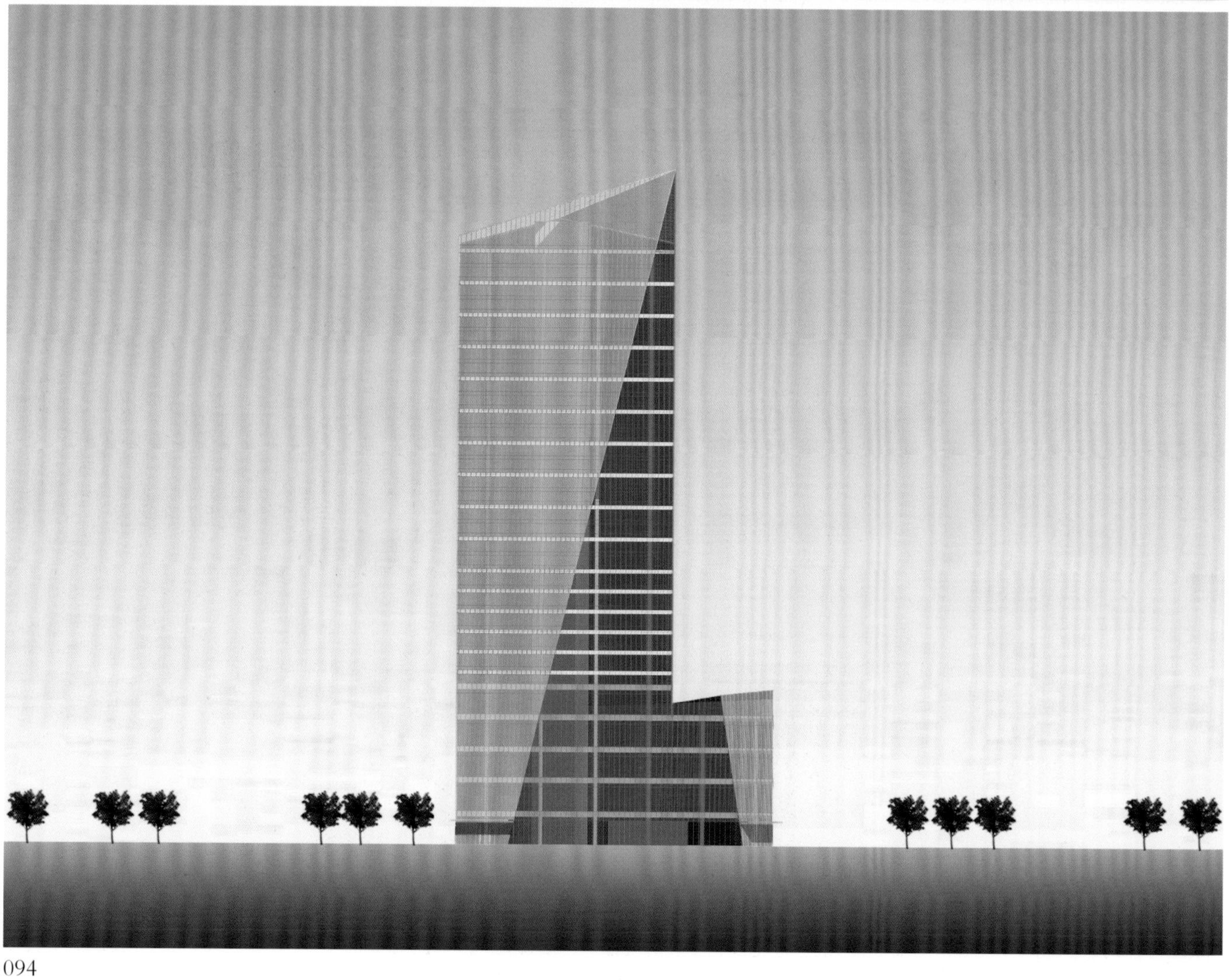

iMac
RETRO!

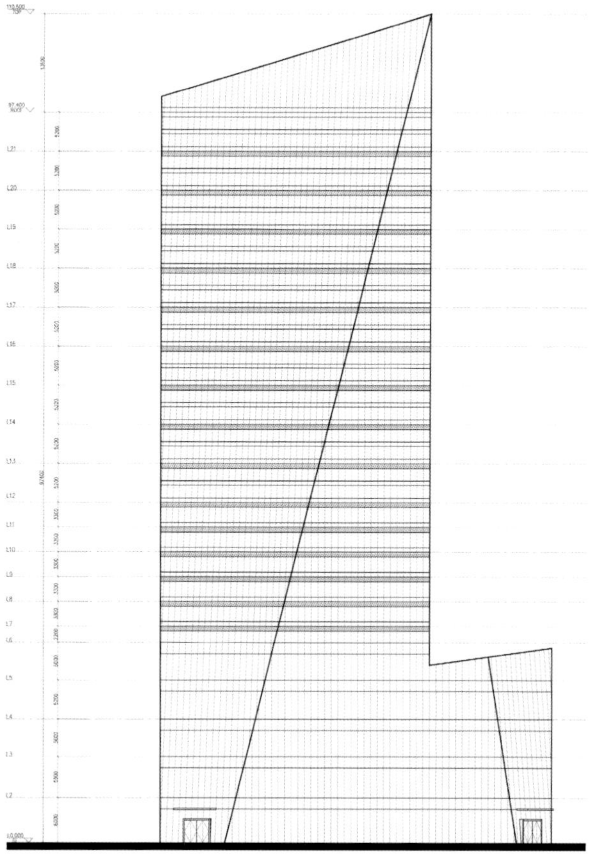

West Elevation 西立面图

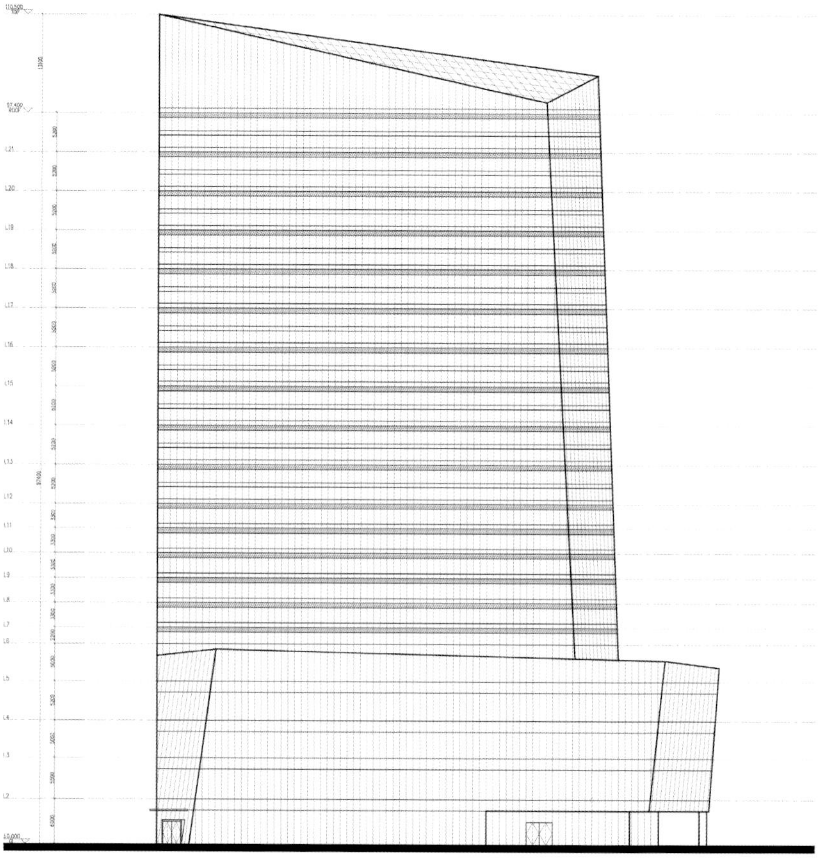

South Elevation 南立面图

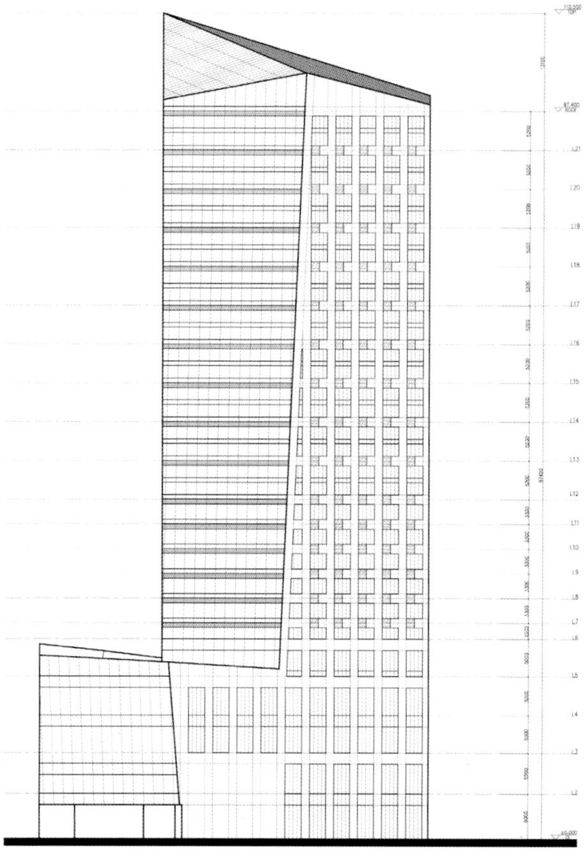

East Elevation 东立面图

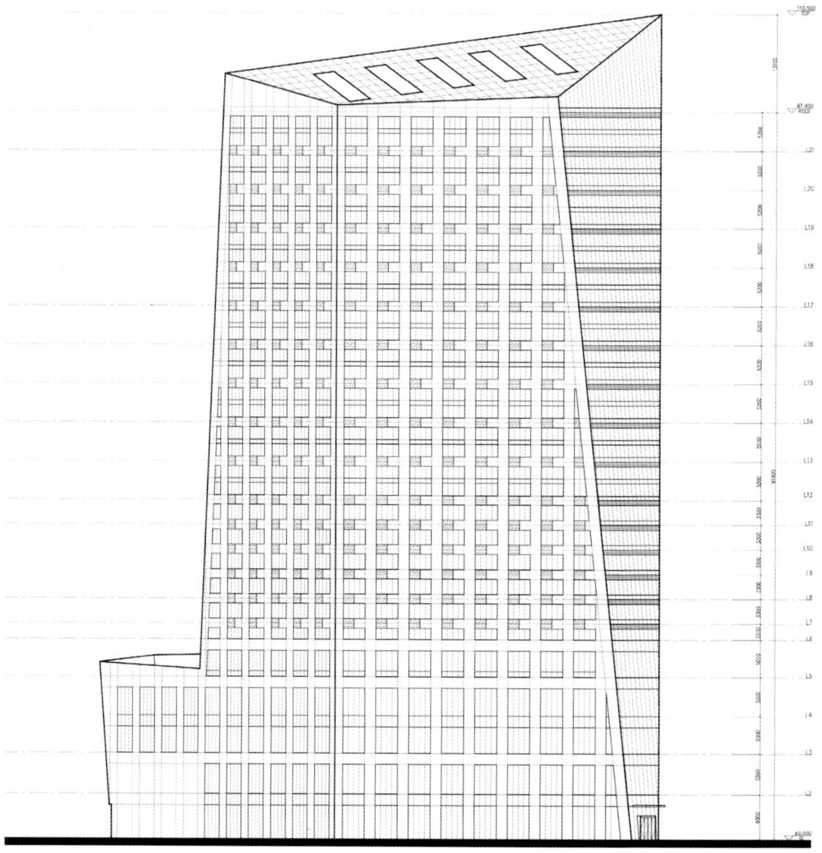

North Elevation 北立面图

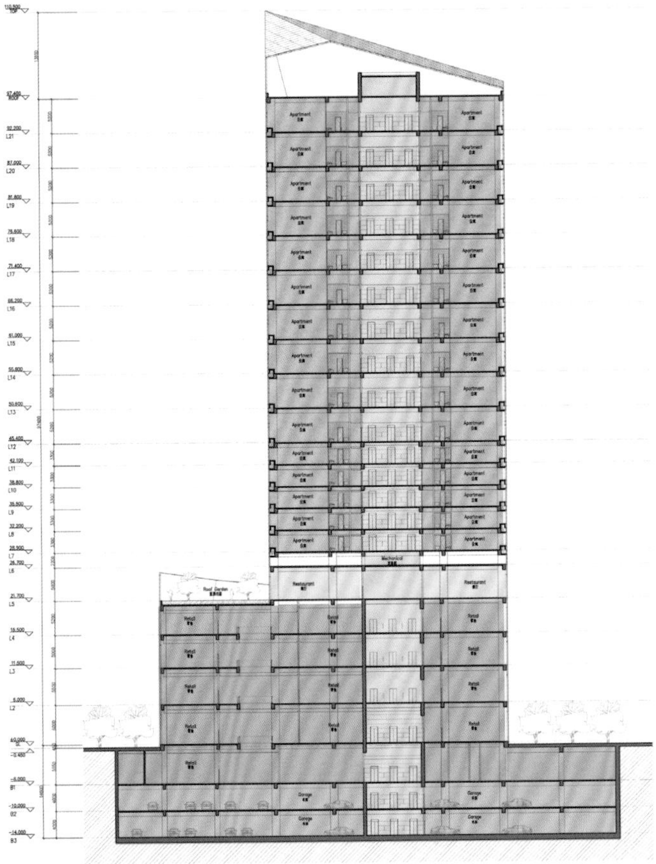

■ Landscape Design

Since most of the plot is covered with buildings and the people flow is large, enough square space should be preserved, the green coverage ratio is relatively low. In order to improve environmental quality, landscape design was fully carried out on the northern square which was cut obliquely, using the similar technique for cutting the facade, to create green view and waterscape. Several diamond-like glass daylighting roofs are established along the street passage of Jianshe Road, turning this green land into a super good place to relax. Meanwhile, daylighting roofs enhance guidance function of the entrance. One can get into the underground shopping mall directly through the underground hall, and the commercial value is promoted.

■ 景观设计

整块基地大部分被建筑物覆盖，且人流量较大，必须留有足够的广场空地，因此相对绿化率较低。为提高环境品质，充分利用基地北面广场进行景观设计，用与立面处理相似的手法将广场多重斜向切割，形成绿化和水景。结合建设路过街通道在地面矗立起几个钻石般造型的玻璃采光顶，使该绿地成为一处极佳的休闲场所。同时采光顶增强了入口的导向性，通过地下大厅还可直接进入商场地下一层，提升了地下商业的价值。

Changhong International R&D Center
长虹国际研发中心

Keywords 关键词

Green Technology
绿色科技

Building Materials
建筑材料

Corporate Image
企业形象

Location: Chengdu, Sichuan, China
Client: Changhong Electronic Technology Co., Ltd.
Architectural Design: KAZIA.LI Design Collaborative
Architects: Clay Vogel, AIA
Size: 170,000 m^2
Height: 150 m
Results: Competition Winner, Under Contract
Status: Construction Started summer 2011

项目地点：中国四川省成都市
业　　主：长虹电子科技有限公司
设计单位：天津凯佳李建筑设计事务所
设 计 师：克雷·沃格尔、AIA
建设规模：170 000 m^2
高　　度：150 m
设计结果：竞标获胜并取得合同
状　　态：2011年夏季开工

Features 项目亮点

Centering on the idea of green and technology, through the use of energy-efficient building materials and equipment, the design greatly reduces the energy consumption of the buildings to achieve sustainable development.

围绕绿色和科技的设计理念，通过使用高效节能的建筑材料和设备，降低建筑的能耗，使建筑具有可持续发展的特征。

■ Design Concept

The architecture curves round forming two opening arms that gesture to welcome guests from all around the world. The buildings are arranged around a plaza, creating a more attractive and pleasant environment. The facade concept was inspired by the digitization of television, an industry in which the client is closely involved. The Changhong International R&D Center design is highly sustainable and uses energy-saving building materials, and incorporates a rainwater-collecting green roof system.

■ 设计理念

该建筑圆形曲线呈现出张开的双臂，是一种欢迎八方来客的寓意。建筑围绕着一个广场，创造出一个更具吸引力和舒适的环境。表皮概念的灵感来自于电视的数字化，在这个行业中，客户是密切参与的。长虹国际研发中心的设计是高度可持续的，使用节能建筑的材料，并采用了雨水收集屋顶绿化系统。

Master Plan 规划总平面图

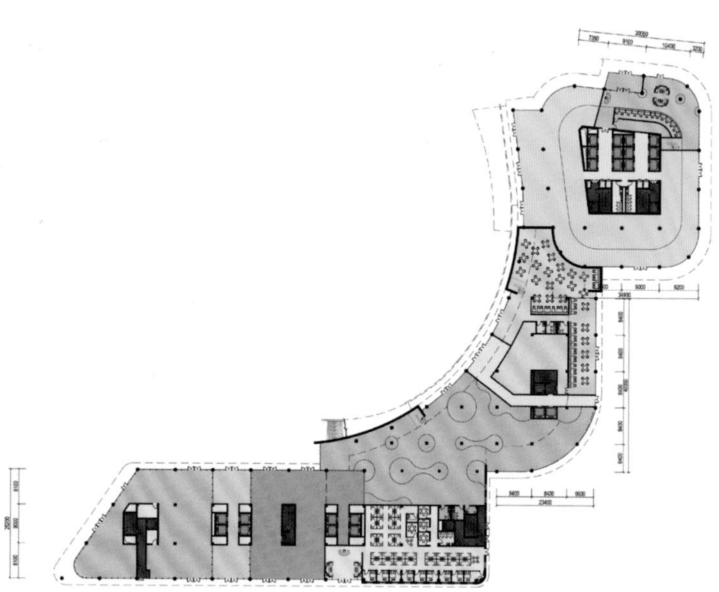

1st Floor Plan 一层平面图

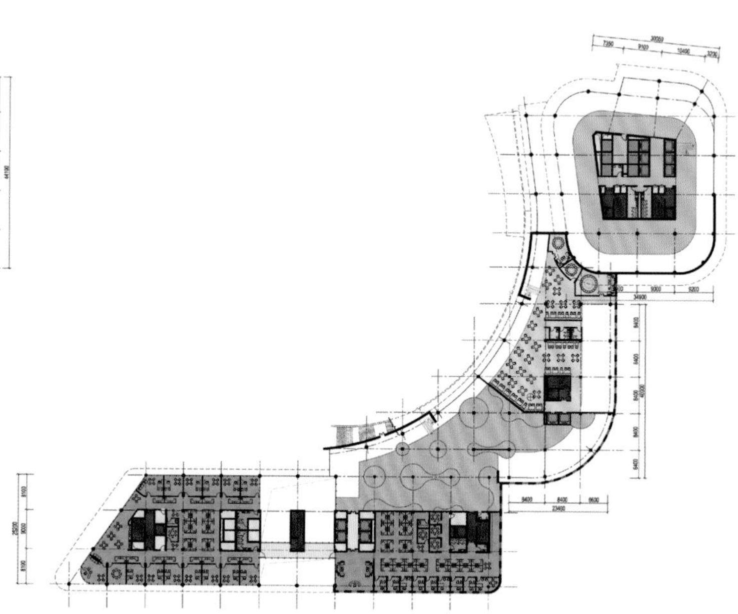

2nd Floor Plan 二层平面图

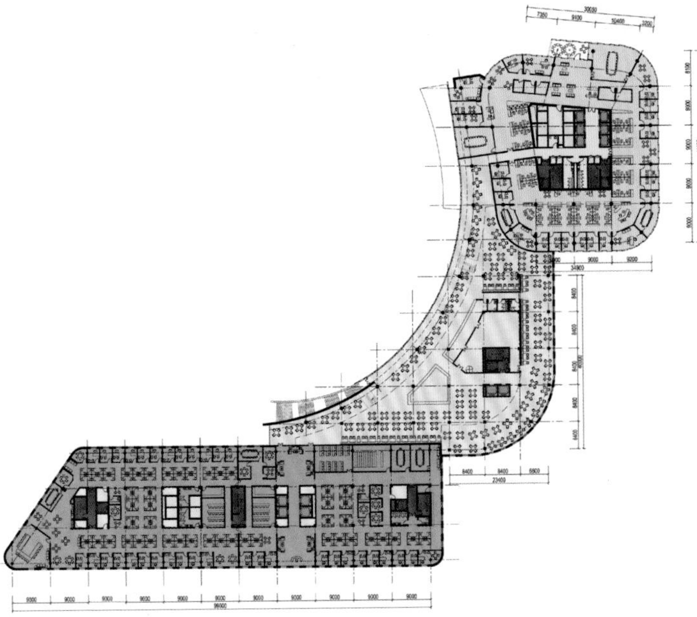

3rd Floor Plan 三层平面图

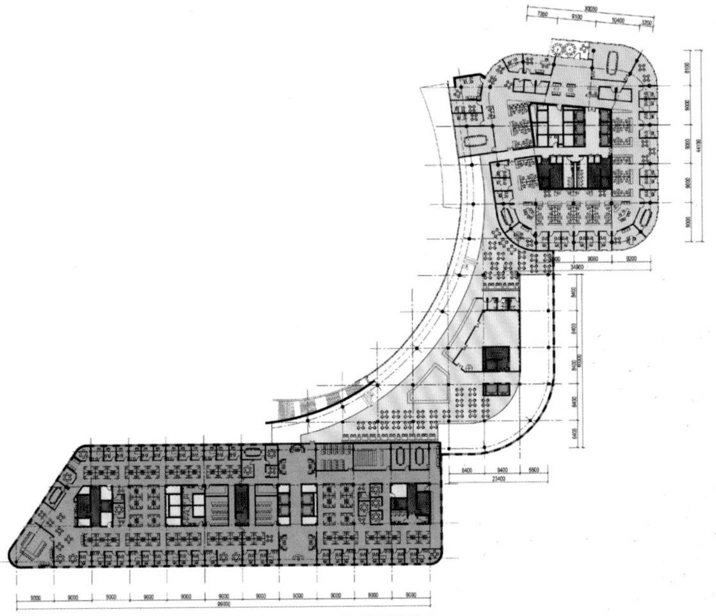

4th Floor Plan 四层平面图

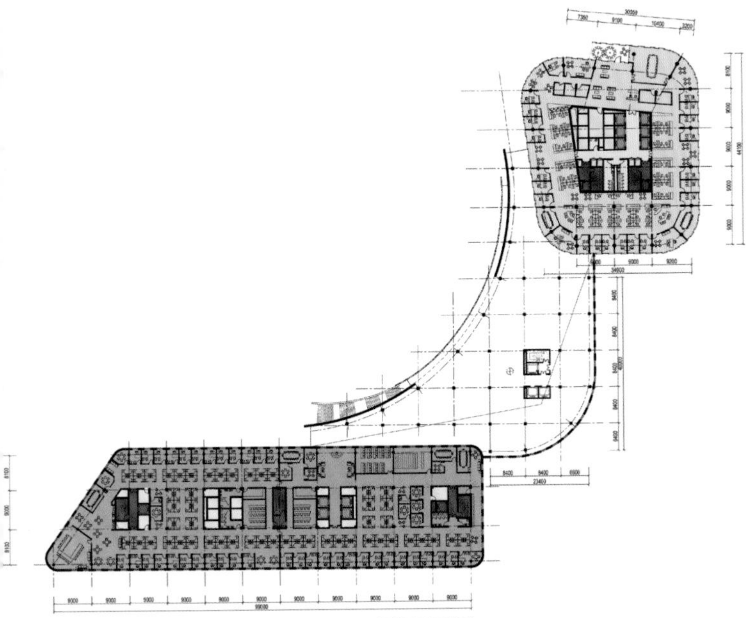

5th Floor Plan 五层平面图

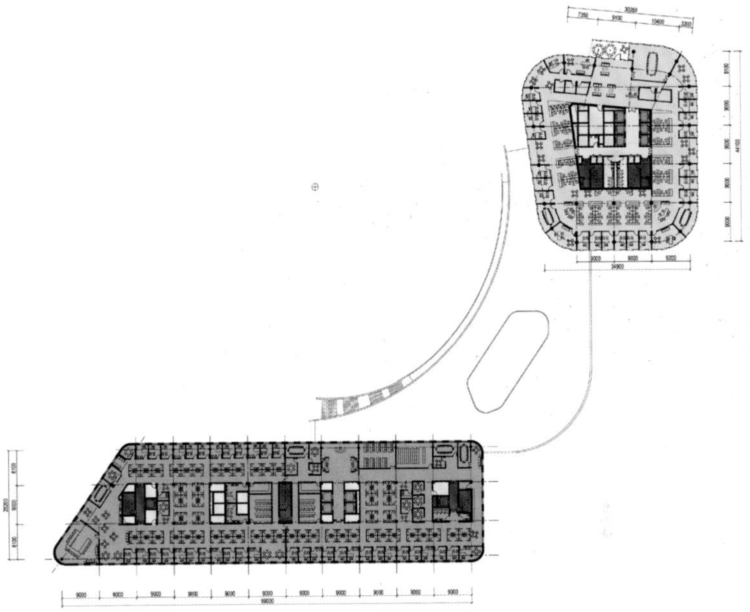

6th Floor Plan 六层平面图

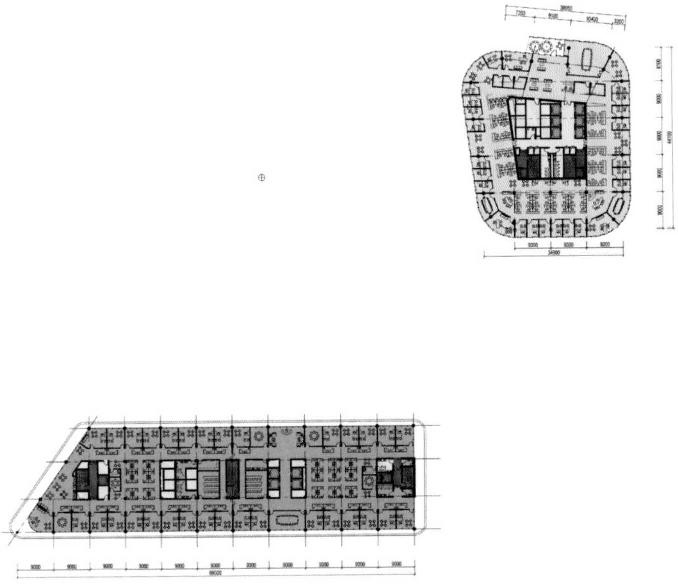

7th~8th Floor Plan 七~八层平面图

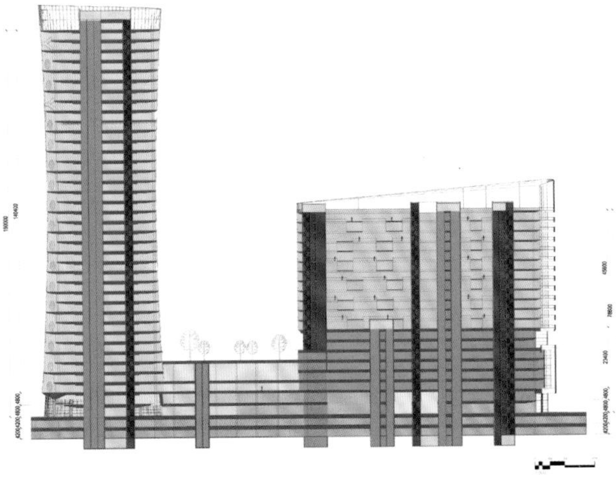

Section 剖面图

East Elevation 东立面图

North Elevation 北立面图

South Elevation 南立面图

West Elevation 西立面图

CHANGHONG

Agricultural Bank of China, Chongqing Branch 中国农业银行重庆分行

Keywords 关键词

Natural Lighting
自然采光

Building Image
建筑形象

Energy Saving
节能环保

Features 项目亮点

According to geographical factors of the project, the design uses as much as possible of the natural light and reduces artificial light use to greatly save energy and protect the environment.

依据项目的地理环境因素，设计过程中尽可能利用充足的自然光，减少使用人造光源，节约能源、保护环境。

Location: Chongqing, China
Client: Agricultural Bank of China
Architectural Design: KAZIA.LI Design Collaborative
Designed By: Clay Vogel, AIA
Floor Area: 92,500 m^2
Height: 160 m
Results: Competition Winner, Under Contract
Status: Construction started March 2011

项目地点：中国重庆市
业　　主：中国农业银行
设计单位：天津凯佳李建筑设计事务所
设 计 师：克雷·沃格尔、AIA
建筑面积：92 500 m^2
高　　度：160 m
设计结果：竞标获胜并取得合同
状　　态：2011年3月开工建设

■ Design Concept

In consideration of the daylighting condition of Chongqing City, the ABC Tower's shape and orientation maximize the use of natural light, reducing artificial light needed to light the building while also minimizing the heat gained with the building. In the situation of global financial unrest, bank needs to shape its image as stability and mightiness. Bank building must deliver the message as the following: elegant and transparent, safe and stable, confident and responsible, reliable and sustainable, and at the same time, welcoming and full of hope.

■ 设计理念

在了解重庆的日照优势及建筑将面临的日照问题（建筑朝向及建筑节能）的前提下，设计过程中尽可能利用充足的自然光，减少使用人造光源，节约能源、保护环境。同时，为避免日照的不利因素对建筑的影响，应选择合适的建筑体型，减少建筑的西朝向。在全球金融动荡的氛围中，银行尤其需要稳固而强大的形象。银行建筑必须传递出这样的信息：简洁而透明，安全而稳定，自信而负责任，可靠而持久；同时，海纳百川，令人充满希望。

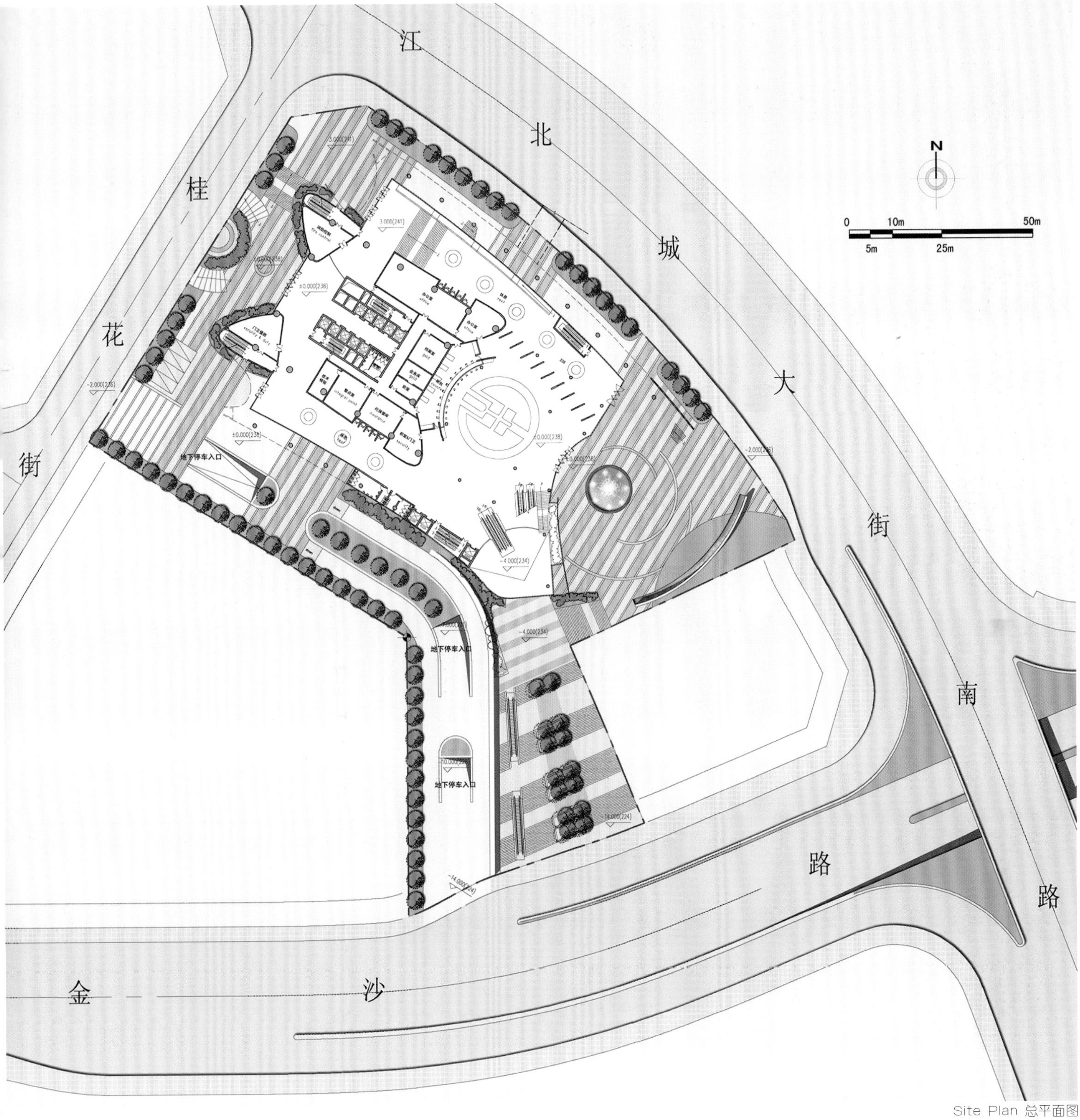

Site Plan 总平面图

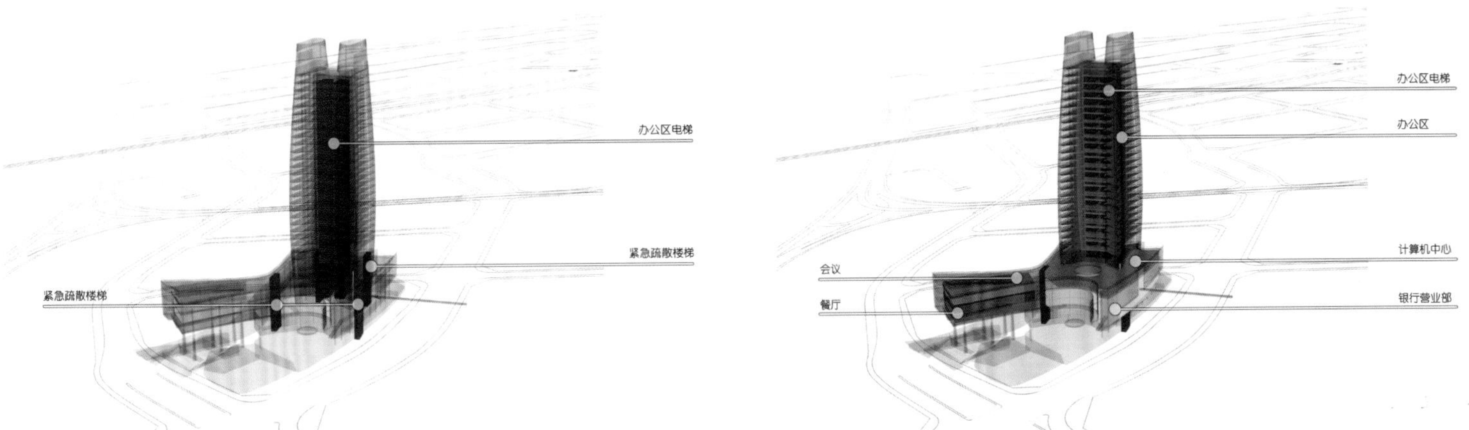

Function Analysis Diagram 功能分析图

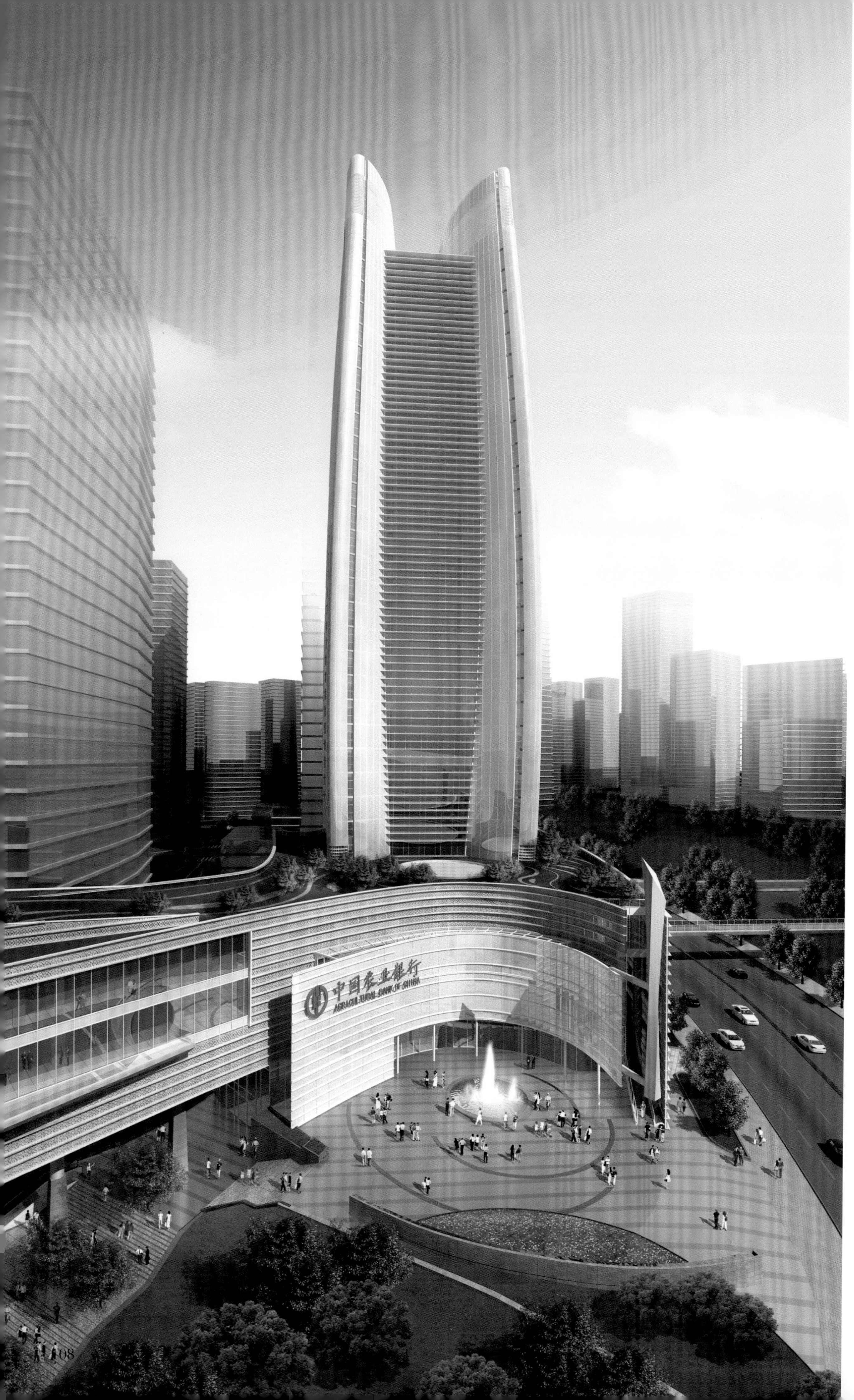
中国农业银行
AGRICULTURAL BANK OF CHINA

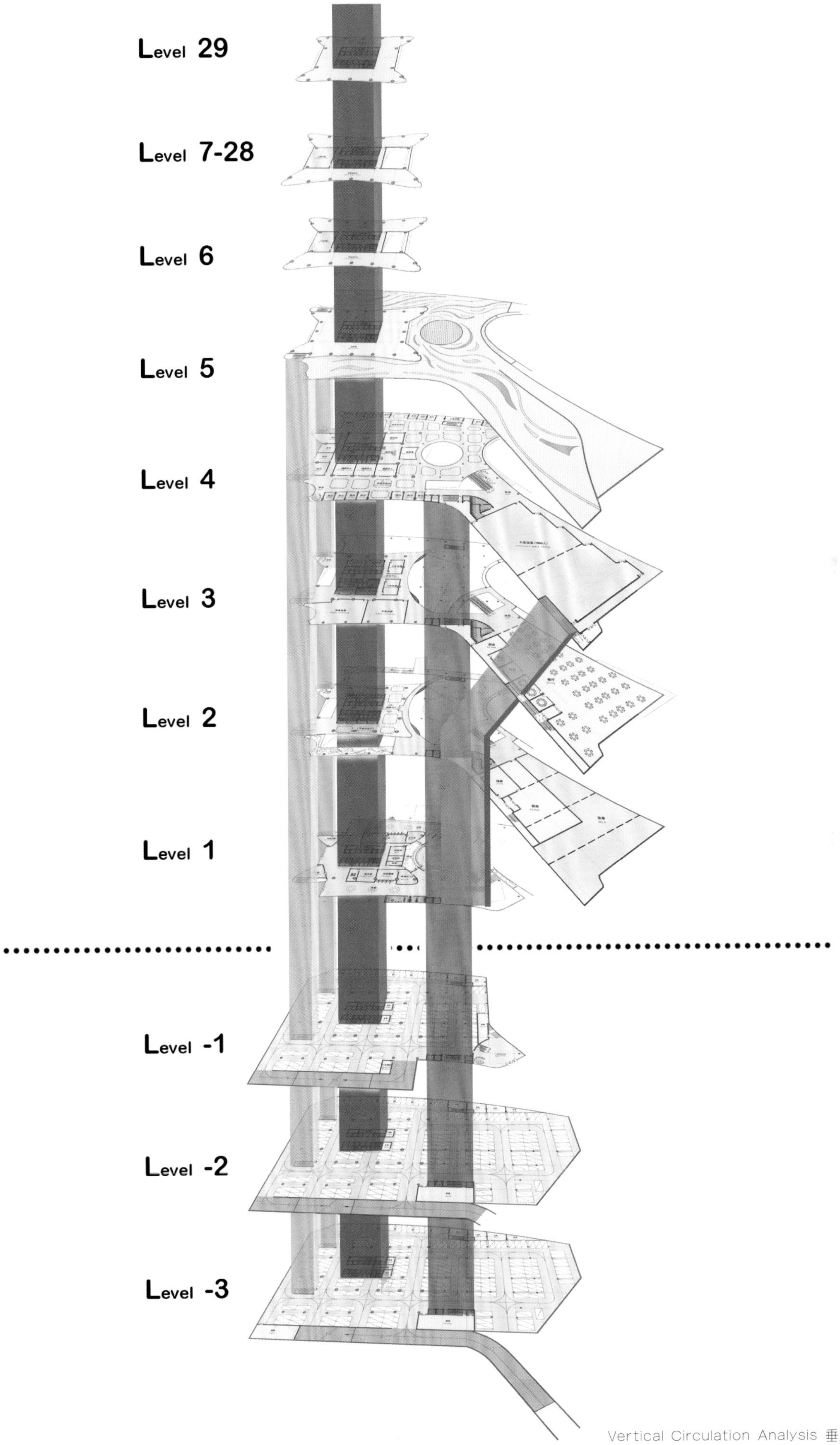

Vertical Circulation Analysis 垂直交通

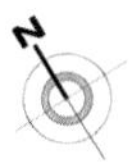

Level 6-29

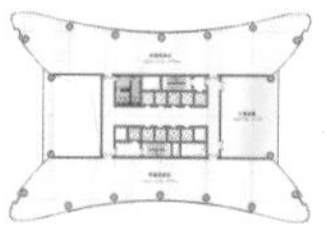

面积： 1390m²
Area: 1390m²

Level 5

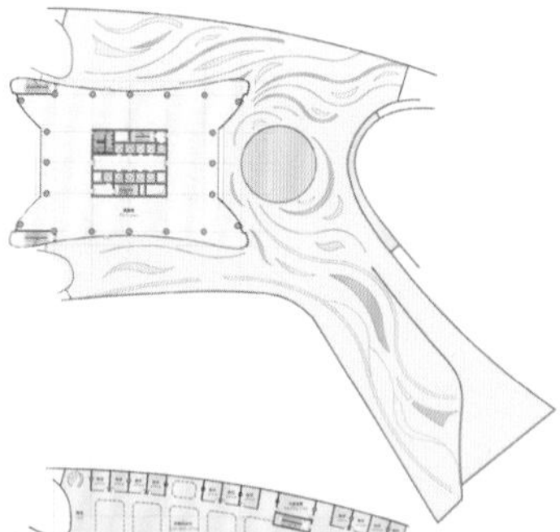

面积： 1500m²
Area: 1500m²

Level 4

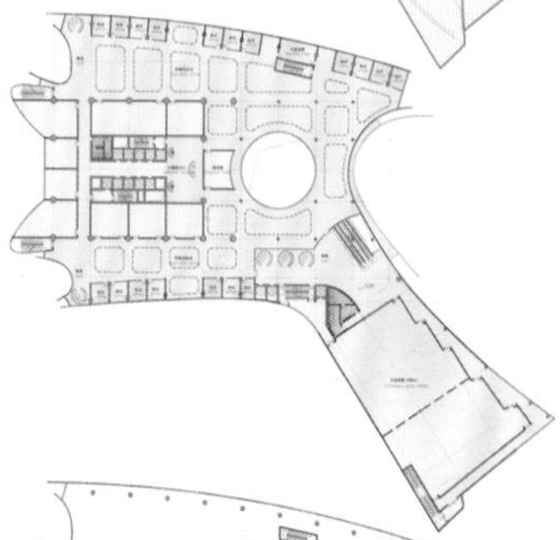

面积： 4950m²
Area: 4950m²

Level 3

面积： 3665m²
Area: 3665m²

Level 2

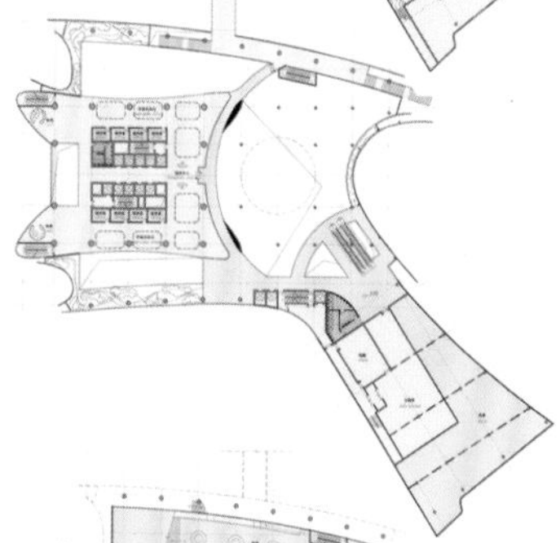

面积： 3812m²
Area: 3812m²

Level 1

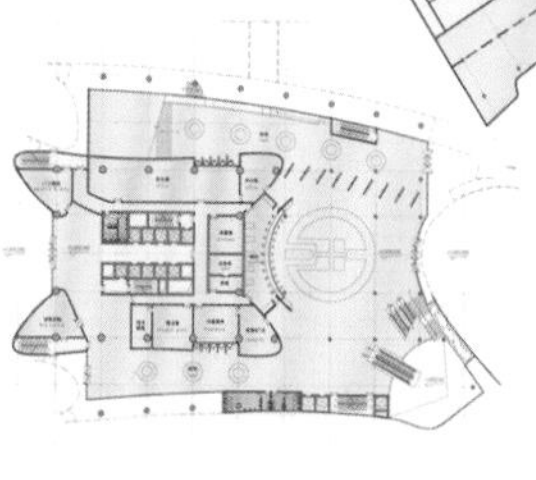

面积： 3816m²
Area: 3816m²

地上总建筑面积： 49 530 m²
total overground bldg Area: 49,530 m²

地下总建筑面积： 21 905 m²
total underground bldg Area: 21,905 m²

Basement

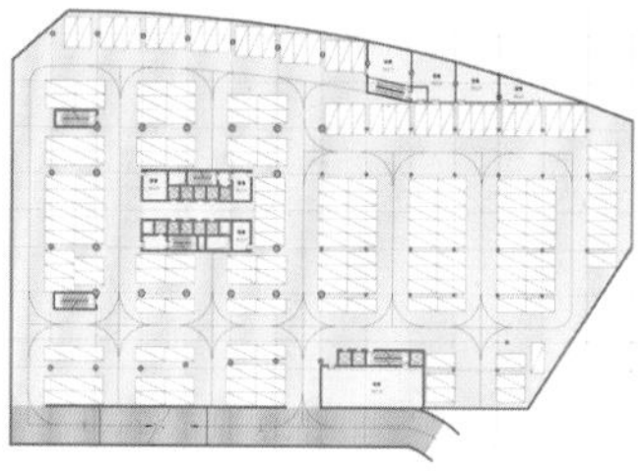

总建筑面积： 71 435 m²
total bldg Area: 71,435 m²

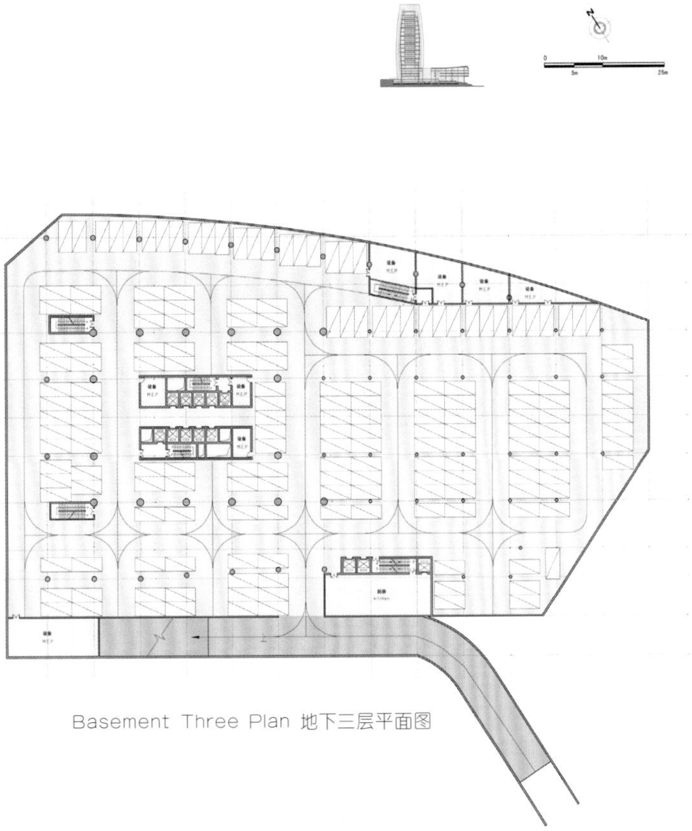

Basement Three Plan 地下三层平面图

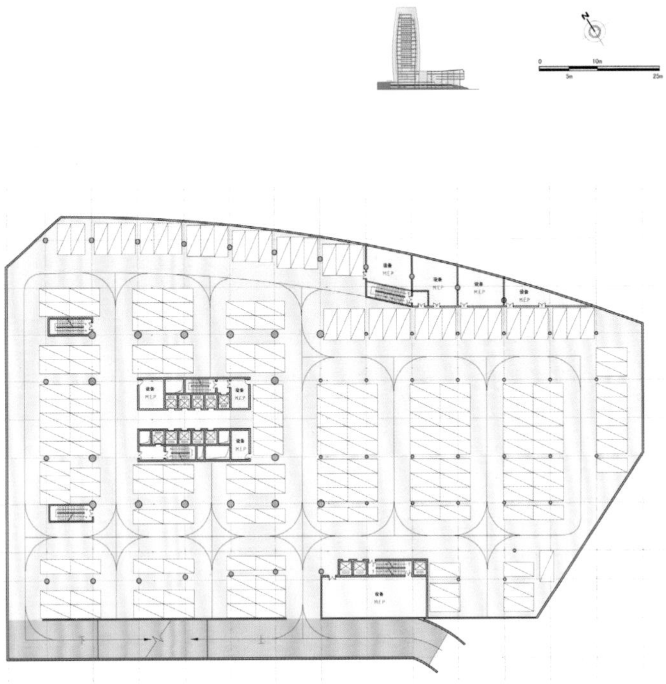

Basement Two Plan 地下二层平面图

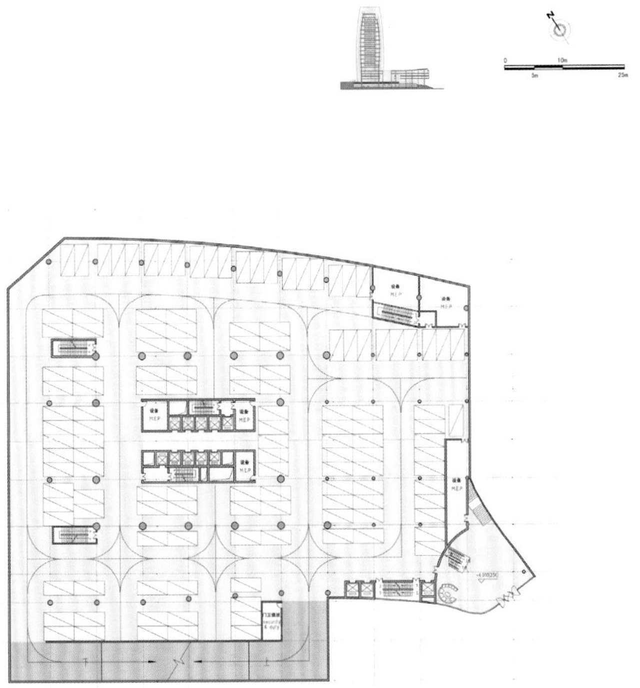

Basement Plan 地下一层平面图

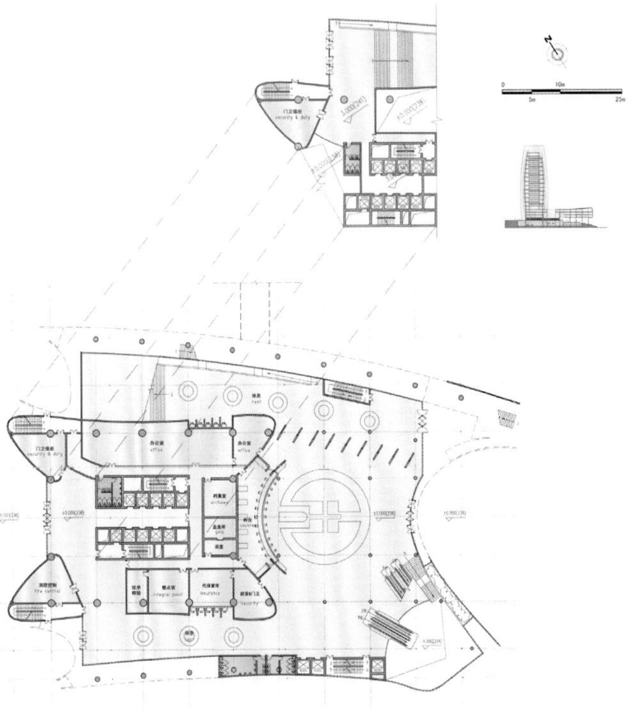

First Floor Plan 一层平面图

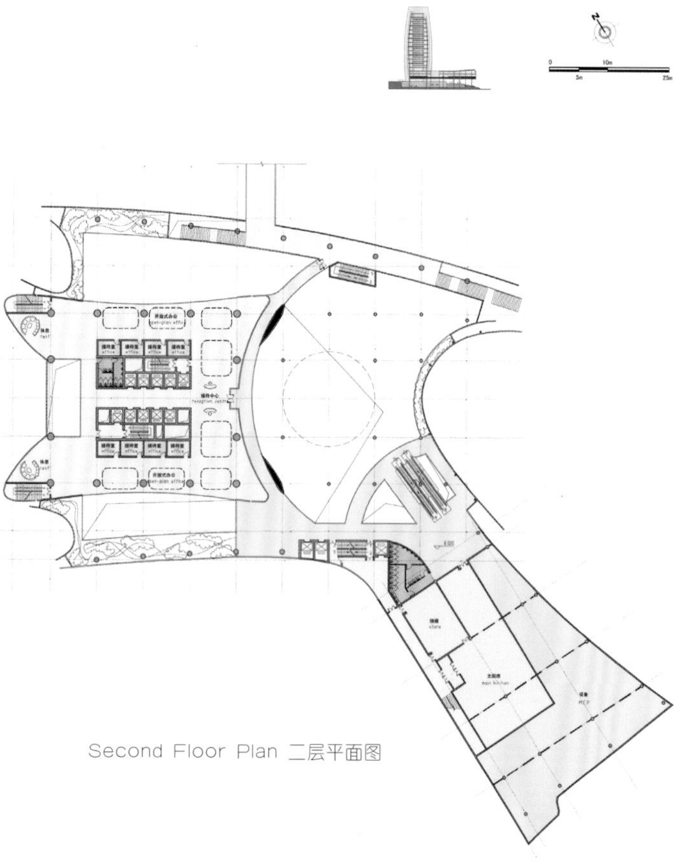

Second Floor Plan 二层平面图

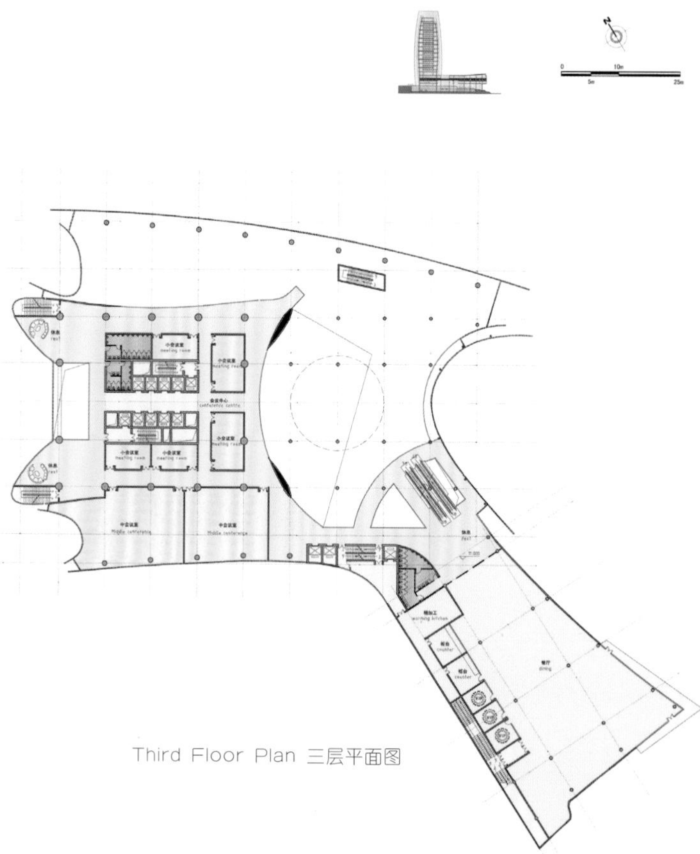

Third Floor Plan 三层平面图

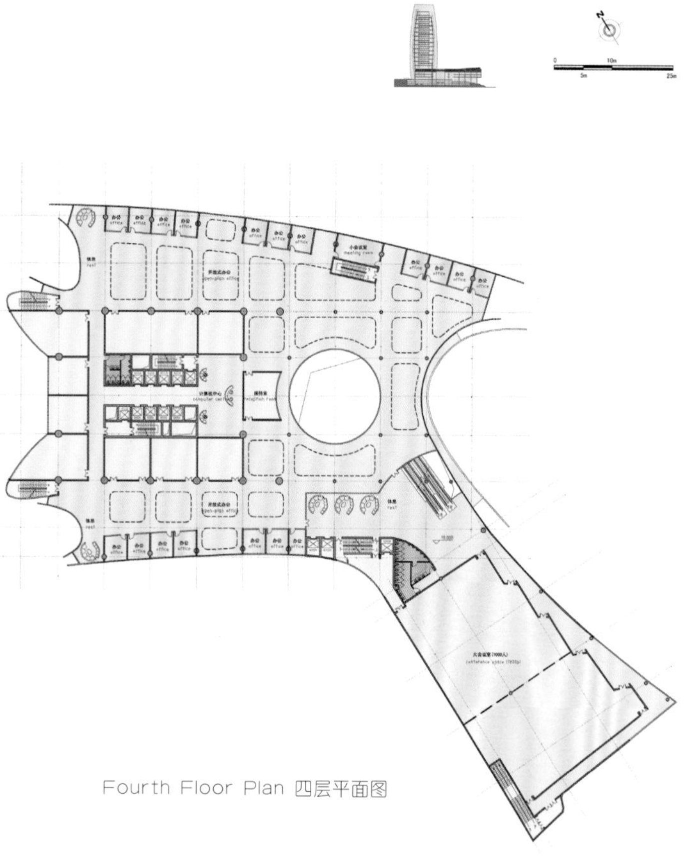

Fourth Floor Plan 四层平面图

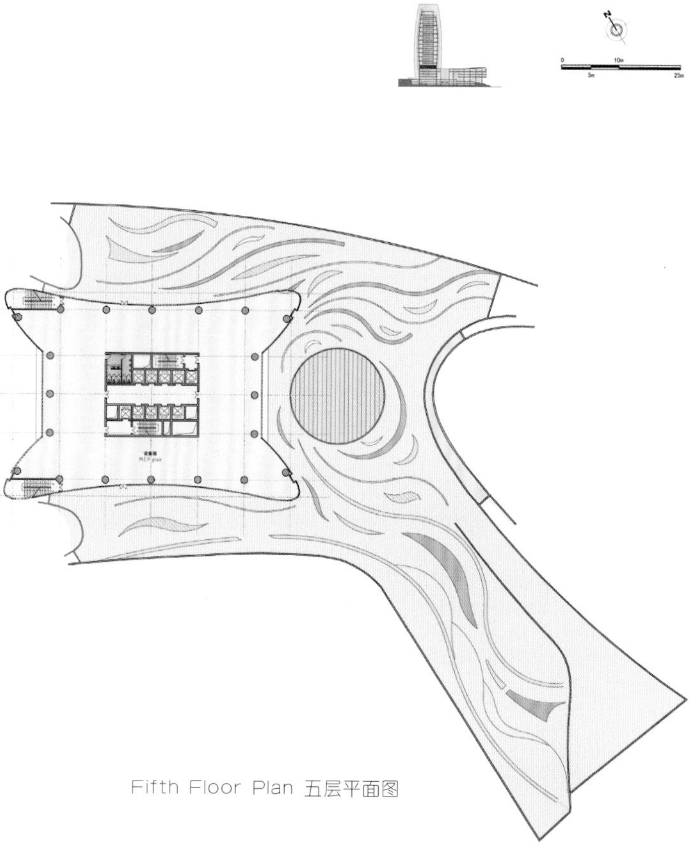

Fifth Floor Plan 五层平面图

Southeast Elevation 东南立面图

Northeast Elevation 东北立面图

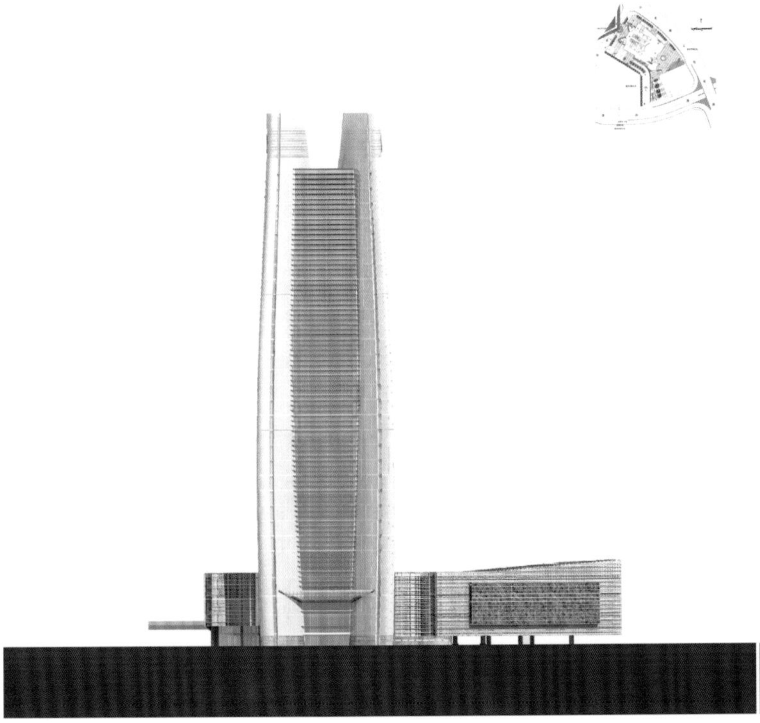

Northwest Elevation 西北立面图

Southwest Elevation 西南立面图

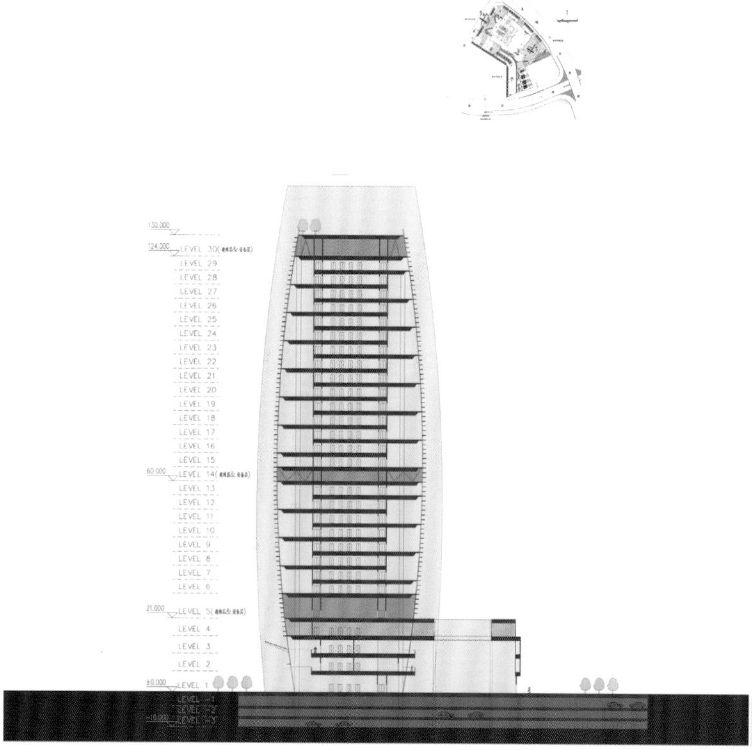

Section 1 剖面图 1

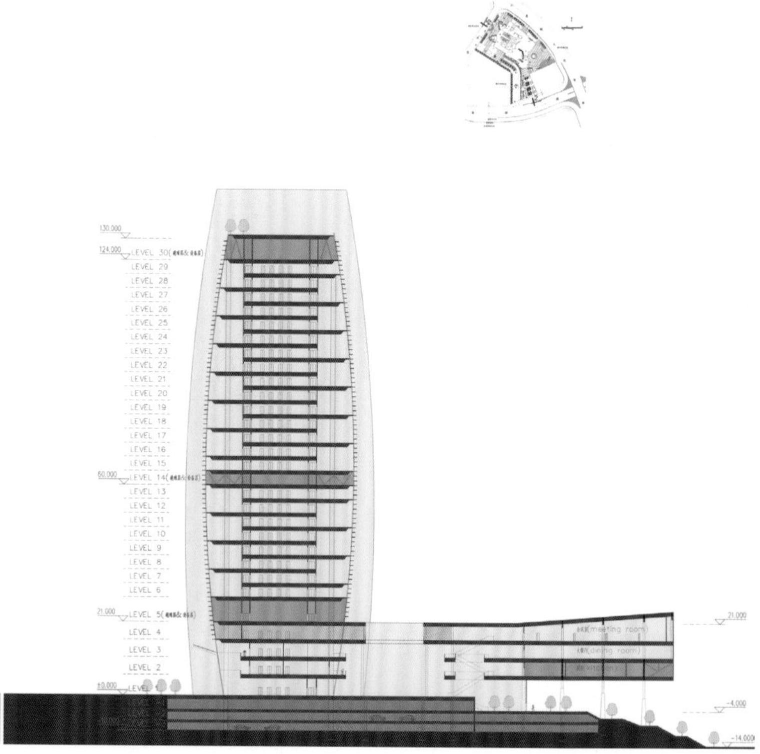

Section 2 剖面图 2

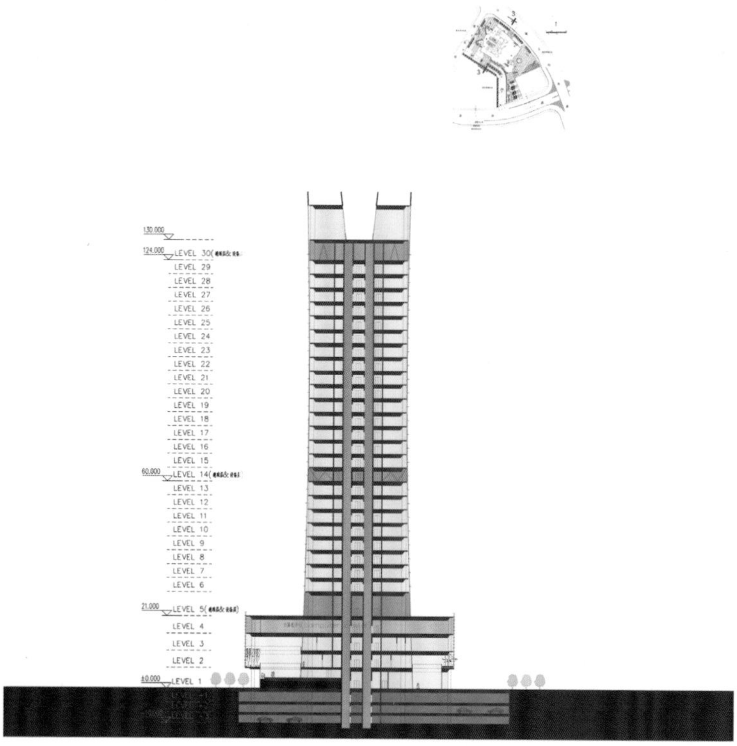

Section 3 剖面图 3

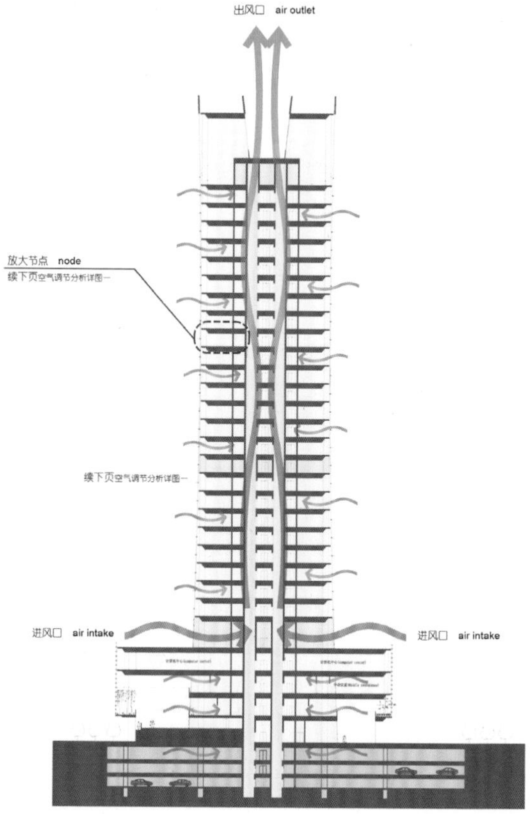

Air Conditioning Analysis Drawing 空气调节分析图

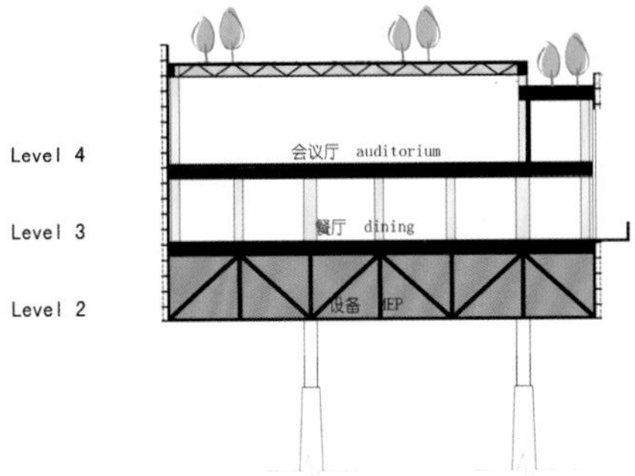

Part Section A-A 局部剖面图 A-A

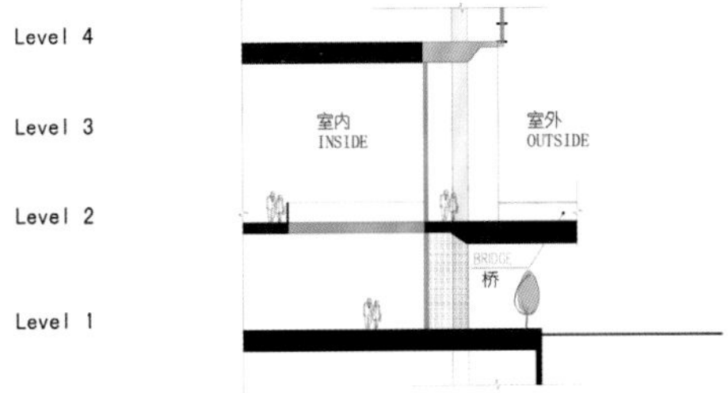

Part Section B-B 局部剖面图 B-B

COMAC Headquarters Base
中国商飞总部基地

Keywords 关键词

Green & Low-carbon
绿色低碳

Mega-frame Structure
巨型框架结构

Vertical Aluminum Plate
竖向铝板

Location: Shanghai, China
Architectural Design: Jiang Architects & Engineers
Land Area: 18,434 m^2
Total Floor Area: 129,170 m^2
Plot Ratio: 5.5
Status: Schematic designed in 2011

项目地点：中国上海市
设计单位：上海江欢成建筑设计有限公司
用地面积：18 434 m^2
总建筑面积：129 170 m^2
容 积 率：5.5
状　　态：2011年方案设计

Features 项目亮点

All the building volumes share the same special facade and giant vertical aluminium plates on east and west sides, which keep from the sunlight efficiently and realize a unified building image.

所有建筑体块均采用同样的立面造型，建筑东西两侧均安装巨型竖向铝板，不仅能起到遮阳作用，还使得建筑形象统一而整体。

■ **Overview**

Located in Shanghai Expo Park, COMAC (short for Commercial Aircraft Corporation of China) Headquarters Base accommodates project management center, financial services center, marketing center, international cooperation center, press center, etc., and is comprised of a 120 m high-rise office building and the annex buildings. As an important component in the lot for central enterprises headquarters, it will interplay with China Pavillion and become the new landmarks for Huangpu River together.

■ **项目概况**

中国商用飞机有限责任公司总部基地位于上海世博园区，规划建设项目管理中心、金融服务中心、市场营销中心、国际交流合作中心、新闻中心等重要机构。项目由120 m的超高层办公楼及其附属裙房组成。作为央企总部地块的重要组成部分，落成后可与中国馆交相辉映，共同构成黄浦江畔的新地标。

■ **Architectural Design**

Since concentrated green land and Huangpu River are on the south and north sides of the base, designers arranged the traditional core tube on the west side of the building, which not only protect the offices from western sun but also leave more space to enjoy the landscape on north and south. Super high-rise tower is in mega-frame structure, and each six office floors form a unit hangs on it, the original compressed vertical components are replaced by the tensioned ones, so that they can minimize the size of indoor vertical components and improve utilization of the plane. All the building volumes share the same special facade and giant vertical aluminium plates on east and west sides, which keep the sunlight efficiently and realize a unified building image.

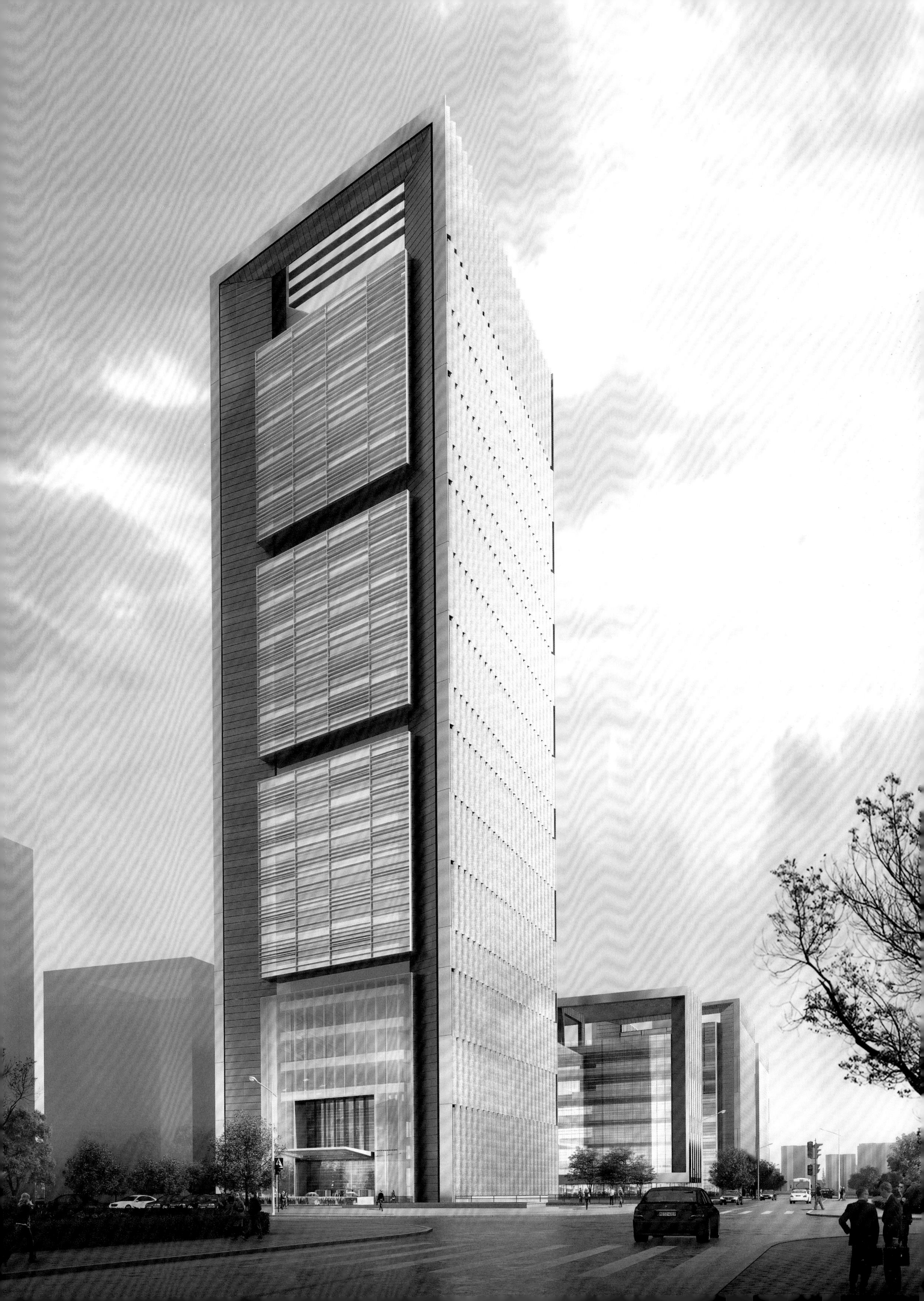

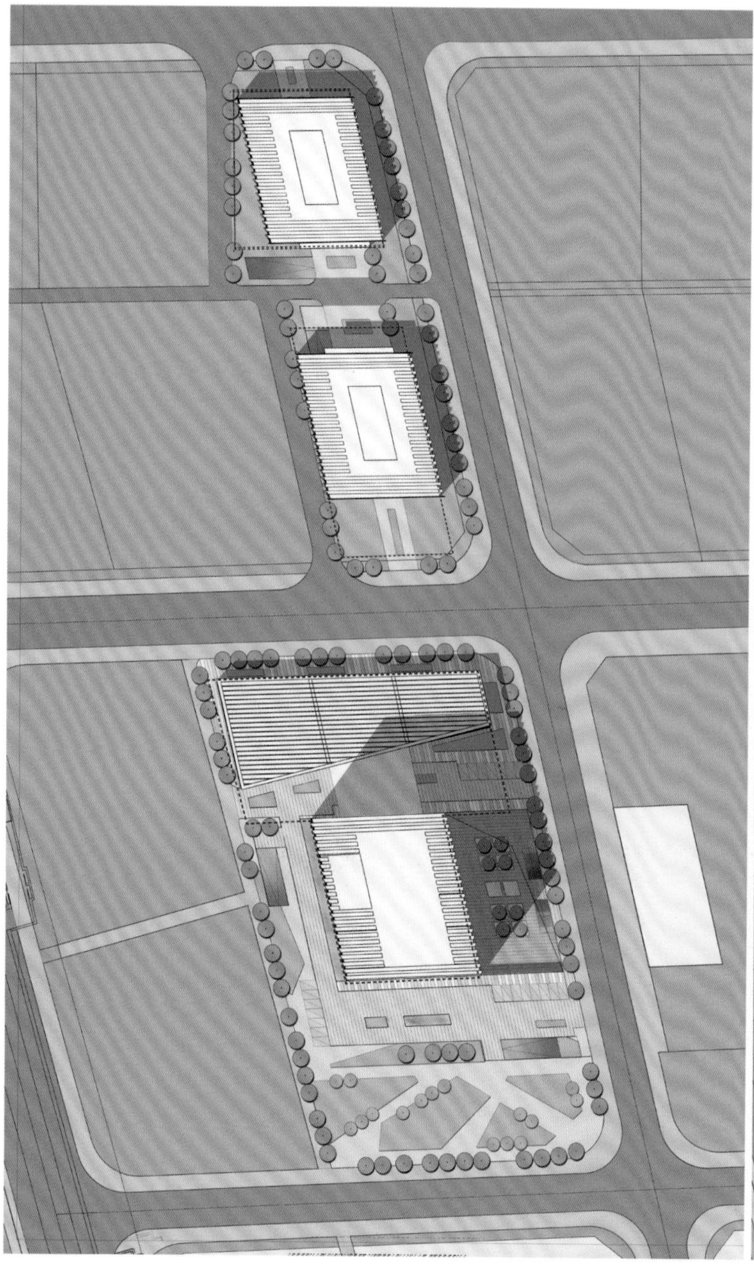

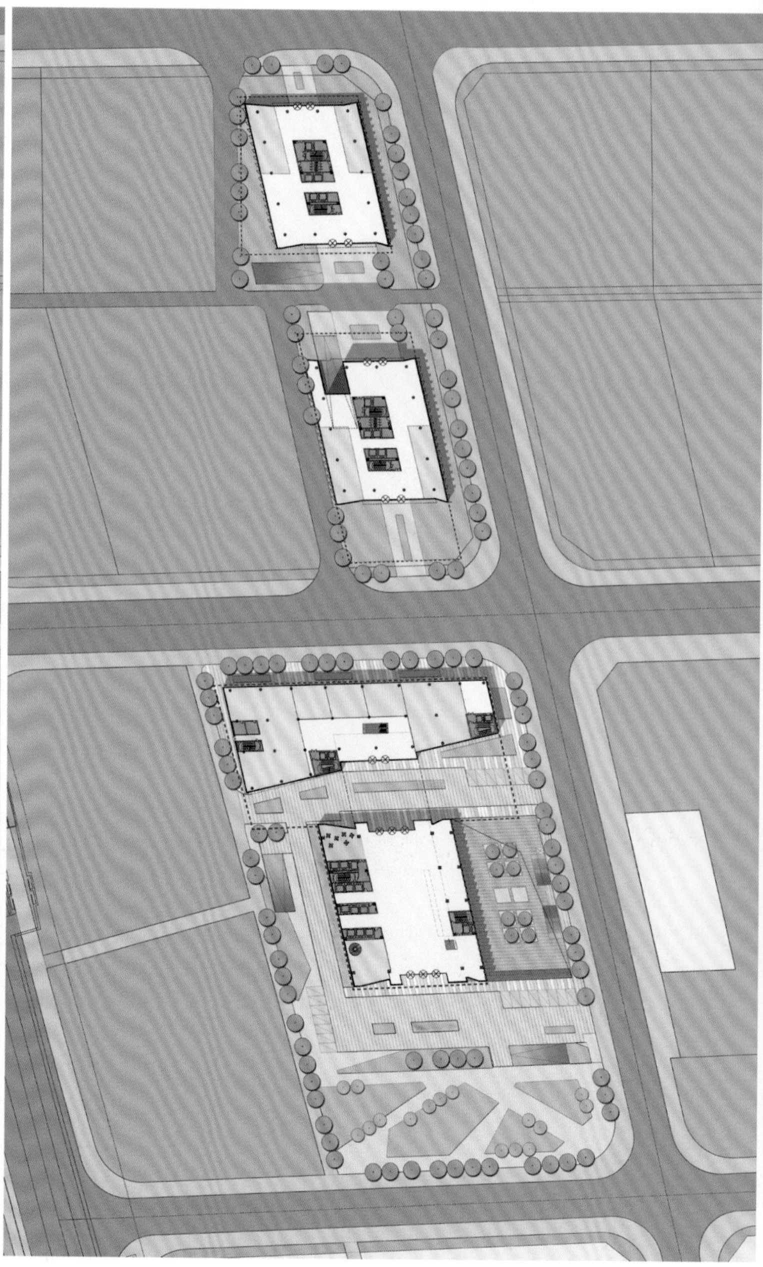

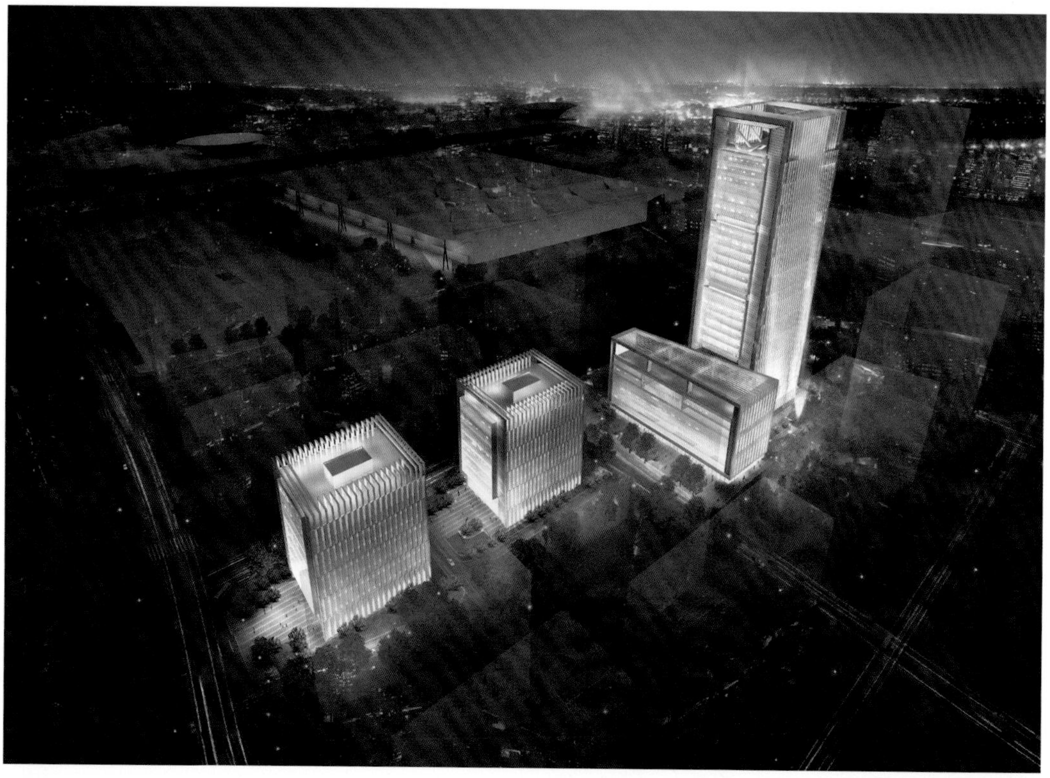

■ 建筑设计

基地南北两侧为集中绿地和黄浦江，所以设计中将传统的中央核心筒放在建筑西侧，既能减少西晒影响又能在中间留出更多的空间尽享南北景观。超高层塔楼选用巨型框架作为结构形式，办公层每六层为一个单元悬挂在巨型框架中，原本受压的竖向构件改为受拉，受力合理，这样就能最大限度减小室内竖向构件尺寸，提高平面使用率。所有的建筑体块均采用同样别致的立面造型，建筑东西两侧均安装巨型竖向铝板，不仅能起到遮阳作用，还使得建筑形象统一而整体。

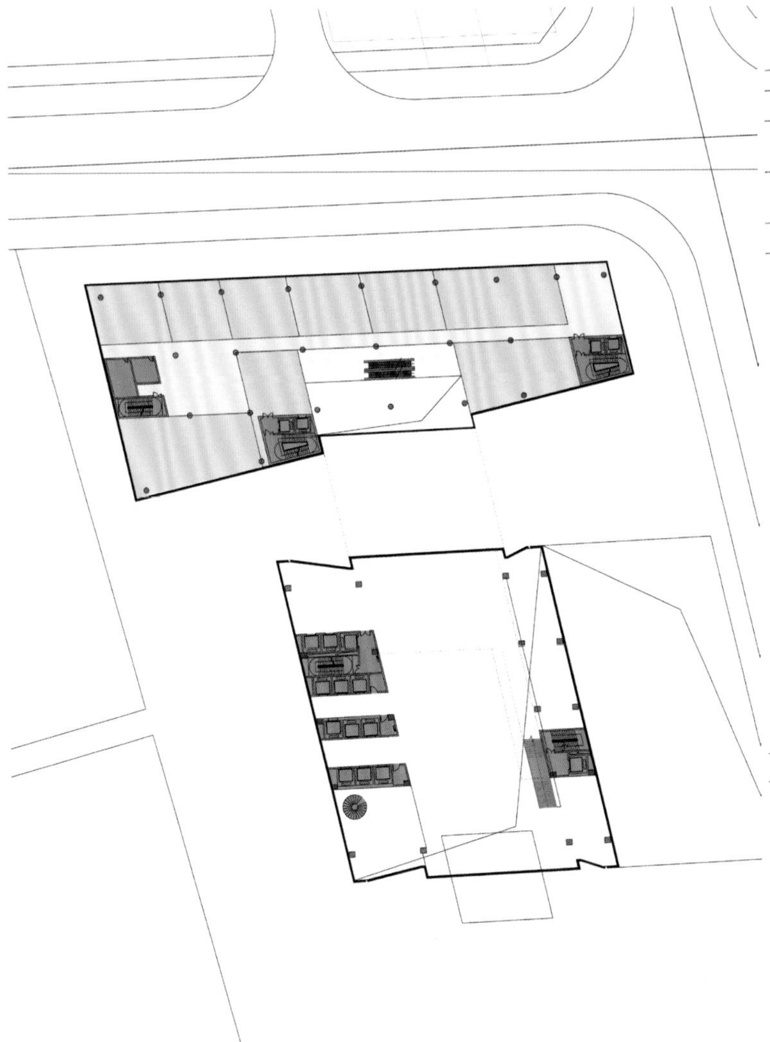

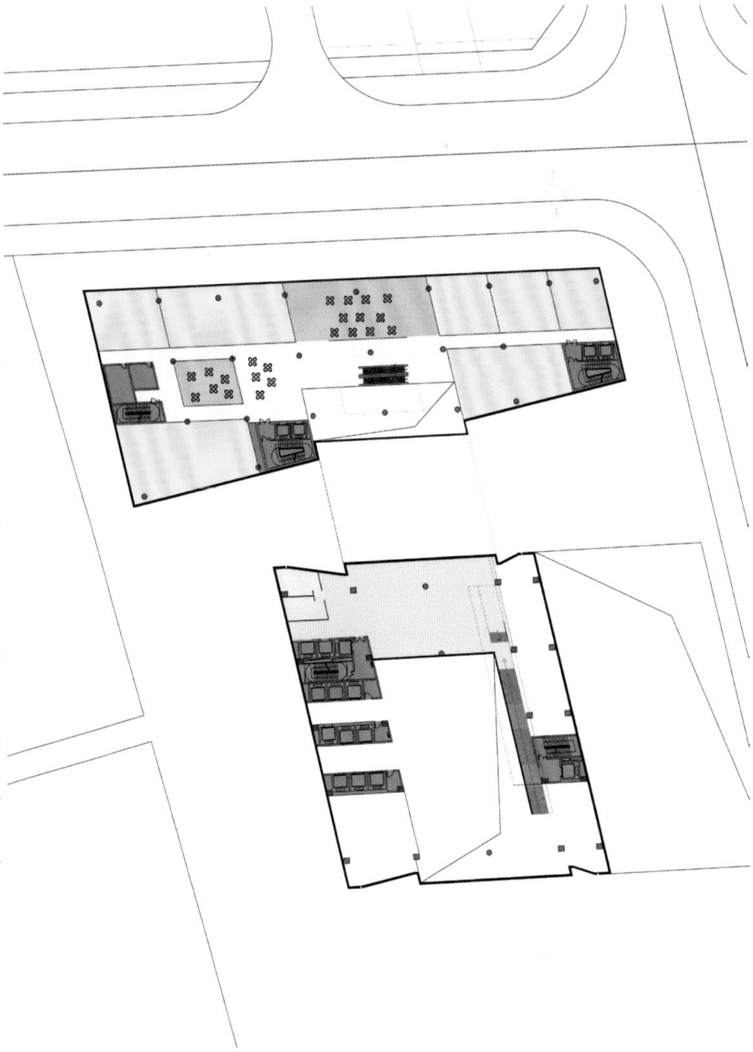

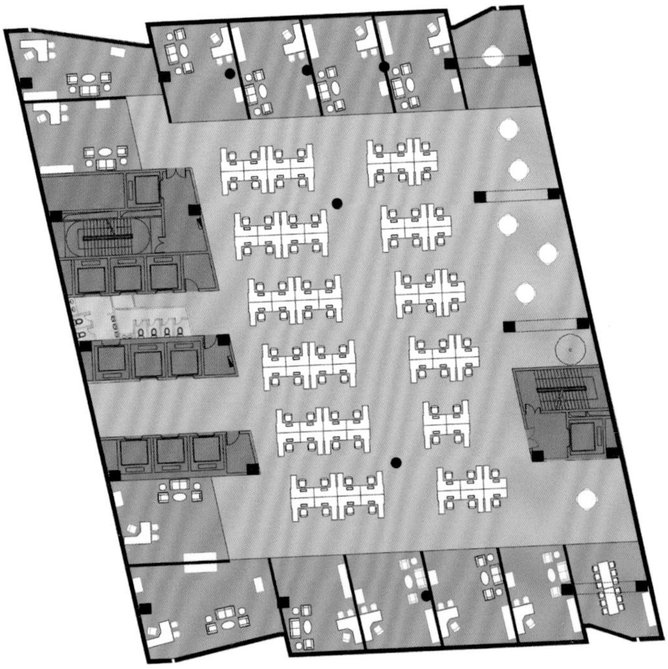

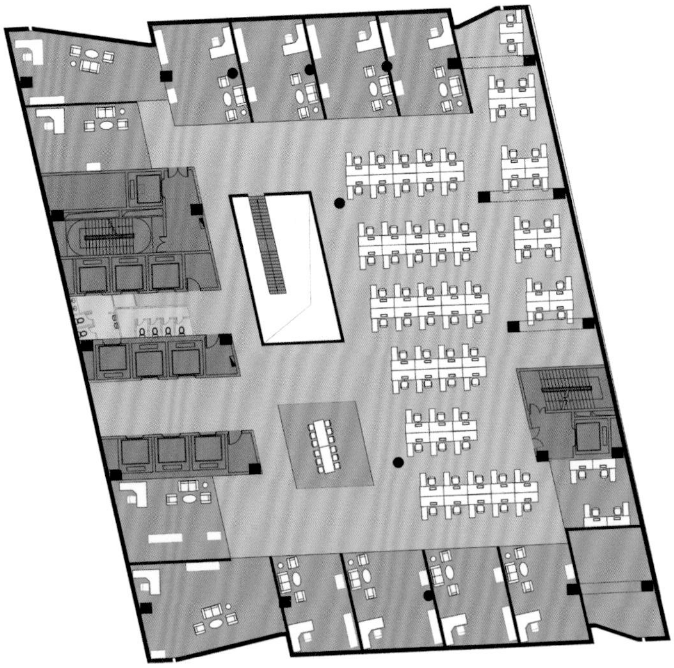

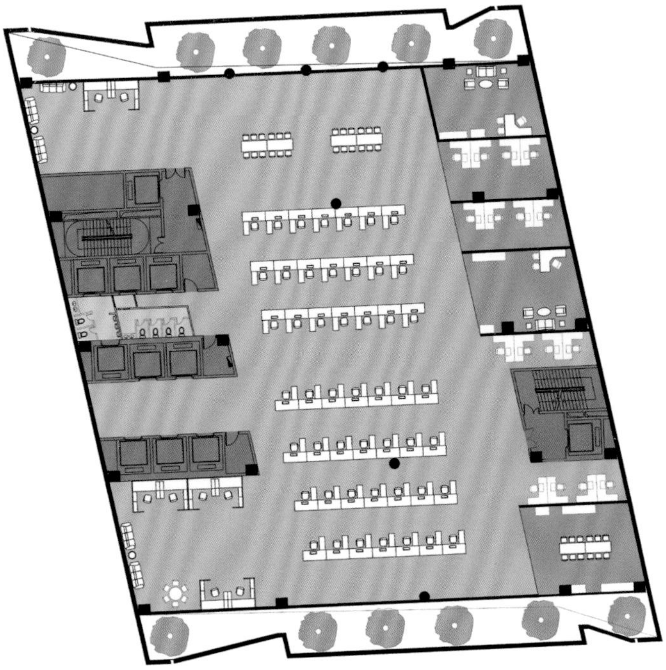

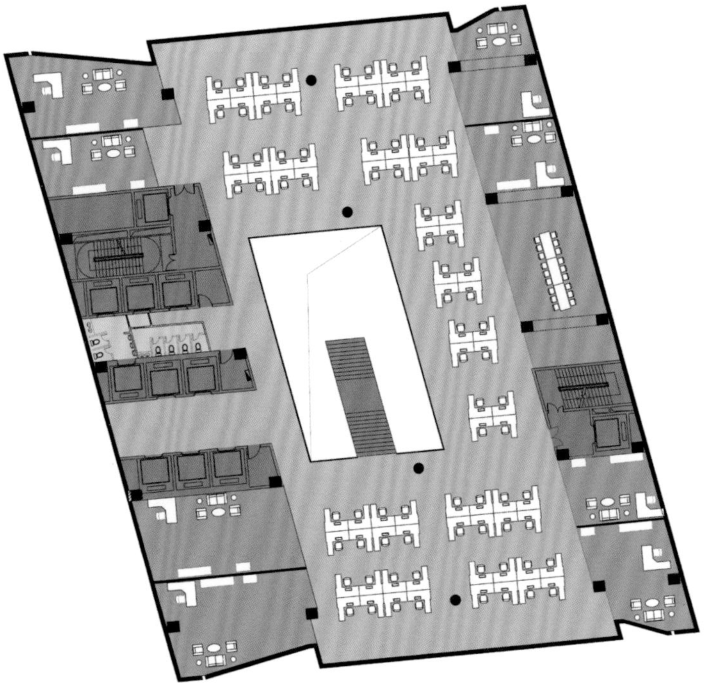

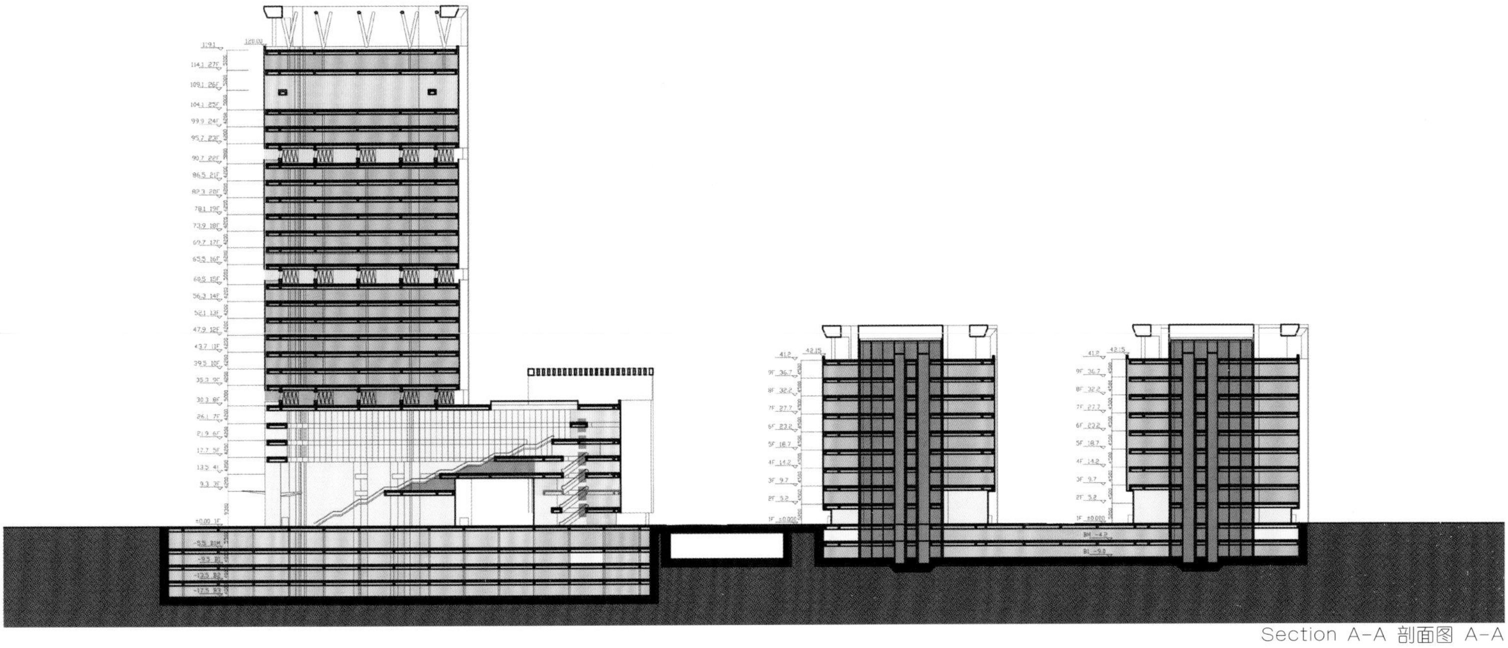

Section A-A 剖面图 A-A

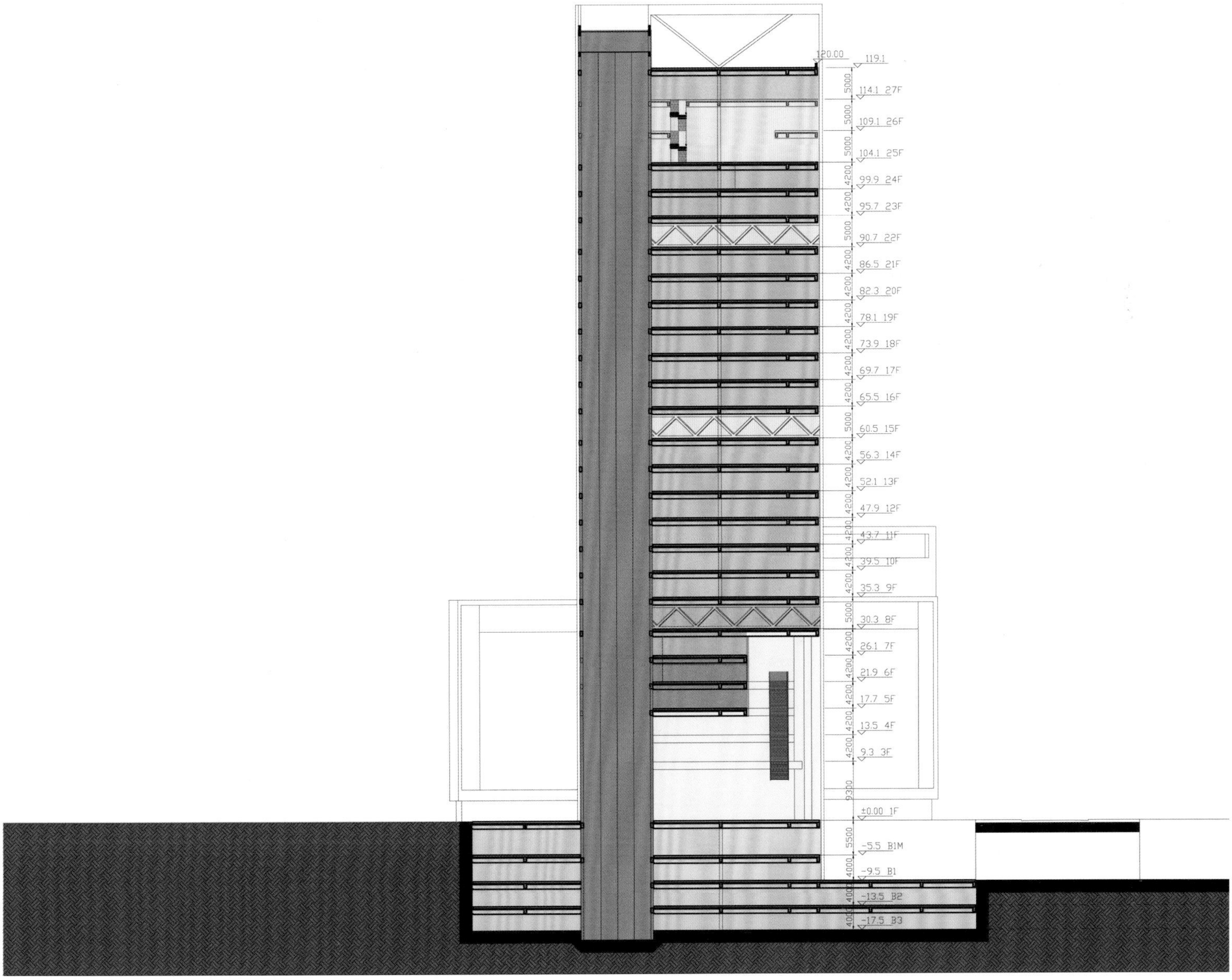

Section B-B 剖面图 B-B

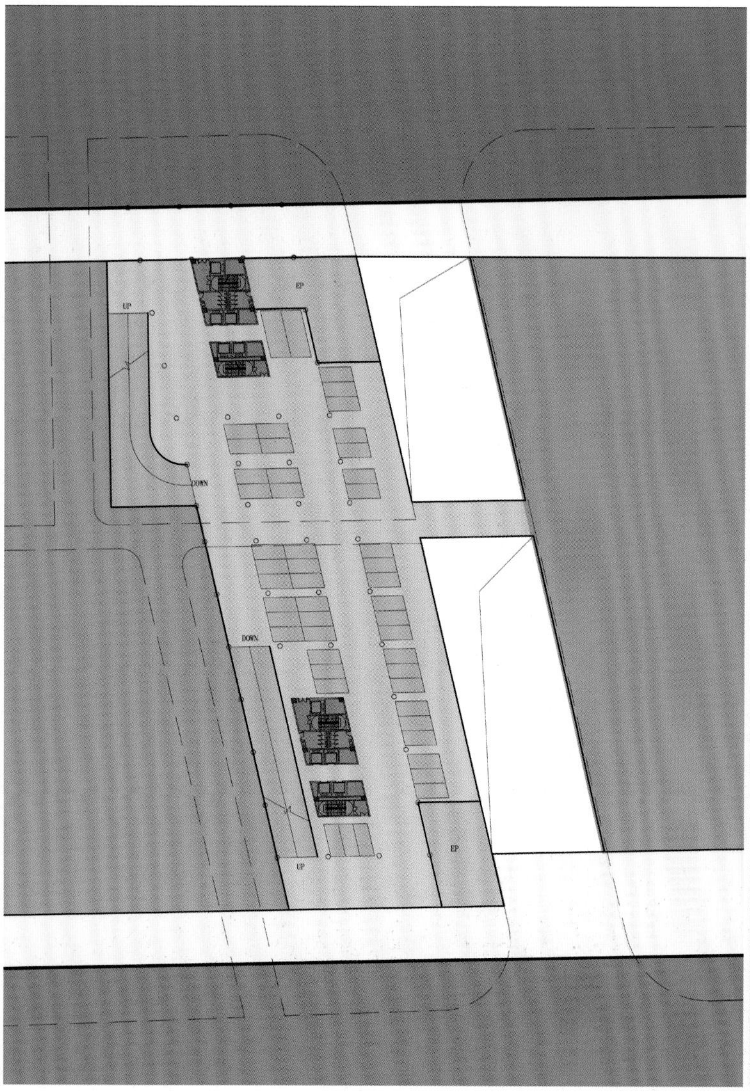
UP
EP
DOWN
DOWN
UP
EP

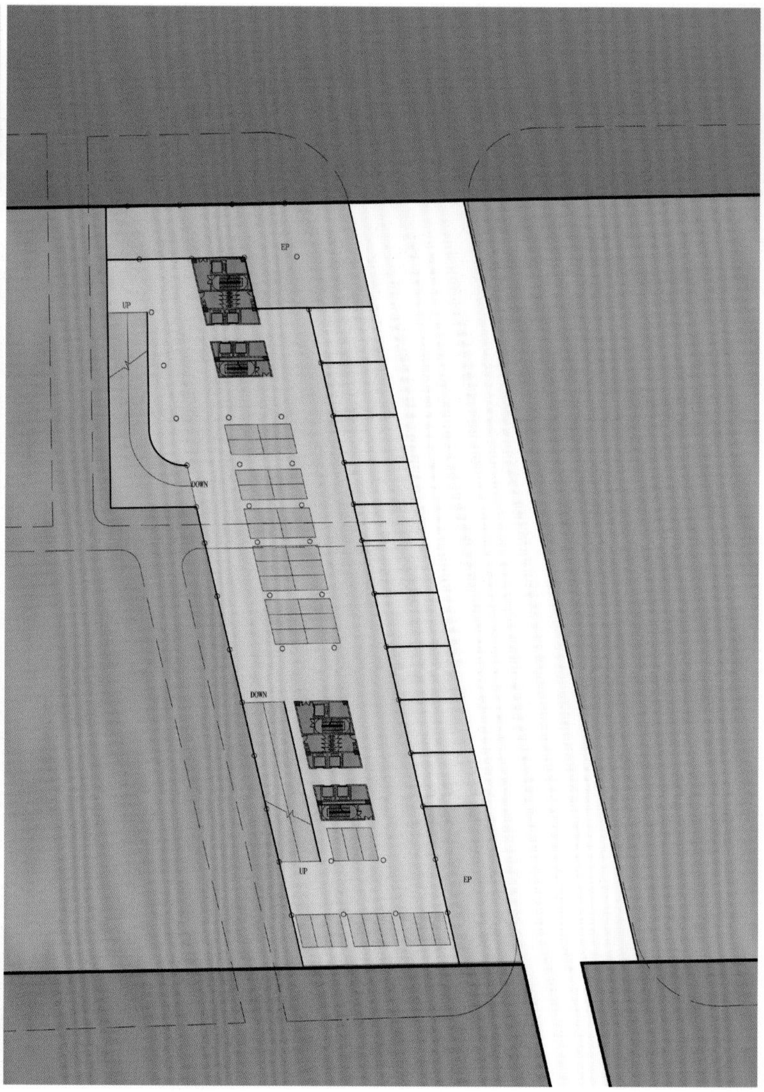
EP
UP
DOWN
DOWN
EP

CSCEC Steel HQ Tower
中建钢构总部大厦

Keywords 关键词

Steel-frame Structure
钢架结构

Huge Truss
巨型桁架

Sky Garden
空中花园

Location: Shenzhen, Guangdong, China
Architectural Design: Jiang Architects & Engineers
Land Area: 2,892.5 m^2
Total Floor Area: 49,727.5 m^2
Plot Ratio: 13.67
Status: Schematic designed in 2013

项目地点：中国广东省深圳市
设计单位：上海江欢成建筑设计有限公司
用地面积：2 892.5 m^2
总建筑面积：49 727.5 m^2
容 积 率：13.67
状　　态：2013年方案设计

Features 项目亮点

The all-steel structure highlights the corporate image of CSCEC Steel and is appropriate for reverse construction method which could speed up the construction progress.

整个建筑采用全钢结构，突现中建钢构的企业形象，同时也易于采用全逆作法，加快施工进度。

■ Overview

CSCEC Steel HQ Tower is located on the west of Zhongxin Road and the south of Dengliang Road in Shenzhen Houhai Business District, which is designed to provide commercial services, covers a land of 2,892.5 m^2 and is comprised of a 150 m super high-rise office building and the annex building.

■ 项目概况

中建钢构总部办公大厦位于深圳市后海中心区中心路西，登良路南。本项目为商业性服务设施用地，总用地面积为2 892.5 m^2。项目由150 m超高办公楼与裙房组成。

■ Architectural Design

Located in Shenzhen Houhai Business District, the base borders a green land to the west and Shenzhen Bay to the east, and is adjacent to high-rise buildings on the south and north, where the sight is blocked. Therefore, the main office space is arranged on the east and west sides with the best view and the central space is dedicated to core tube, atrium and meeting space. The office building is arranged to segment and segment, and refuge storey and sky garden are inserted, which divides the office space into three parts for the headquarters, rent and sale respectively. In addition, the floor slab is staged accordingly and the sectional area of the column is half reduced. Vertical trusses support the east and west, dual core tube + huge truss support the south and north, and the thickness of core tube wall is reduced by half and the size of structural elements is decreased significantly, which increase the comfort level and utility rate for the office. Besides, the all-steel structure highlights the corporate image of CSCEC Steel and is appropriate for reverse construction method which could speed up the construction progress.

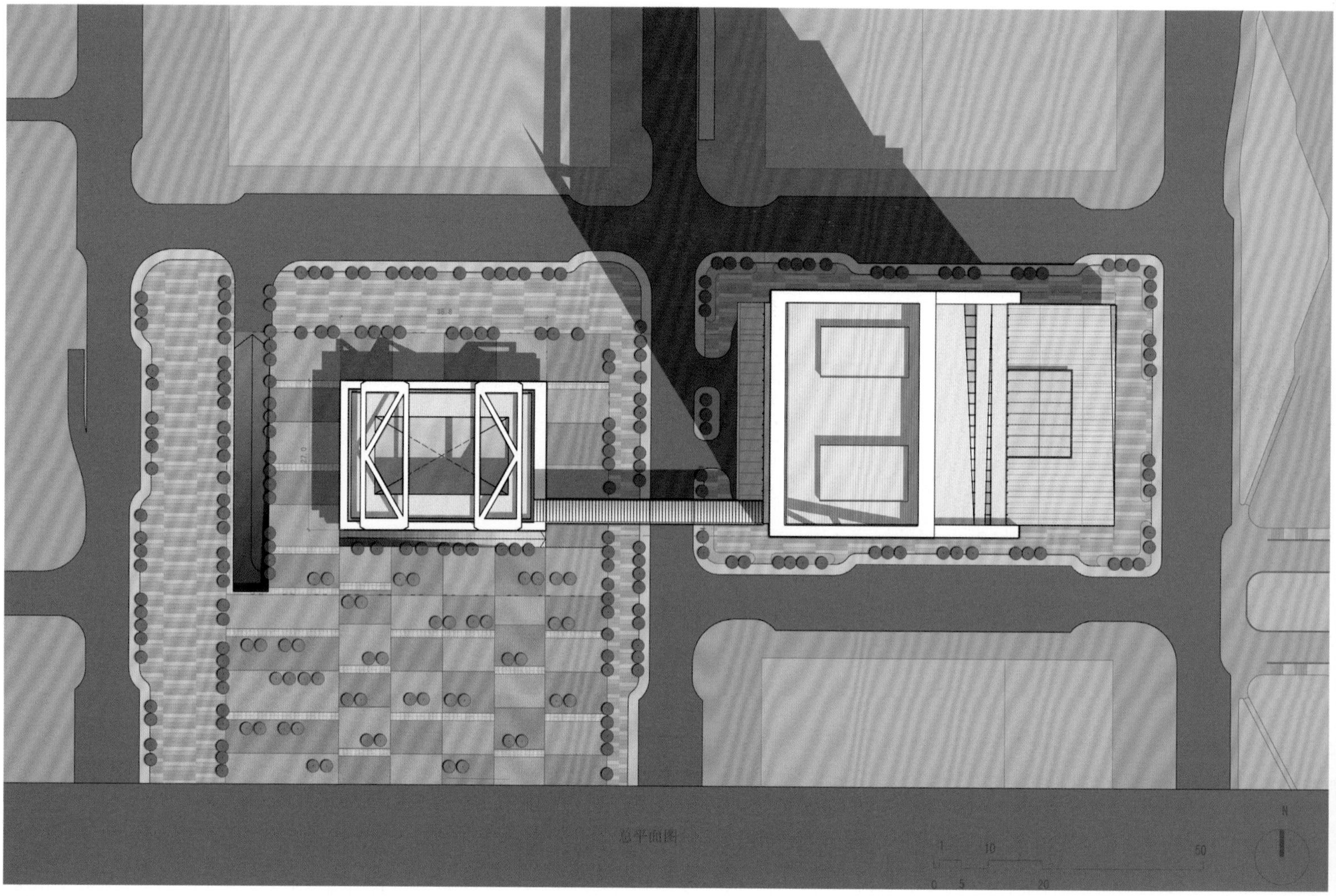

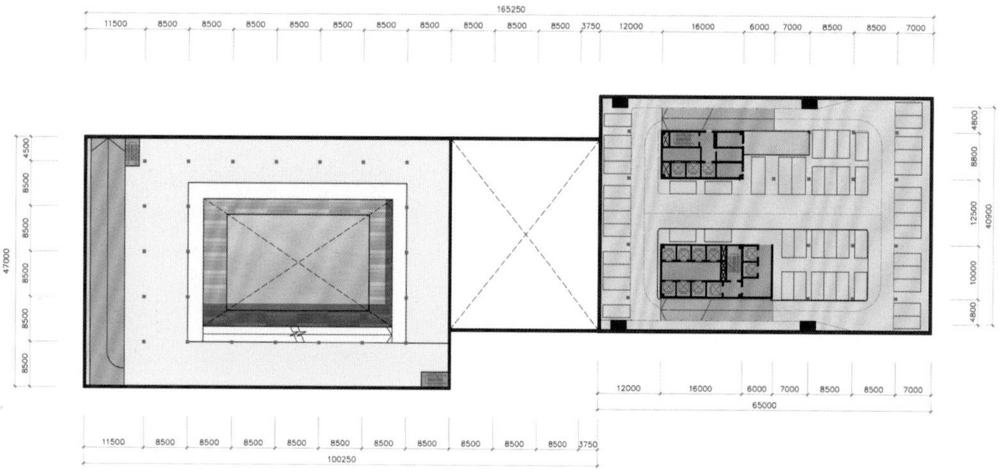

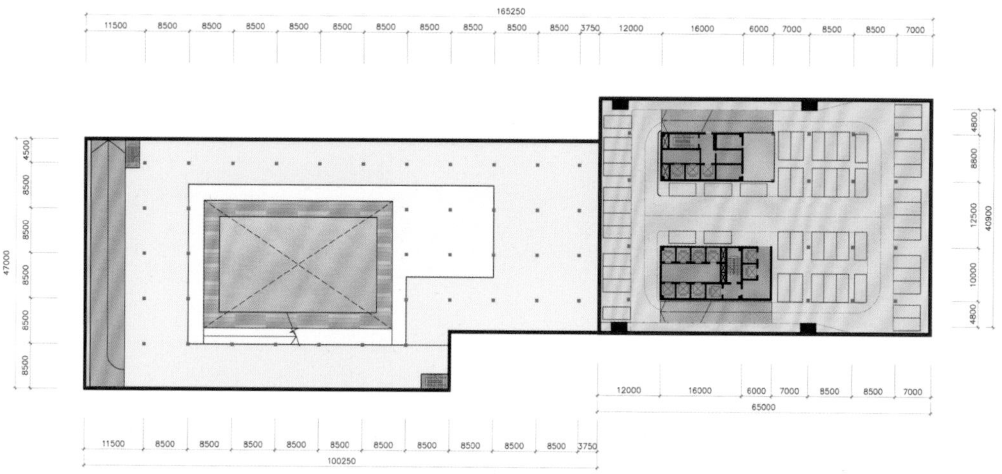

■ 建筑设计

基地位于深圳市后海金融区，西侧地块规划为绿地，东侧为深圳湾，而地块南北侧都有高层建筑阻隔视线，因此设计中将主要办公空间布置在景观最好的东西两侧，中部空间作为核心筒、中庭以及会议空间。将办公楼分段布置，其间设置了避难层和空中花园，这样能将主要办公空间按总部自用、出租、出售三大功能区分开来。将办公楼分段布置后，结构上也采用楼板分段支撑，立柱截面积至少减少一半。东西向依靠两侧垂直桁架，南北向依靠中间双核心筒+巨型桁架，核心筒壁厚减少一半，结构构件尺寸又能大幅减少，增加了办公舒适度及实用率。整个建筑采用全钢结构，突现中建钢构的企业形象，也易于采用全逆作法，加快施工进度。

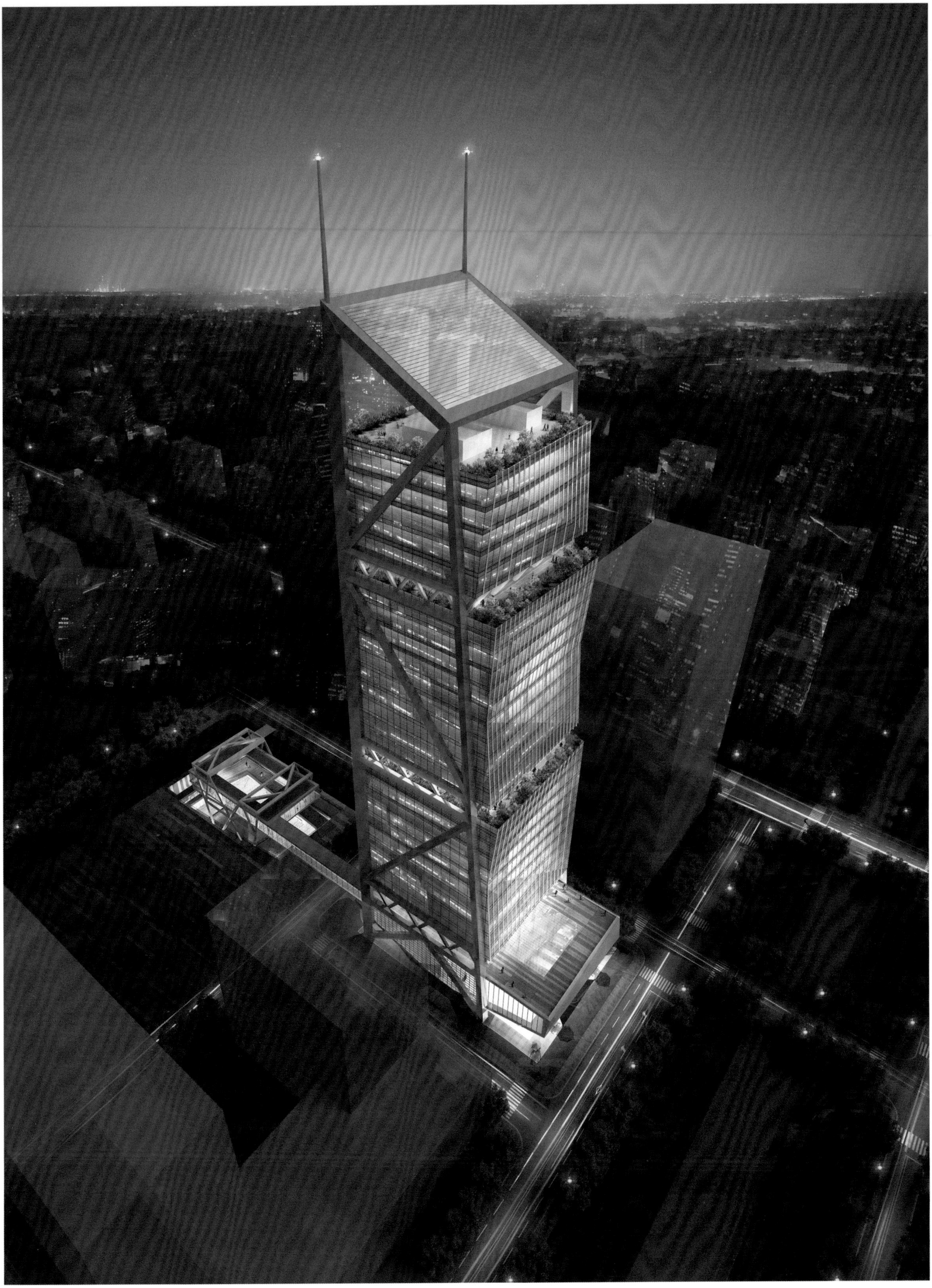

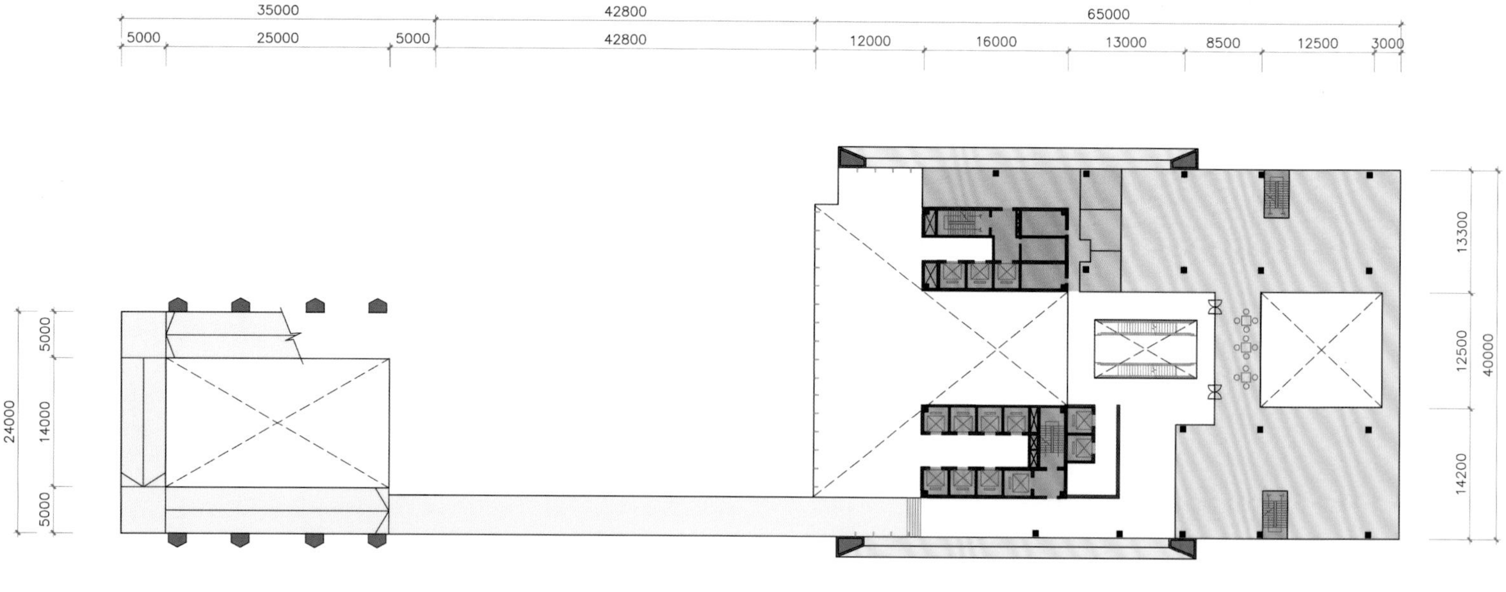
35000
42800
65000
5000
25000
5000
42800
12000
16000
13000
8500
12500
3000
24000
5000
14000
5000
13300
12500
14200
40000

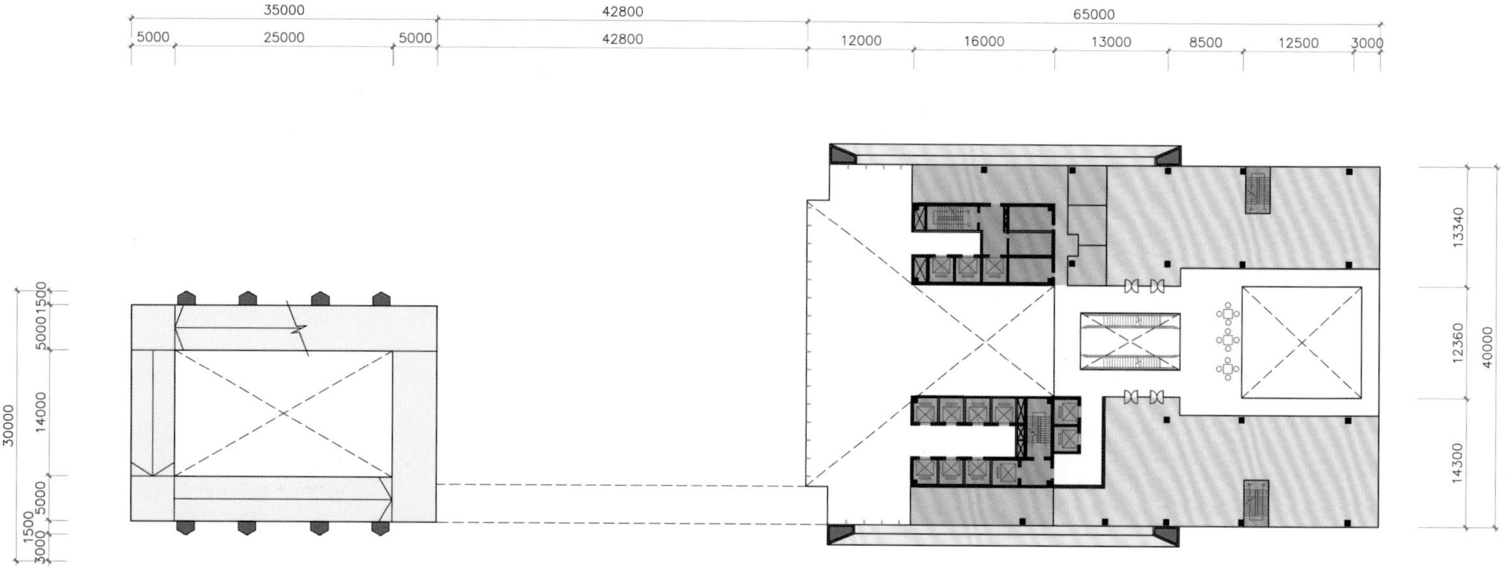
35000
42800
65000
5000
25000
5000
42800
12000
16000
13000
8500
12500
3000
30000
1500
5000
14000
5000
1500
3000
13340
12360
14300
40000

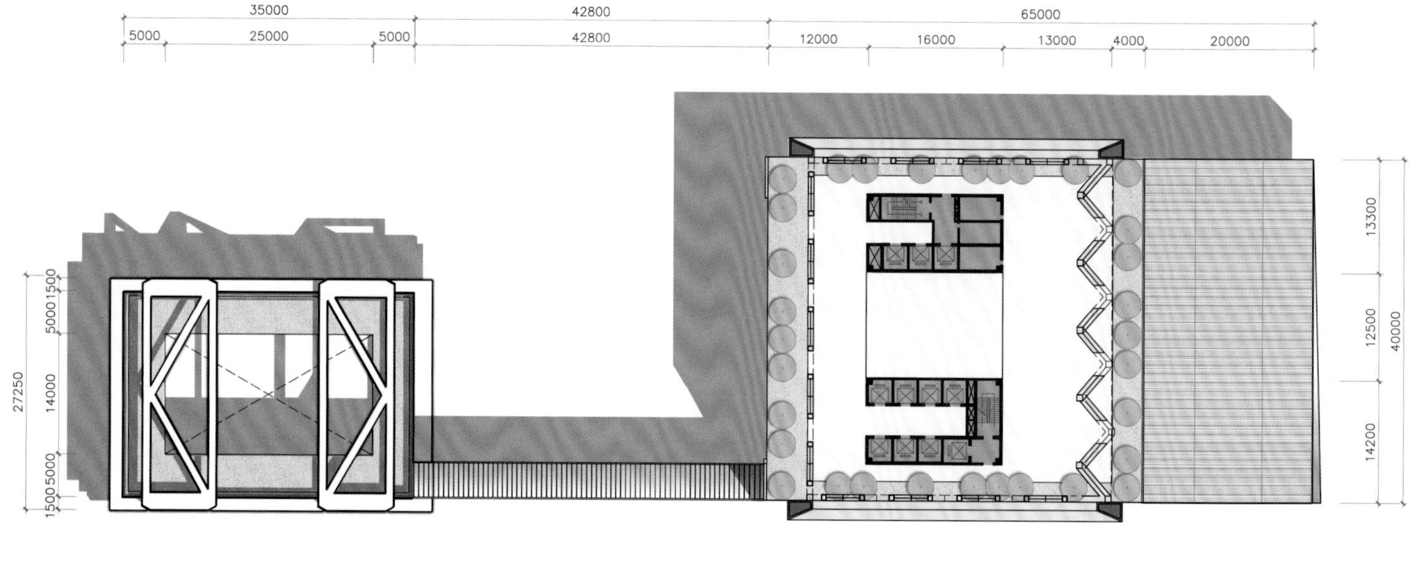
35000
42800
65000
5000
25000
5000
42800
12000
16000
13000
4000
20000
27250
1500
5000
14000
5000
1500
13300
12500
14200
40000

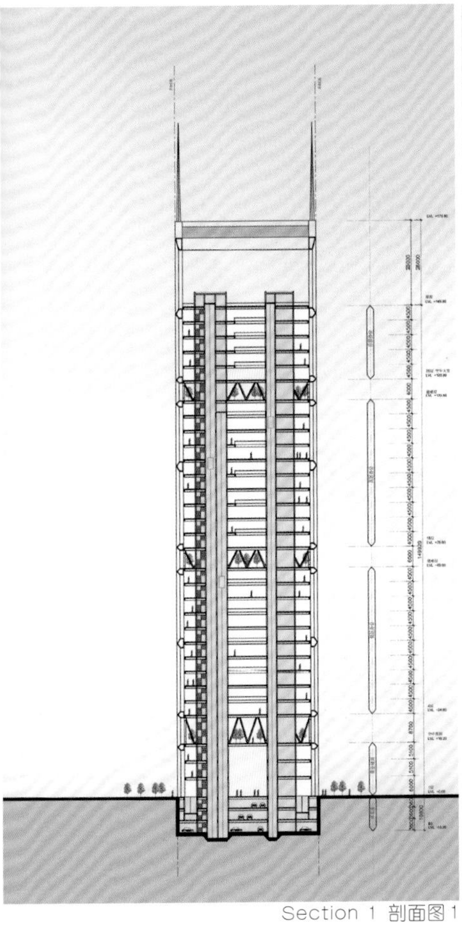
Section 1 剖面图1

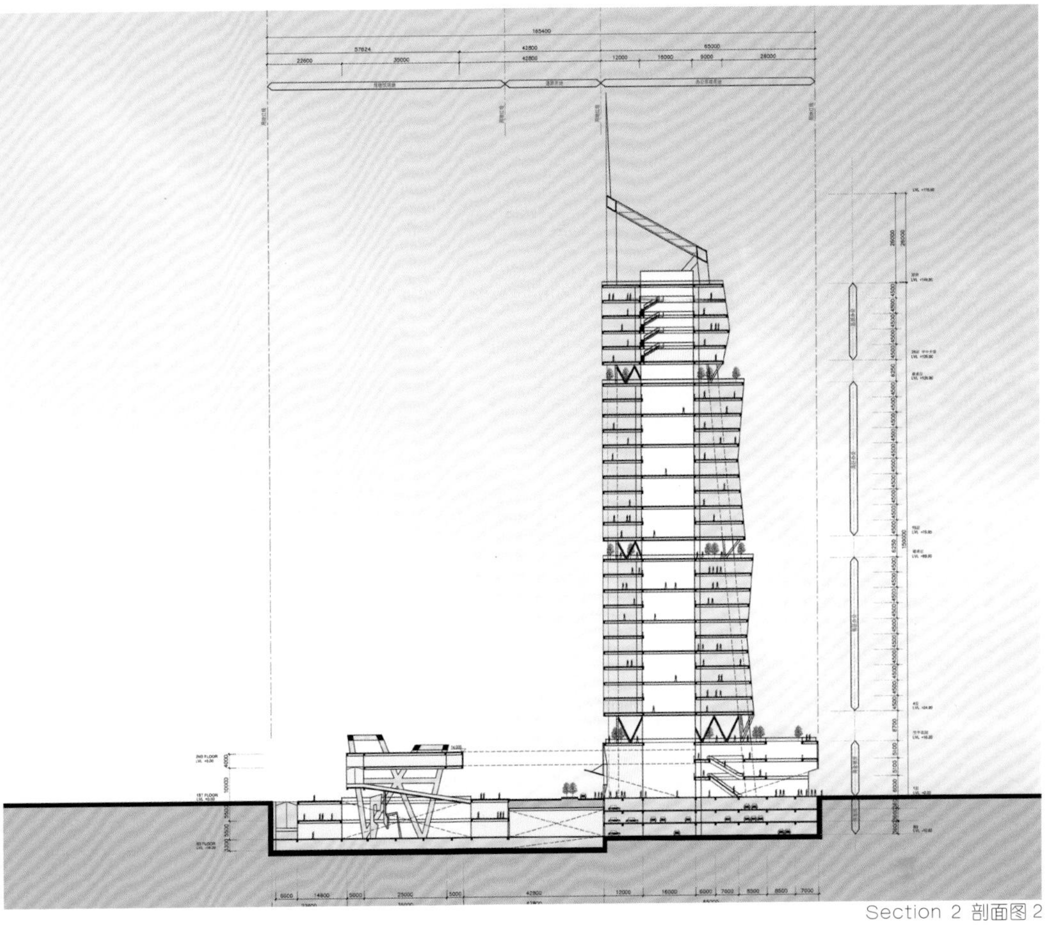
Section 2 剖面图2

New Jiangwan Office Park
新江湾城办公园区

Keywords 关键词

Flexibility and Efficiency
灵活高效

Creativity
创意性

Facade Design
立面设计

Location: Shanghai, China
Owner: Shanghai Chengtou Group
Architectural Design: 10 DESIGN
Design Partners: Barry Shapiro, Scott Findley, Adrian Boot
Land Area: 160,000 m^2
Floor Area: 400,000 m^2
Date of Construction Begins: March 2012
Date of Completion: 2014

项目地点：中国上海市
业　　主：上海城投
设计单位：10 DESIGN (拾稼设计)
设计合伙人：夏太白 葛芬利 魏智仁
用地面积：160 000 m^2
建筑面积：400 000 m^2
施工日期：2012年3月
完工日期：预计2014年

Features 项目亮点

Key objectives of the master plan are to create a good sense of scale, neighborhood and community, which is meant to be clean, modern and sophisticated.

项目规划以一系列大型庭院和一座中央社区公园为核心，旨在建造一座大型的办公园区，园区整体形象干净、现代、精致。

■ Overview

The New Jiangwan Office Park is a 16 hm^2 technical business park that is organized around a series of large courtyards and a central community garden. Key objectives of the master plan are to create a good sense of scale, neighborhood and community, a great environment for people to work in and a flexible design that meets the client's future market expectations featuring for example innovative 3,000 m^2 to 6,000 m^2 office headquarters.

■ 项目概况

新江湾城科技园位于上海市杨浦区新江湾城，是一个约16万m^2大的商务园区发展项目。项目规划以一系列大型庭院和一座中央社区公园为核心，旨在建造一座大型的办公园区，为在此上班的人们提供一个优良的工作环境，并创造一个符合客户未来市场发展需要和期望的灵活设计，如具有创意的3 000~6 000 m^2的办公总部。

■ **Architectural Design**

The building facades are designed to be cost effective, flexible and have a strong architectural character. The depth of the facades and the orientation of windows take solar radiation into consideration to help to reduce operating costs and create a better working environment within the buildings. The overall image of the Technical Park is meant to be clean, modern and sophisticated to attract future tenants and create a strong working environment. It will be an important addition to the New Jiangwan District of Shanghai.

■ **建筑设计**

建筑立面的设计符合成本效益，灵活且具有很强的建筑风格。立面深度和窗户的方向将考虑到利用太阳光的照射，以降低运营成本，创造一个更好的室内工作环境。该办公园的整体形象是秉承干净、现代和精致的标准来吸引租户，创造具有强大吸引力的工作环境。新江湾科技园区将成为上海新江湾城区的一个重要组成部分。

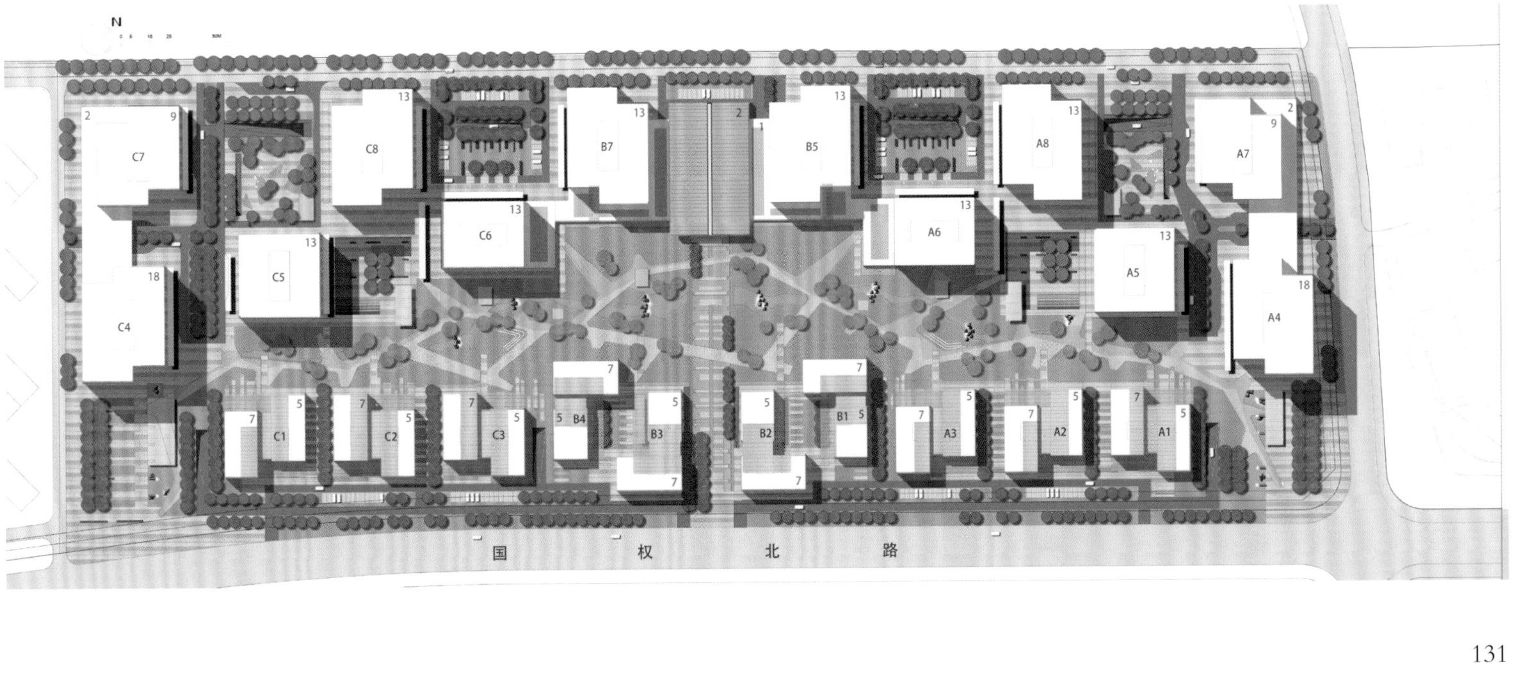

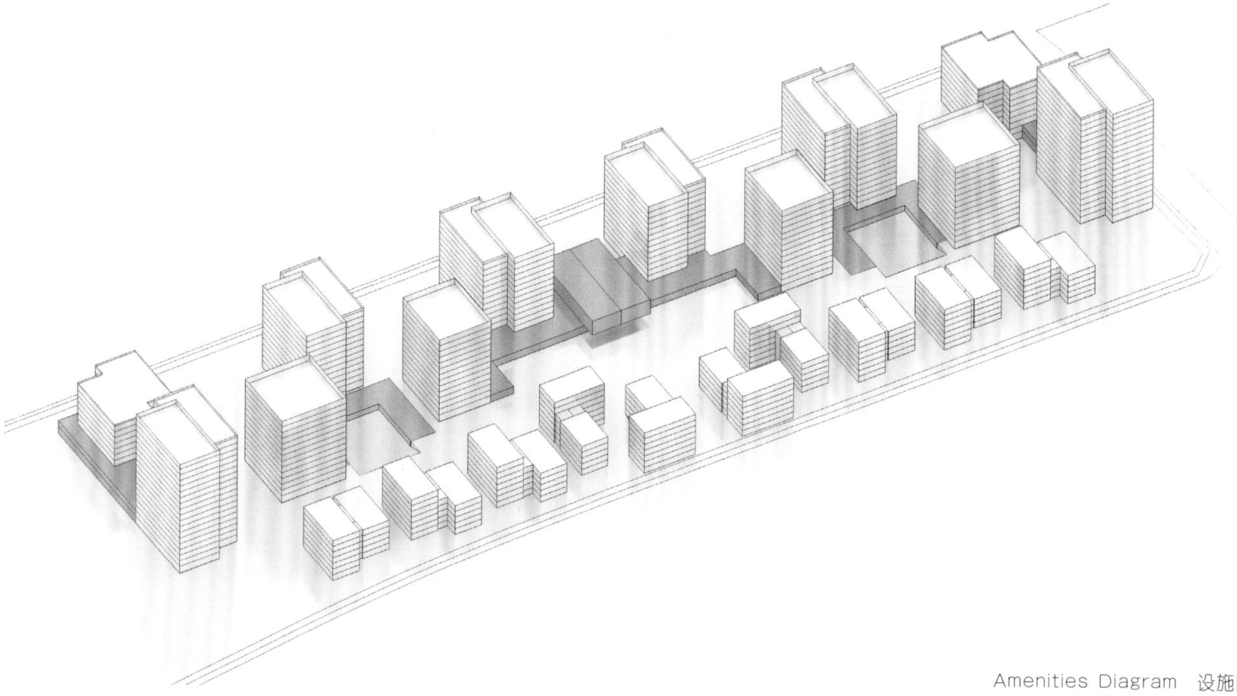

Amenities Diagram 设施图

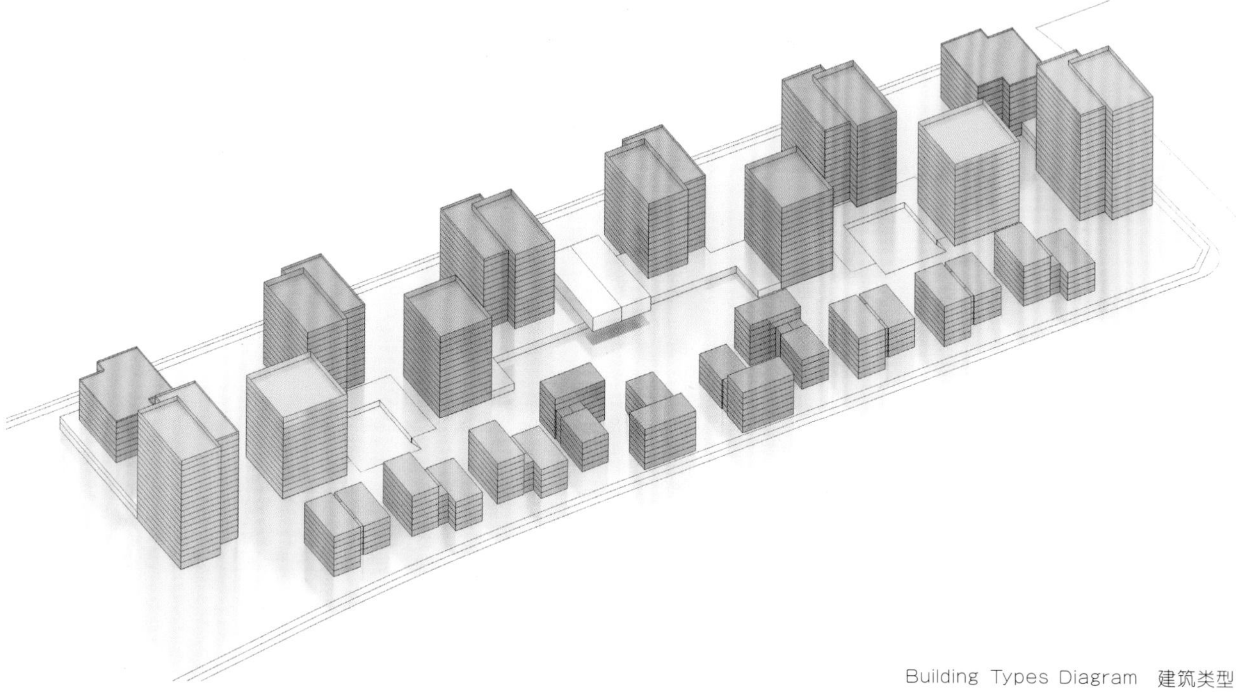

Building Types Diagram 建筑类型图

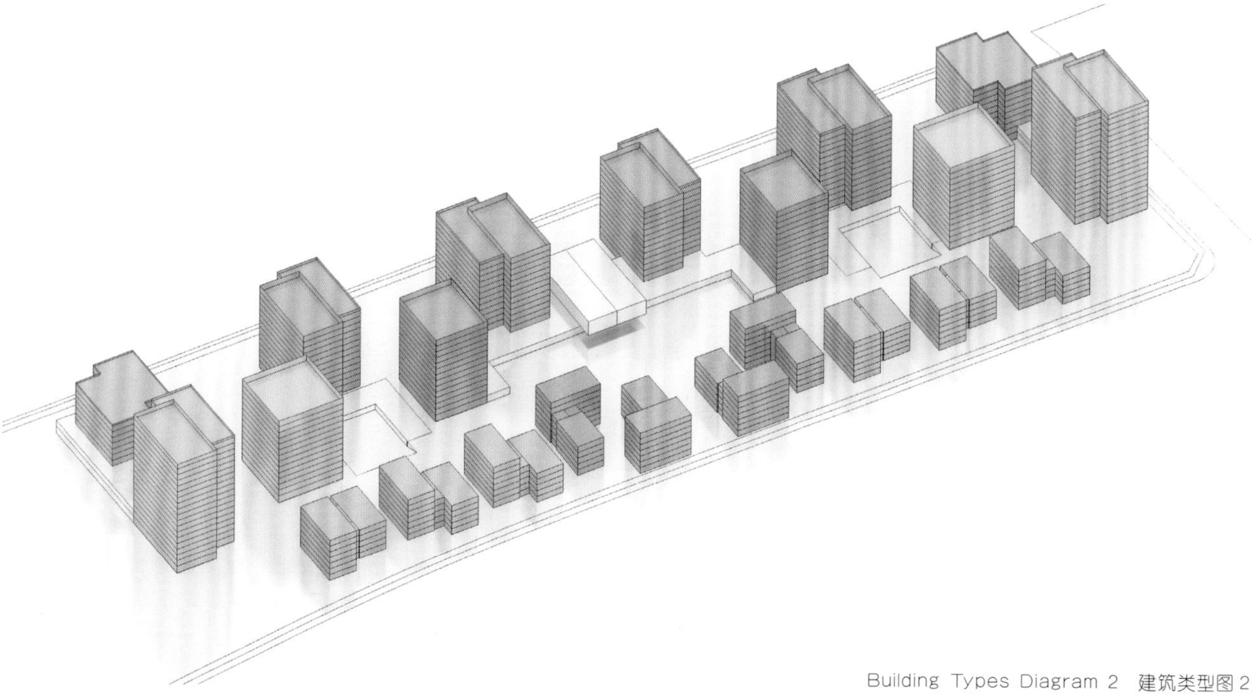

Building Types Diagram 2 建筑类型图 2

Concept Idea Feature Elements 概念特征元素图

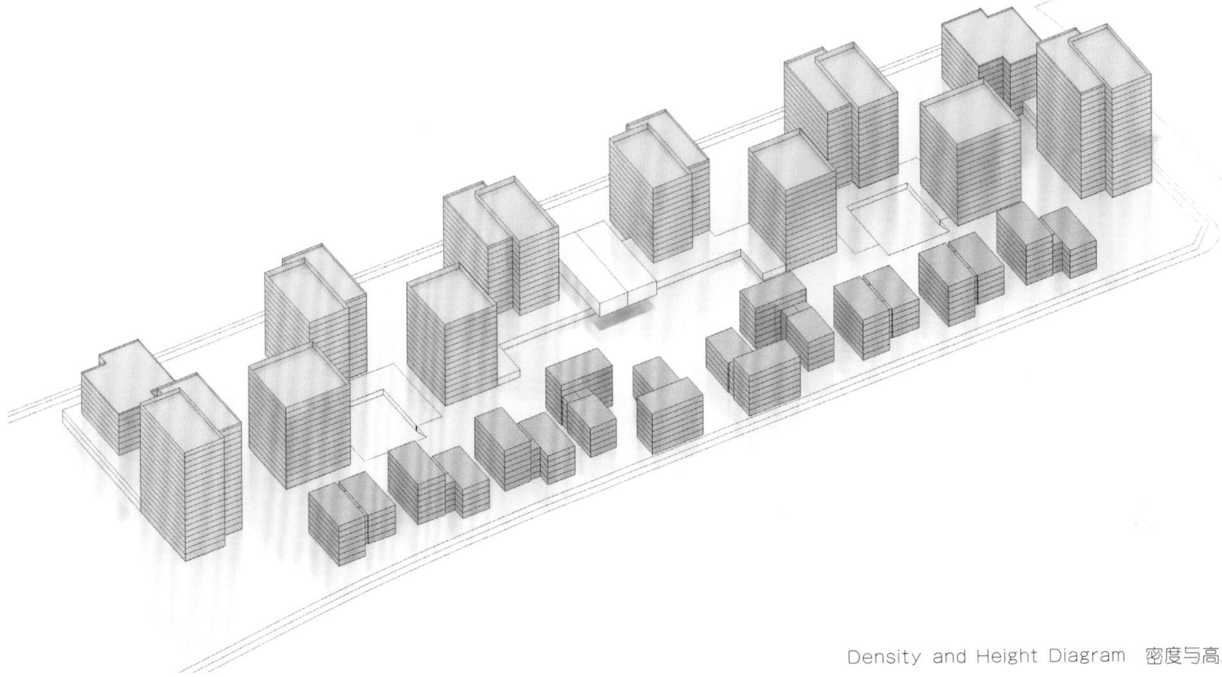

Density and Height Diagram 密度与高度图

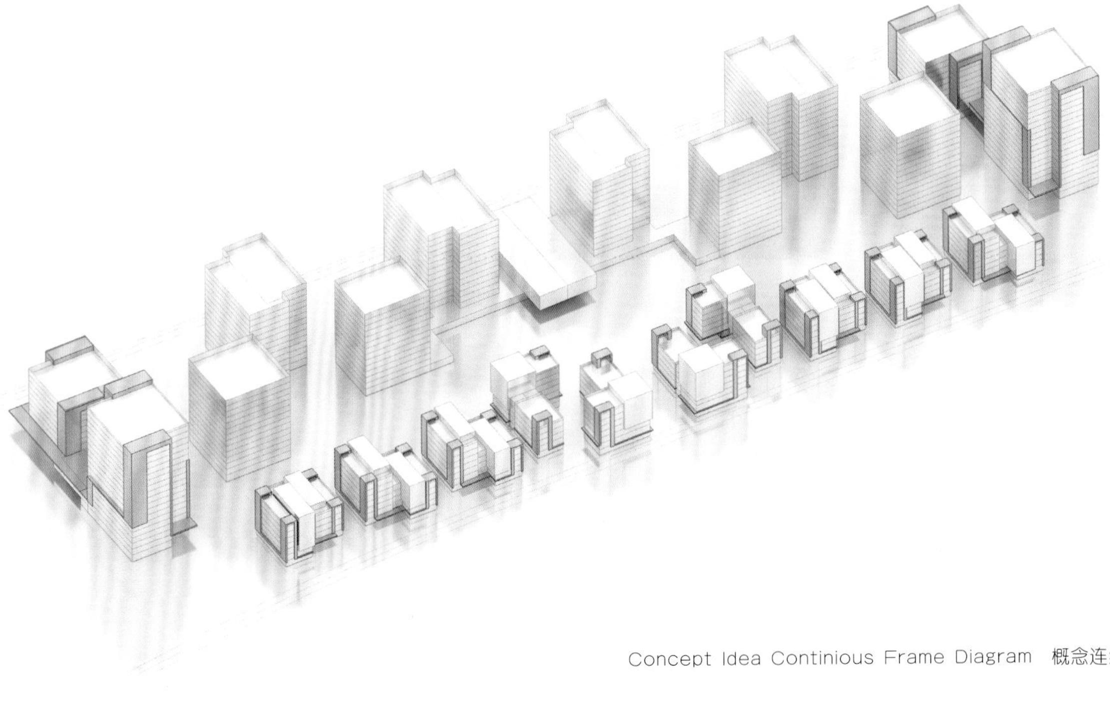

Concept Idea Continious Frame Diagram 概念连续性框架图

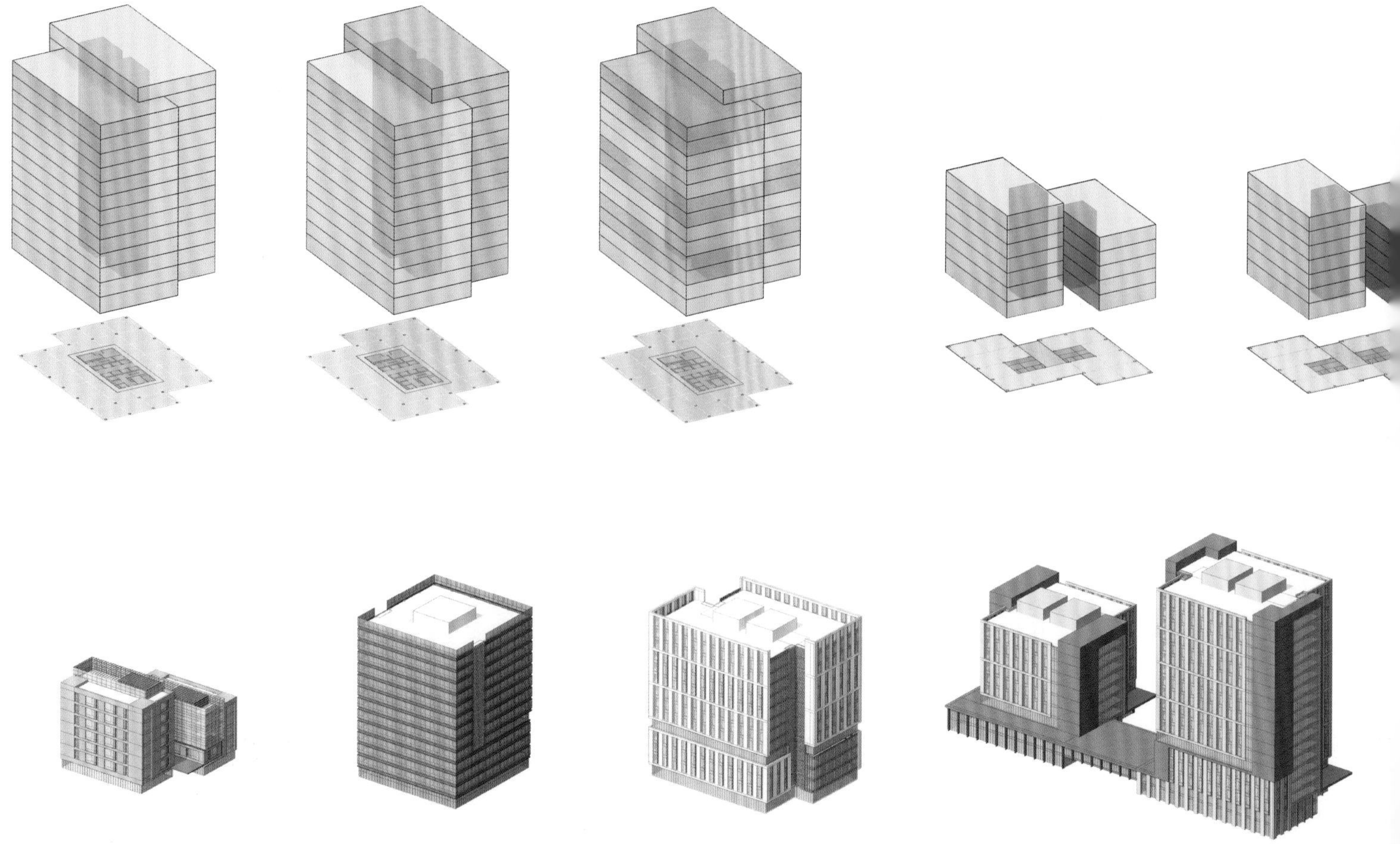

AXOs Summary Diagram　AXOs 总体图

Section 1 剖面图1

Section 2 剖面图2

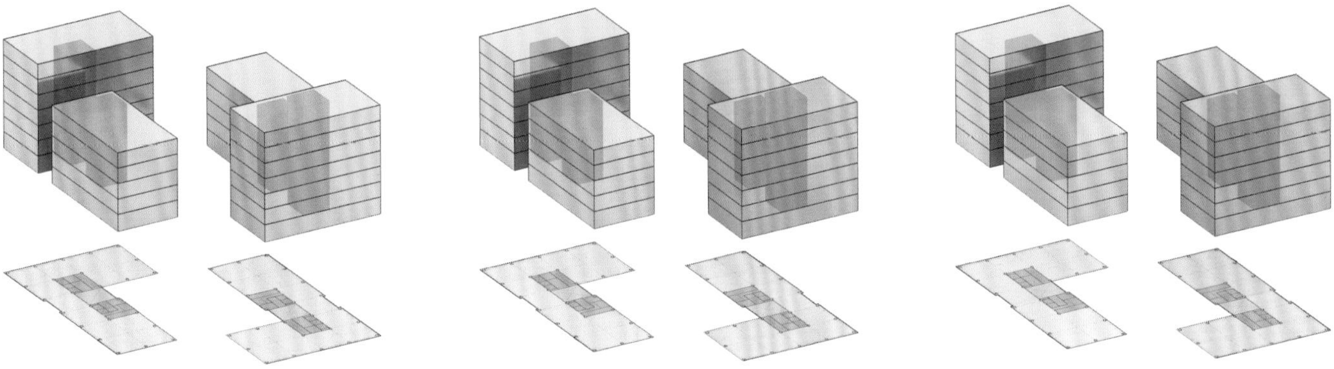

Flexibility Diagram 弹性图

Beijing New Scitech Plaza Renovation Scheme 北京新赛特大厦改造方案设计

Keywords 关键词

Functional Integration
功能整合

Commercial Atmosphere
商业氛围

Landmark Building
地标形象

Location: Chaoyang District, Beijing, China
Client: Scitech Group
Architectural Design: a+a Anderloni Associates
Total Floor Area: 248,100 m^2
Plot Ratio: 9.54

项目地点：中国北京市朝阳区
客　　户：赛特集团
建筑设计：a+a 建筑设计公司
总建筑面积：248 100 m^2
容 积 率：9.54

Features 项目亮点

The design of the building is inspired by "dragon and phoenix". By combining tradition with modernity, oriental culture with western culture, and using some fashionable and dynamic elements, it will create a new oriental and futuristic landmark for Beijing.

建筑设计以“龙凤呈祥”为设计根基，融入时尚、动感等元素，将古现代、东西方文化融合，打造具有东方韵味、人文前卫的北京新地标。

■ Overview

It is the renovation scheme for Beijing Scitech Plaza which was firstly built in the beginning of 1990s. As a leader of high-end shopping centers in China, it is ideally located in the famous CBD of Beijing, on east of Chang'an Road, 10 minutes drive to the Tian'anmen Square. It stands among foreign embassies and consulates, top residential districts, restaurants and hotels, enjoying an advantaged commercial atmosphere.

■ 项目概况

该项目为北京赛特购物中心改造工程， 北京赛特购物中心开业于90年代初，是中国高端购物中心的领跑者，它的地理位置得天独厚，位于长安街的东侧，地处北京著名的中央商务区，距天安门近十分钟的车程，其周围外国领事馆、高档住宅区、高档饭店林立，具备得天独厚的商业环境。

■ Design Concept

Before renovation, there were four independent buildings: a shopping center, an hotel building and two office buildings. These functions were distributed with low efficiency and commercial value. The solution to these problems is to integrate different functions and increase commercial spaces to promote its commercial value. Moreover, F&B, entertainment and recreation facilities are added to cater to modern lifestyle.

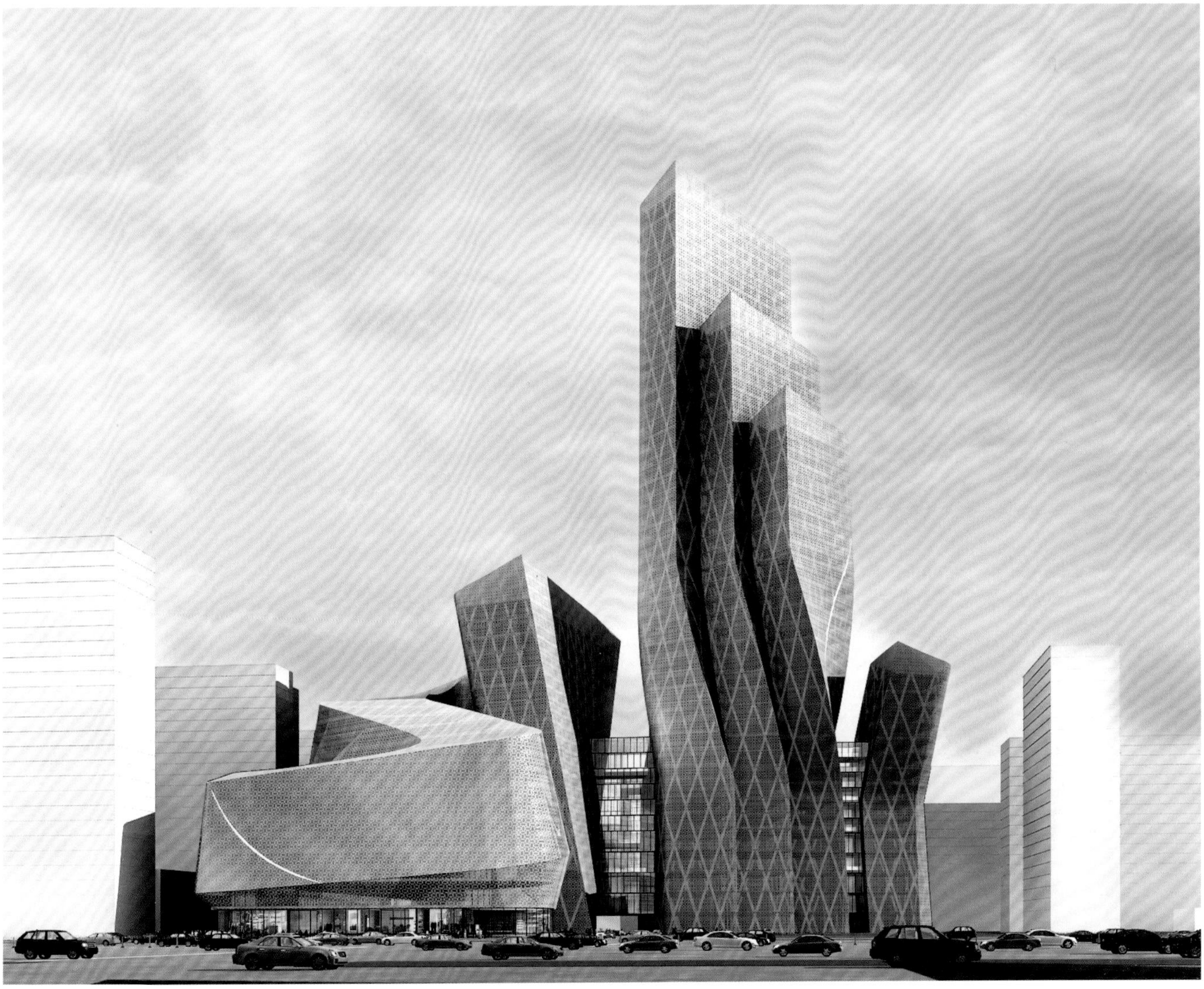

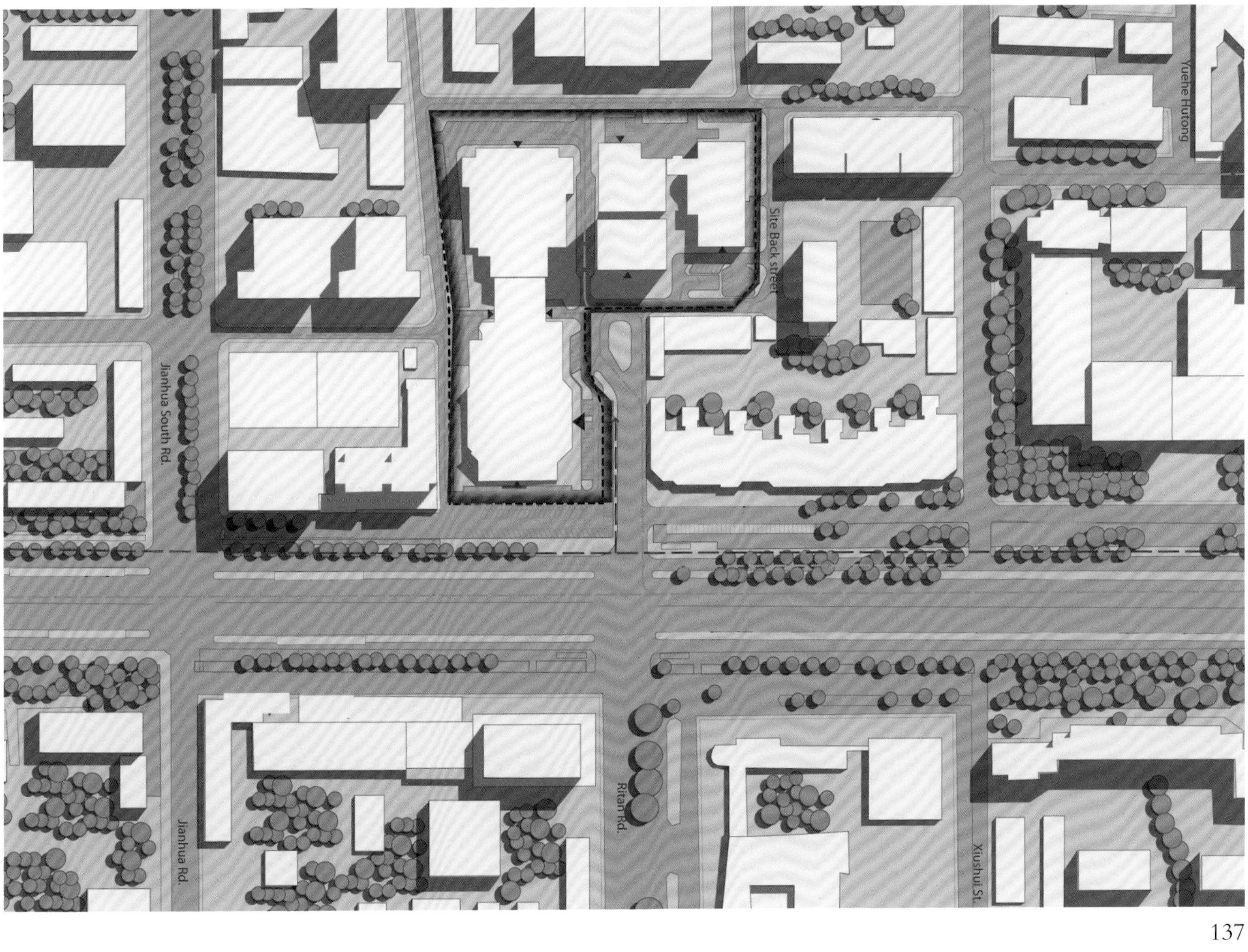
Yuehe Hutong
Site Back street
Jianhua South Rd.
Jianhua Rd.
Ritan Rd.
Xiushui St.

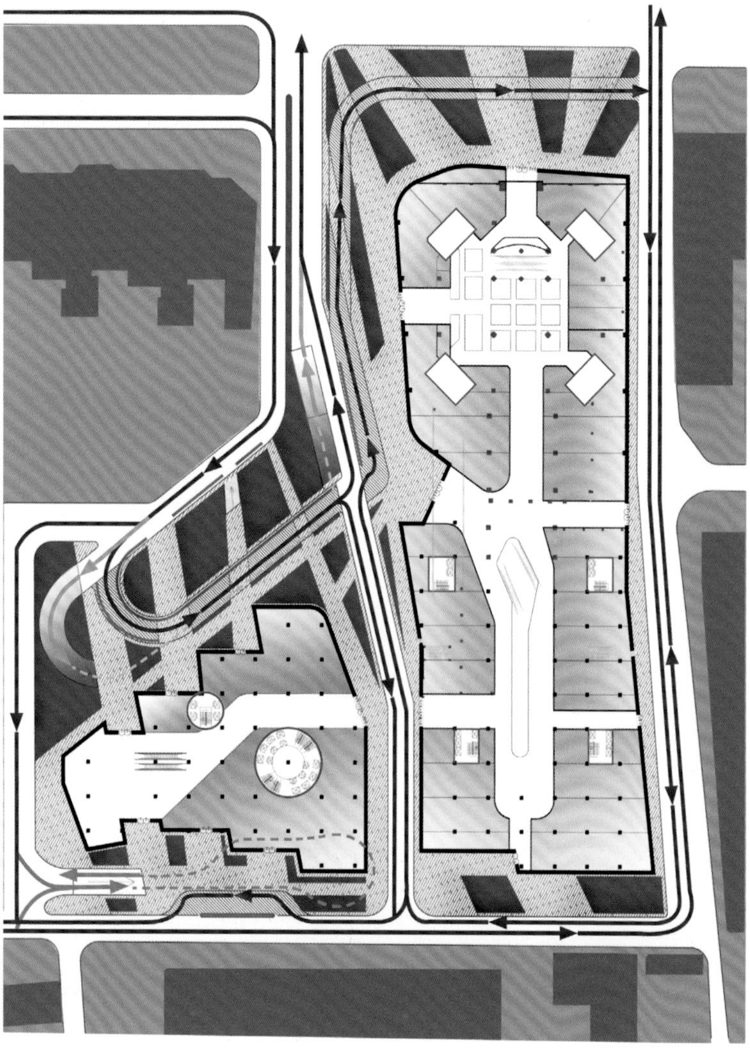

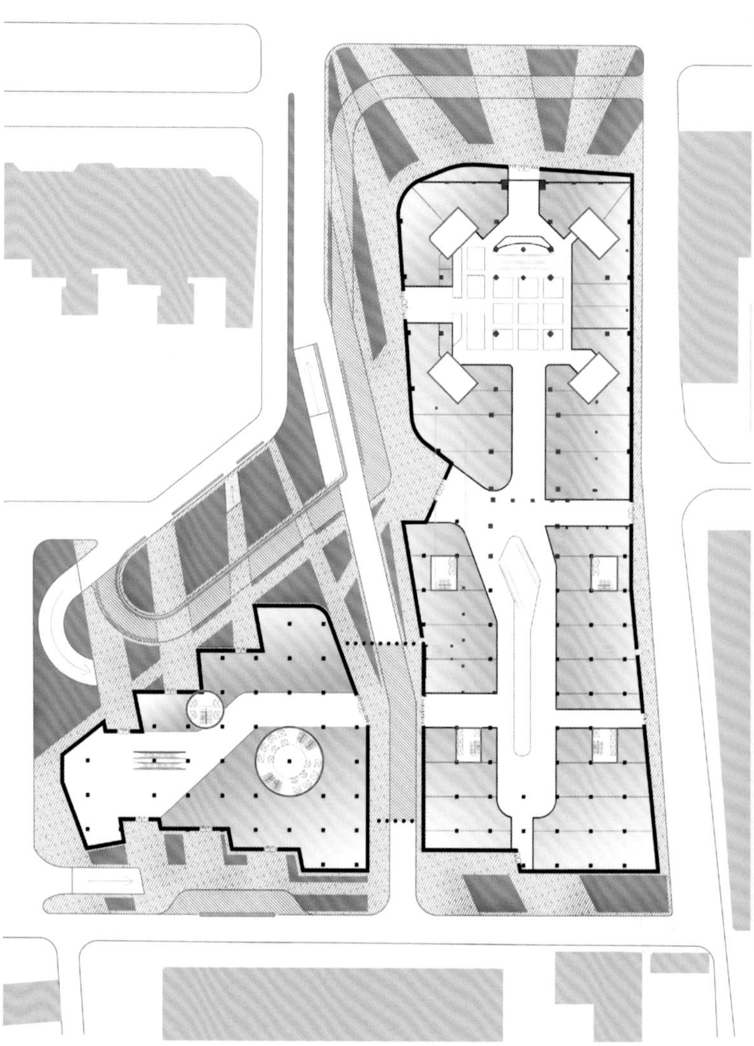

■ 设计理念

改造前该区域有四栋独立的建筑，分别为购物中心、酒店及两栋办公楼，建筑功能分散，使用效率不高，从周边的整体商业氛围来看商业价值较低，针对这些问题设计团队进行了一系列的改造。建筑方案将商业、办公、酒店等功能融于一体，将各个功能重新整合并合理分配，增加商业面积，提高商业价值，更穿插设置餐饮、娱乐、休闲等功能，使其成为符合现代人生活方式、人性化的建筑综合体。

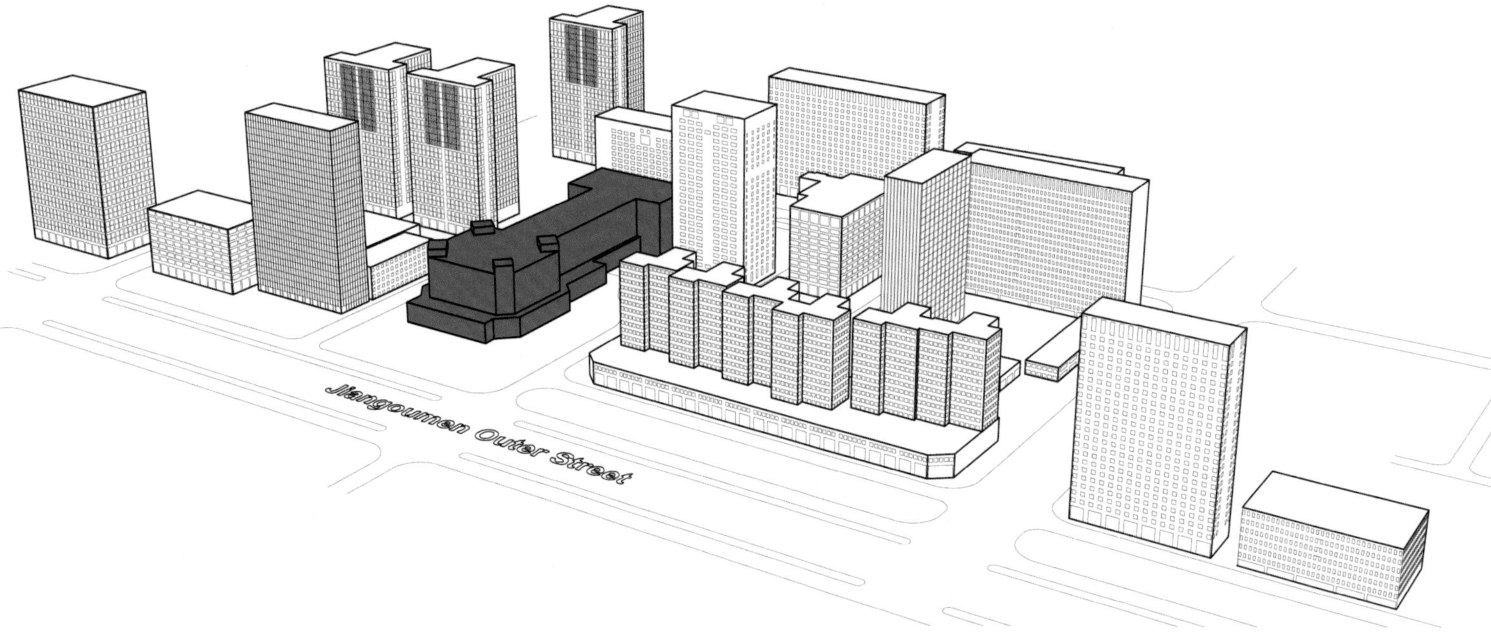

■ Planning and Layout

The plaza is 220 m high with 50 floors over ground and 5 floors underground. The first basement floor and Floor 1 to 6 are commercial spaces for the department store and shopping center which cater to different consumers. Floor 7 to 35 are offices, and Floor 36 to 50 are hotel rooms which enjoy a panoramic view of the Imperial Palace. In terms of landscape planning, the existing road extend to the southwest corner, and two curves will be added to present a clear traffic system.

■ 功能配置

建筑高220m，共50层，地下5层。地下一层到6层为商业空间，商业空间分为百货与购物中心两个区域，满足不同类型消费者的消费需求。7~35层为办公区，36~50层为酒店，在酒店里可以将故宫风采一览无余。建筑景观规划上，为了缓解该区域的交通压力，交通流线交叉严重等问题，将现有的道路拉伸到西南角，增加道路长度，并设置两条弯道，将现有交通流线进行了梳理。

■ Architectural design

As the plaza located in Beijing, just east to the Tian'anmen Square, it should embody the essence of traditional Chinese culture. Thus the design of this building is inspired by "dragon and phoenix". By combining tradition with modernity, oriental culture with western culture, and using some fashionable and dynamic elements, it will create a new oriental and futuristic landmark for Beijing.

■ 建筑设计

建筑位于历史悠久的古都北京，并位于天安门东侧，建筑设计势必要传承中国五千年历史文化的精髓。建筑设计以“龙凤呈祥”为设计根基，融入时尚、动感等元素，将古现代、东西方文化融合，打造具有东方韵味、人文前卫的北京新地标。

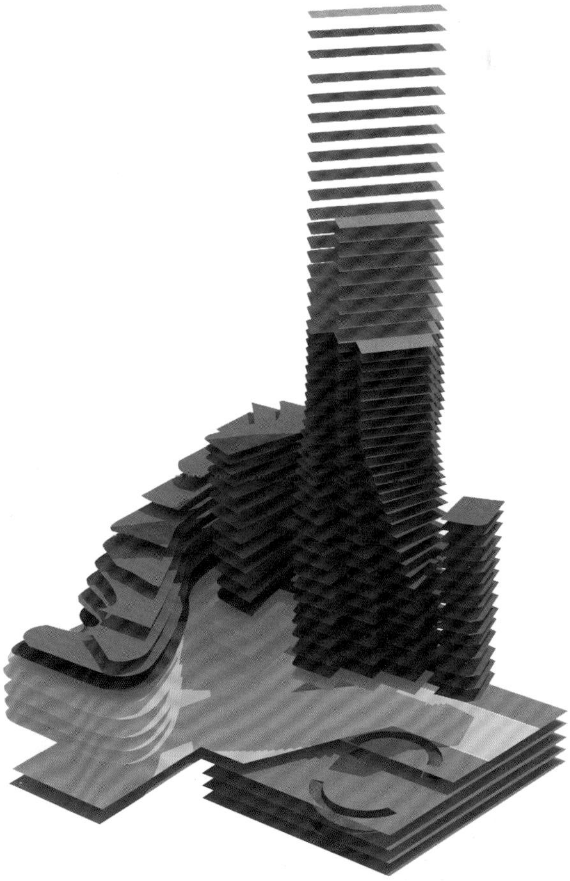

赛特大厦
SCITECH TOWER
赛特饭店 SCITECH HOTEL
赛特集团
BURBERRY
GIORGIO ARMANI
a. testoni

Guangzhou Nansha Administration Center Main Buildings Phase I

广州市南沙区行政中心主体建筑一期

Keywords 关键词

Ecological Measure
生态措施

High Arcade
挑高拱廊

Facade Design
立面设计

Location: Nansha District, Guangzhou, China
Architectural Design: PES Architects
Main Government Building: 31,500 m^2
Assistant Functions: 20,000 m^2
Citizen Center: 14,500 m^2
Service Functions: 44,600 m^2
Total Floor Area: 110,600 m^2
Green Coverage Ratio: 55.8%

项目地点：中国广东省广州市南沙区
建筑设计：芬兰PES建筑事务所
主体行政楼：31 500 m^2
辅助功能：20 000 m^2
市民中心：14 500 m^2
服务功能：44 600 m^2
总建筑面积：110 600 m^2
绿 化 率：55.8%

Features 项目亮点

The designers have chosen a compact, ecological architecture form, and the even, slightly curving facade surface gives people a composed and even monumental impression, which is needed in the architecture of administration building.

设计采用紧凑、生态的建筑形式，建筑外立面平整而略带弧度，给人一种沉着稳重甚至不朽的印象，这正是行政建筑所需要的品质。

■ Overview

For Nansha Administration Center, the designers have created a round central square, around which a 13 meters high arcade connects all buildings together. This kind of arcade with colonnade is a well-known urban element in creating a protecting shelter against rain and sun heat.

■ 项目概况

设计师为南沙行政中心一期主体建筑设计了一个圆形中心广场，由13 m高的拱廊走道将所有单体建筑连接在一起。拱廊走道与柱廊的搭配是一种众所周知的城市设计元素，可以形成一道保护屏障，避免建筑遭受雨水的和太阳带来的损害。

■ Structure Design

All the first phase buildings will be constructed by using a high-quality column-beam reinforced concrete system. Only the staircase, elevator and technical shafts are built by using load bearing concrete walls in order to stabilize the structure horizontally. This structure has such important advantage that all inner walls can be placed according to their functional needs and can even be rearranged later at any time.

■ 结构设计

所有一期建筑都将采用高质量的钢筋混凝土梁柱结构。只有楼梯、电梯和技术轴使用承重混凝土墙来稳定其水平结构。这种结构具有明显的优势，即所有内部的立面都可以根据其功能需求进行设置，甚至可以在任何时间重新排列组合。

■ Facade Design

The designers have chosen a compact, ecological form for separating the first phase buildings. Each facade has even, slightly curving surface without any bends, hollows or protrusions. This kind of facade is ideal, since it is both ecological and economical. The slightly curving forms, however, will give people a composed and even monumental impression, which is needed in the architecture of administration building.

■ 立面设计

设计师选择了一种紧凑、生态的建筑形式将一期工程的每一栋建筑独立分开。每个建筑立面外观都是平整而略带弧度，没有任何弯曲、凹陷或突起。这种外观设计很完美，既生态又经济。稍带弧度的外形给人一种沉着稳重甚至不朽的印象，这正是行政建筑所需要的品质。

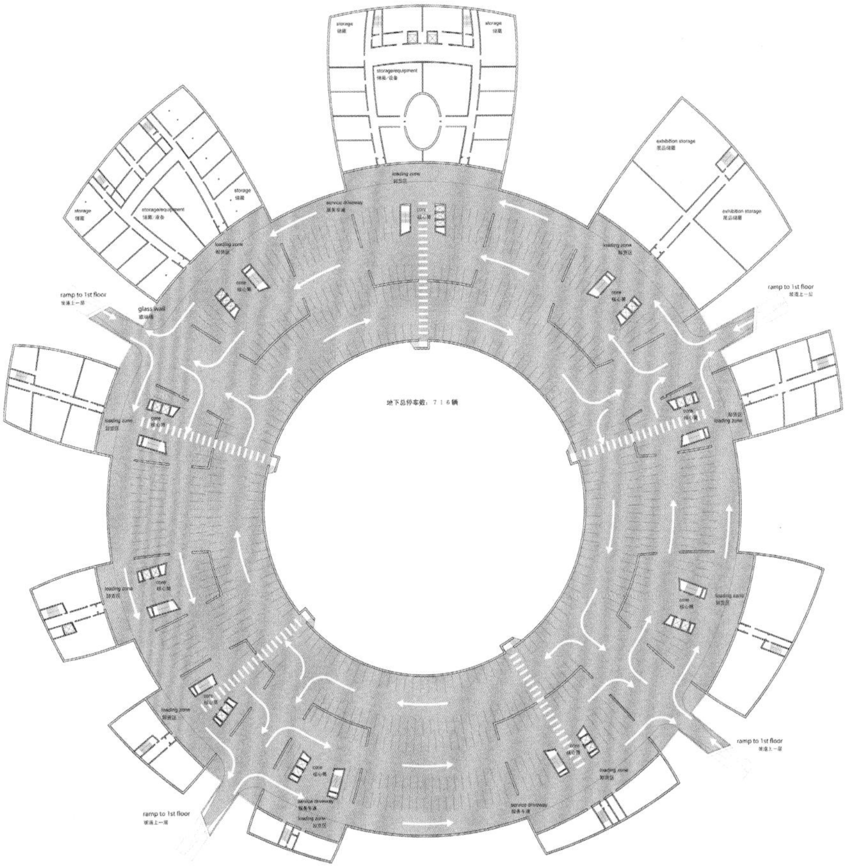

Basement Plan 地下层平面图

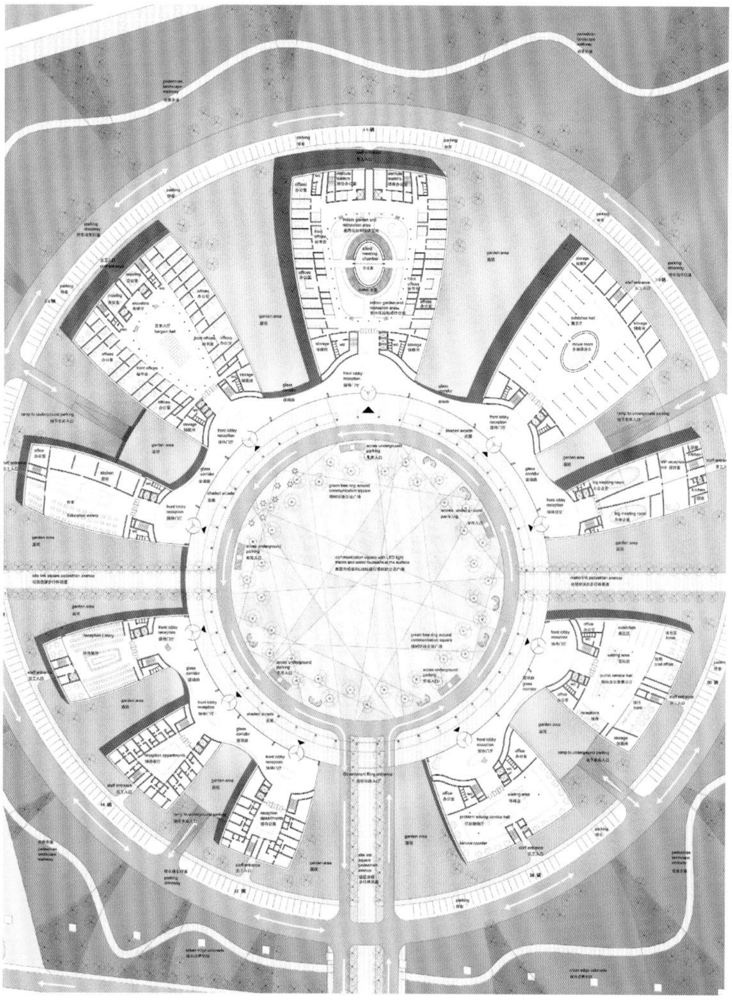

First Floor Plan 一层平面图

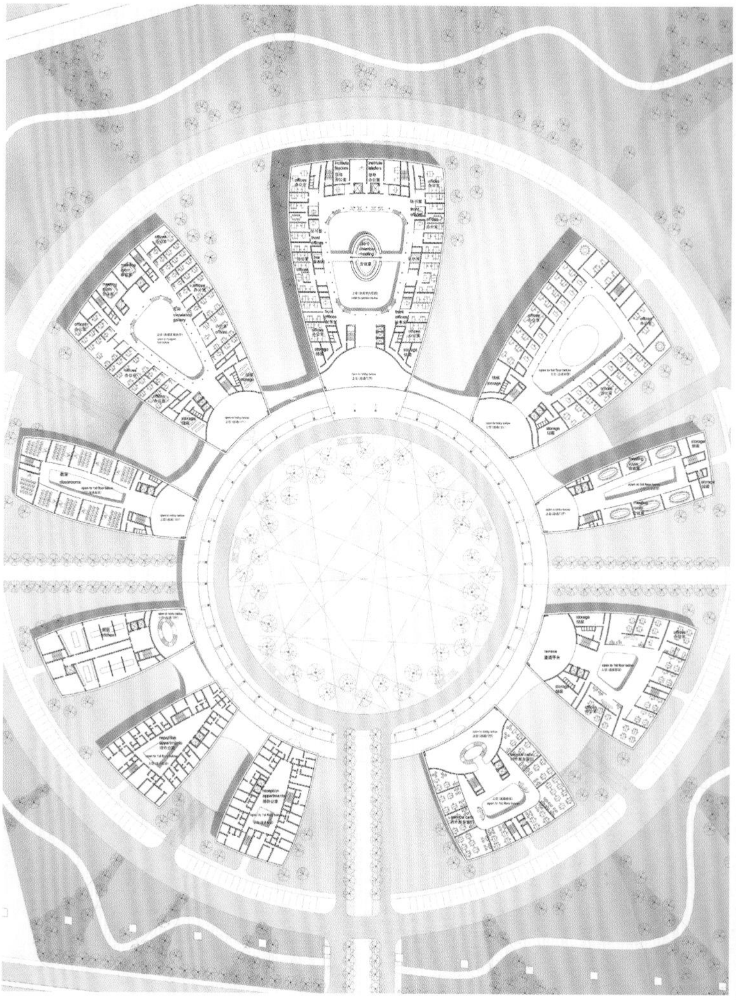

Second Floor Plan 二层平面图

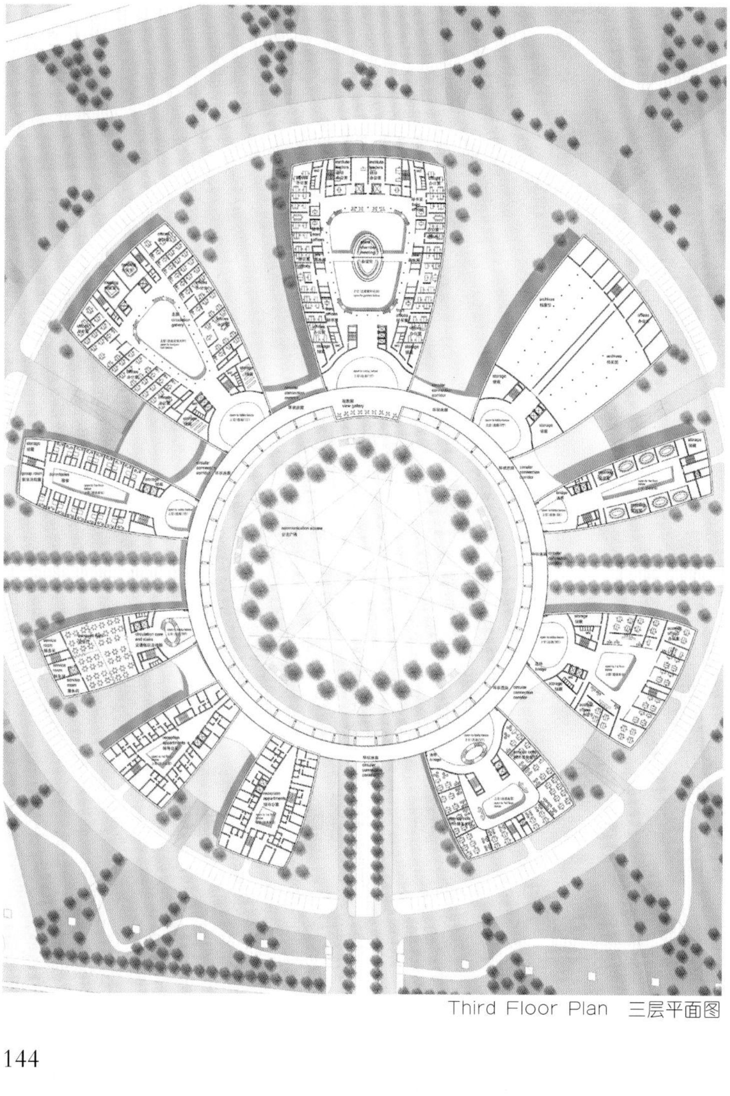

Third Floor Plan 三层平面图

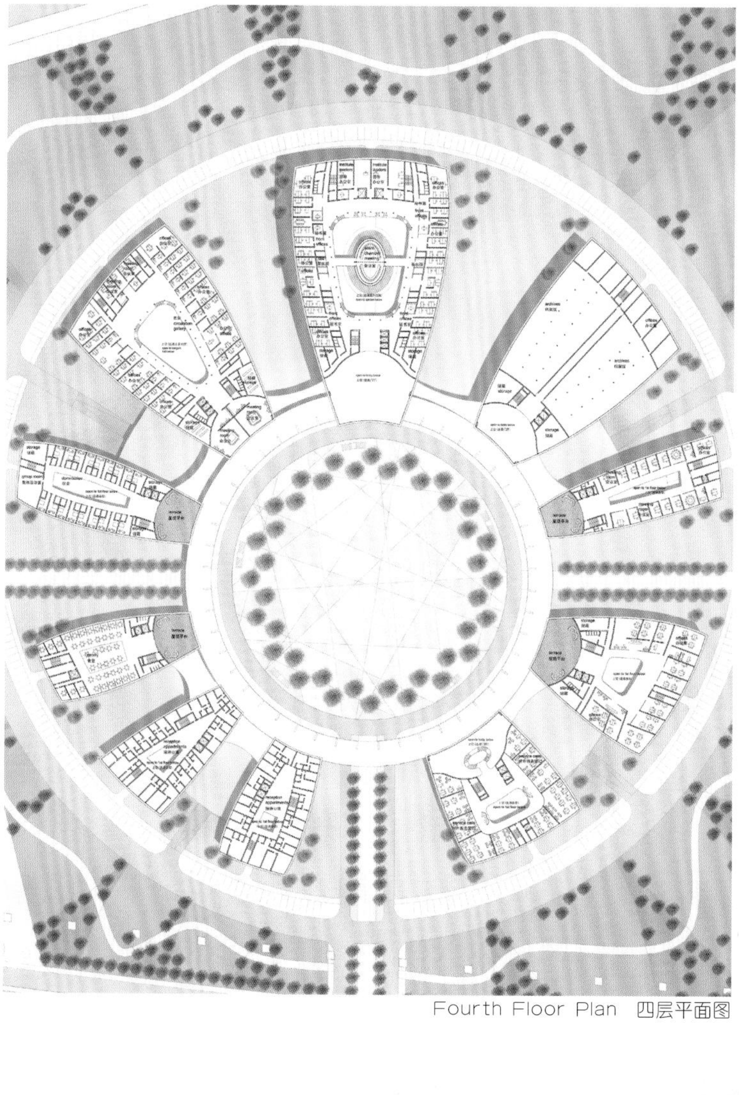

Fourth Floor Plan 四层平面图

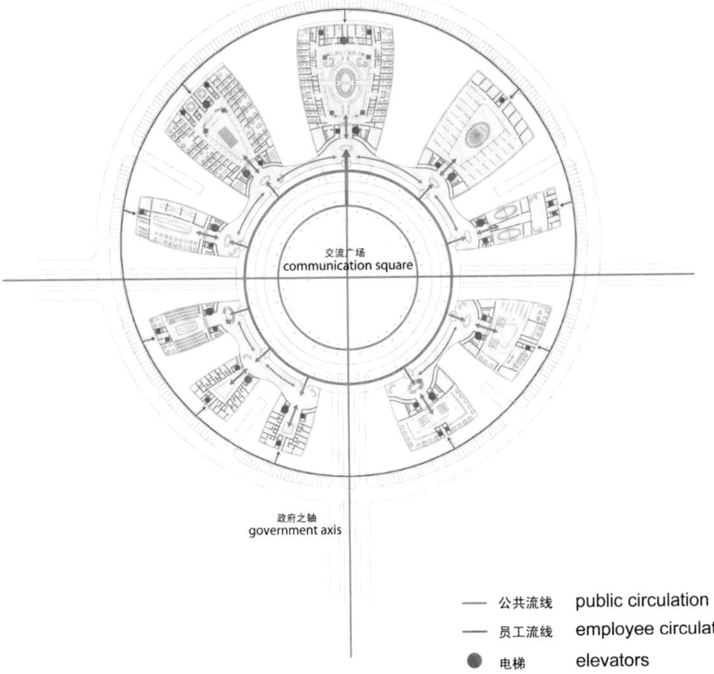

— 公共流线 public circulation
— 员工流线 employee circulation
● 电梯 elevators
■ 疏散楼梯 emergency stairs
■ 开放楼梯 open stairs
□ 建筑 buildings

1st Floor Pedestrian Circulation 一层步行流线

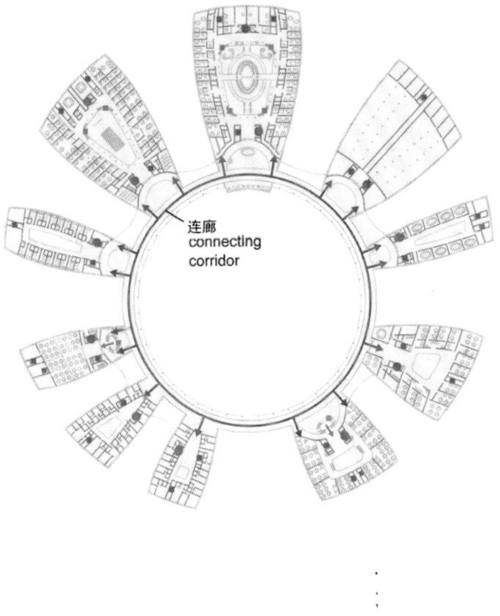

— 公共流线 public circulation
— 员工流线 employee circulation
● 电梯 elevators
■ 疏散楼梯 emergency stairs
■ 开放楼梯 open stairs
□ 建筑 buildings

3rd Floor Pedestrian Circulation 三层步行流线

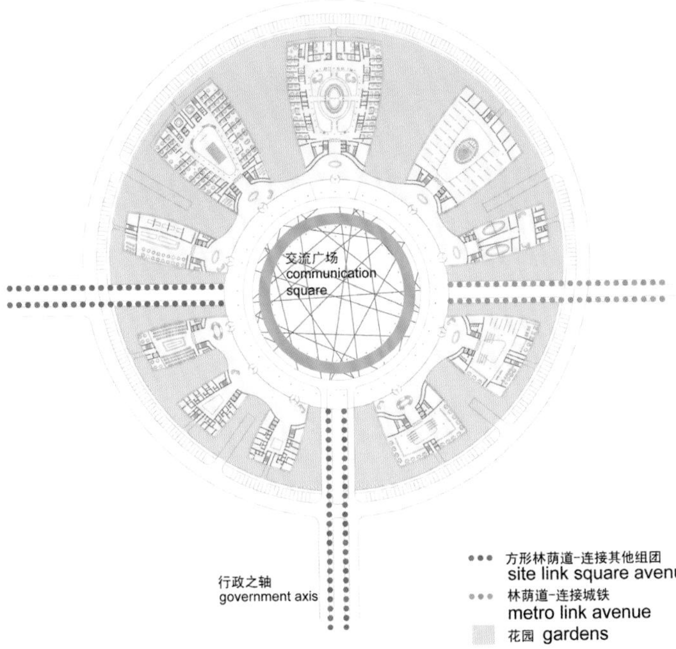

••• 方形林荫道-连接其他组团
site link square avenues
••• 林荫道-连接城铁
metro link avenue
花园 gardens
交流广场绿带
comunication square green
— 广场地面的LED灯
LED lights in square floor
建筑 buildings

Green Area 绿化区域

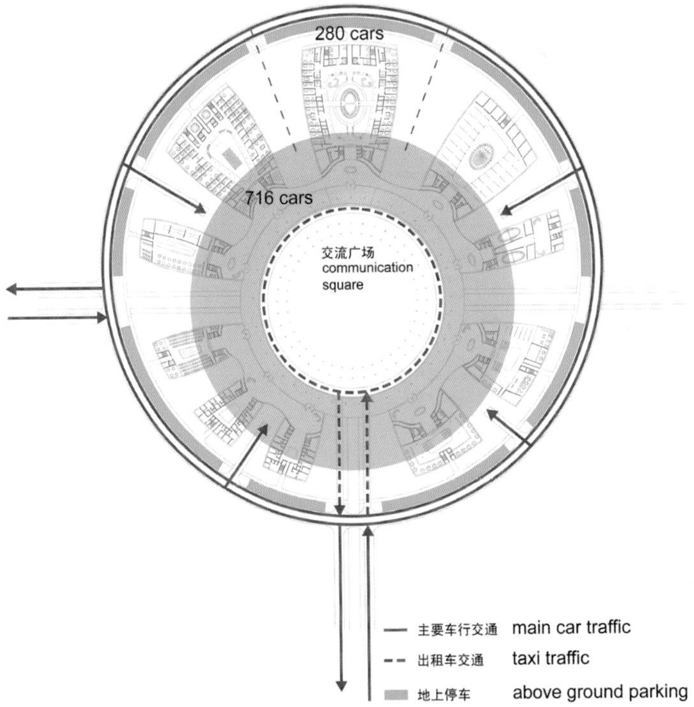

— 主要车行交通 main car traffic
- - 出租车交通 taxi traffic
地上停车 above ground parking
地下停车 underground parking

Traffic and Parking 交通与停车

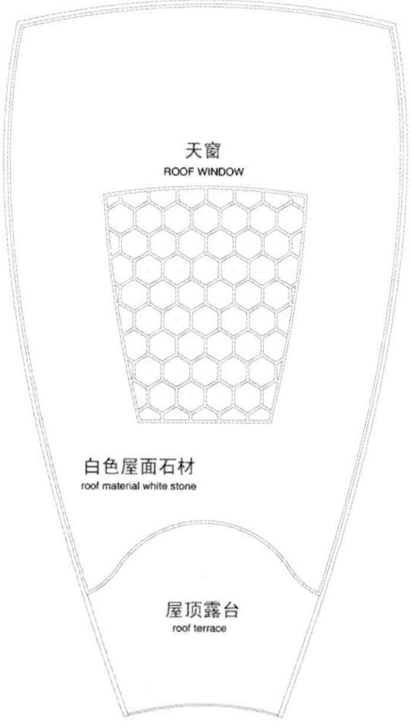

Roof Plan 屋顶平面

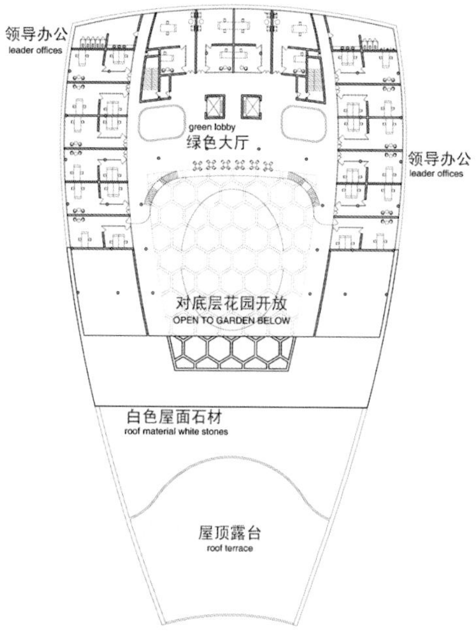

10th Floor 10层平面

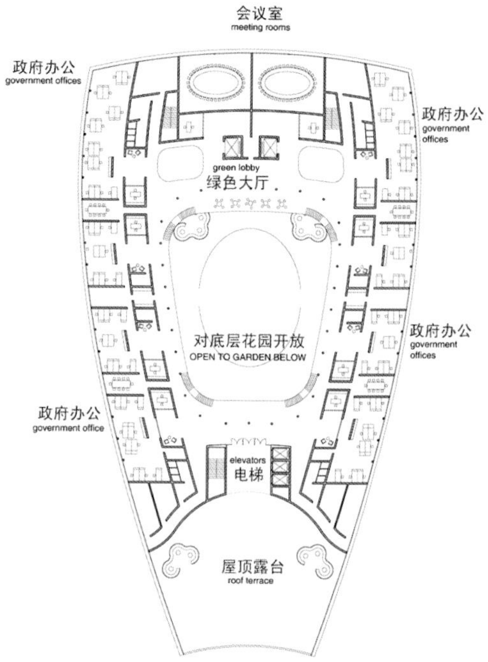

9th Floor 9层平面

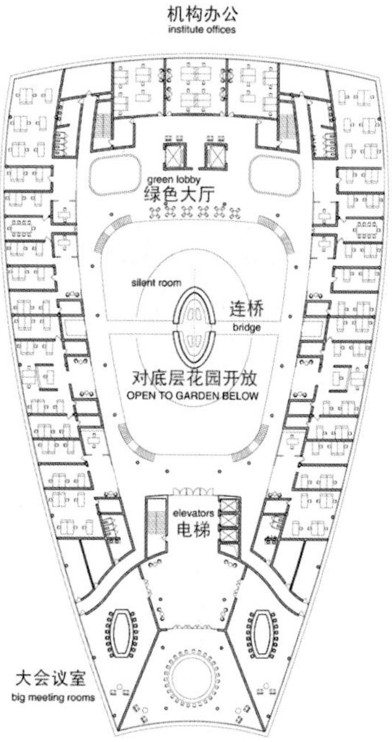

5th Floor 5层平面

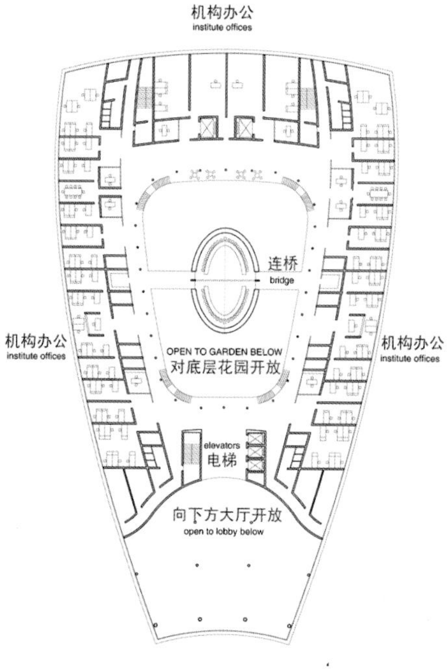

4th Floor 4层平面

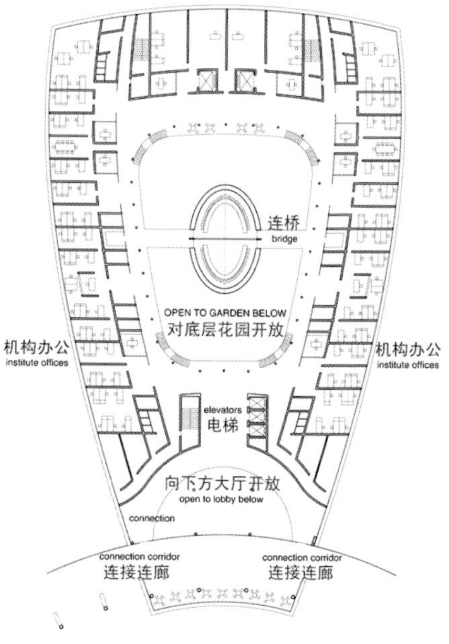

3rd Floor 3层平面

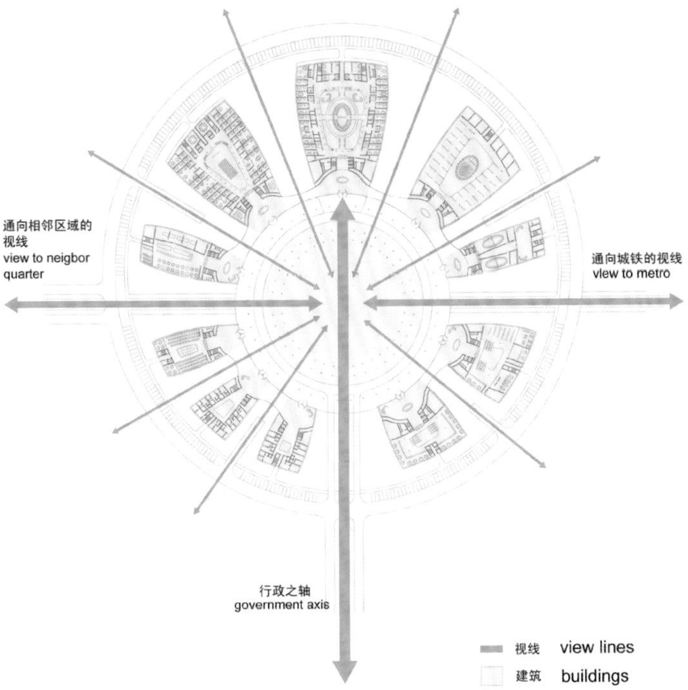

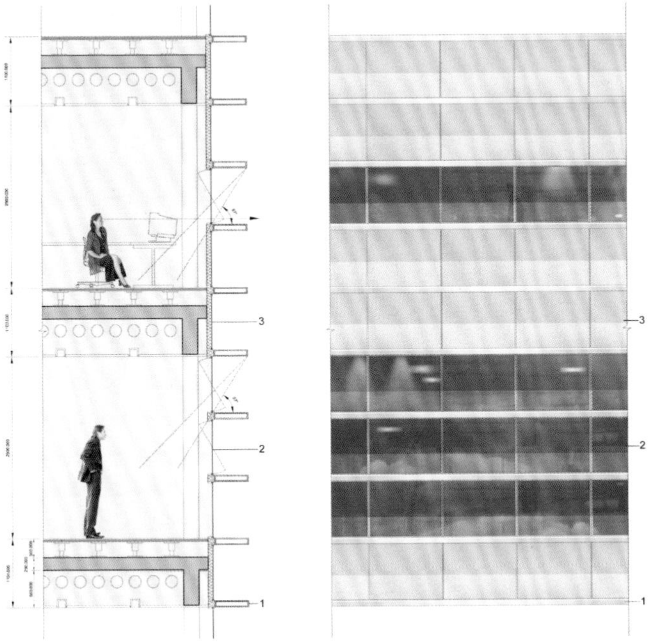

1. 遮阳金属百叶
Metal Louvers For Sun Protection
2. 用于自然通风的可开启玻璃窗
Windows To Open For Natural Ventilation
3. 用于覆盖外墙实体部分的磨砂玻璃
Silk Printed Glass To Cover The Solid Parts Of The Facade

Facade Detail 立面细部

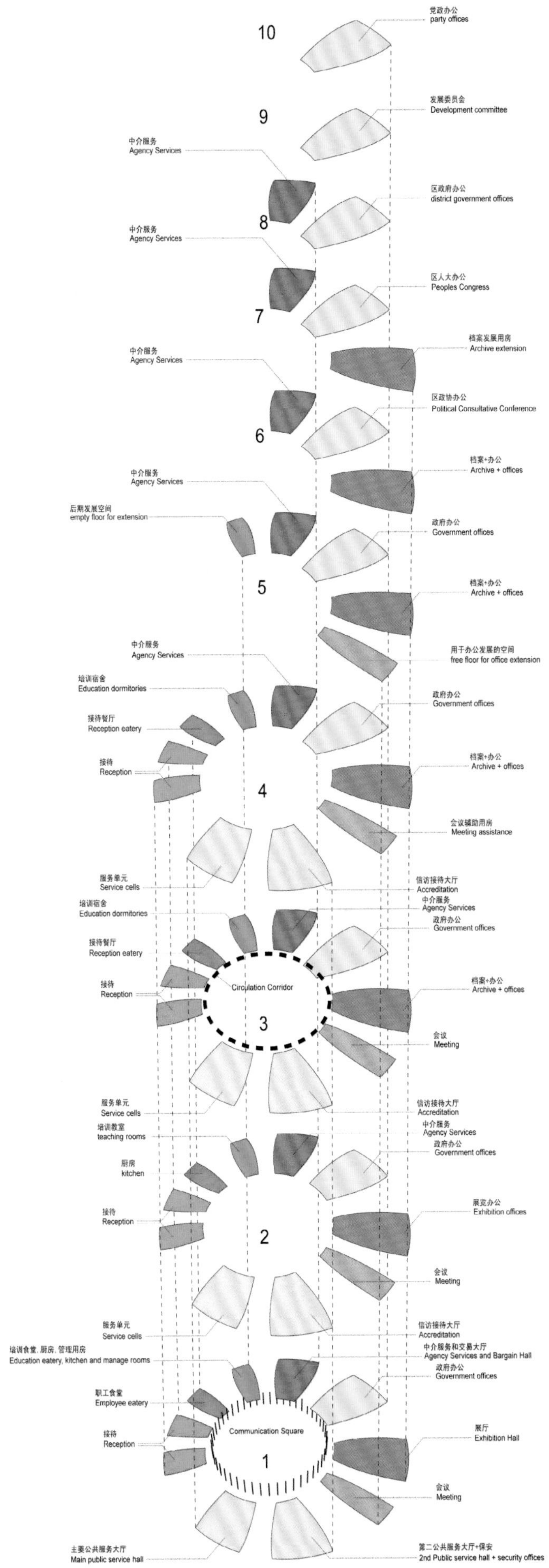

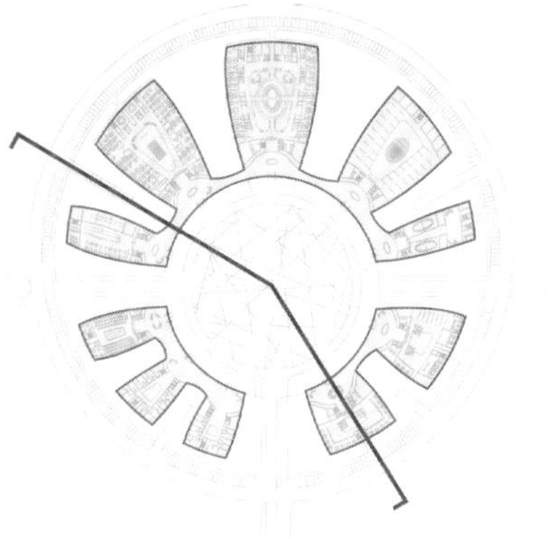

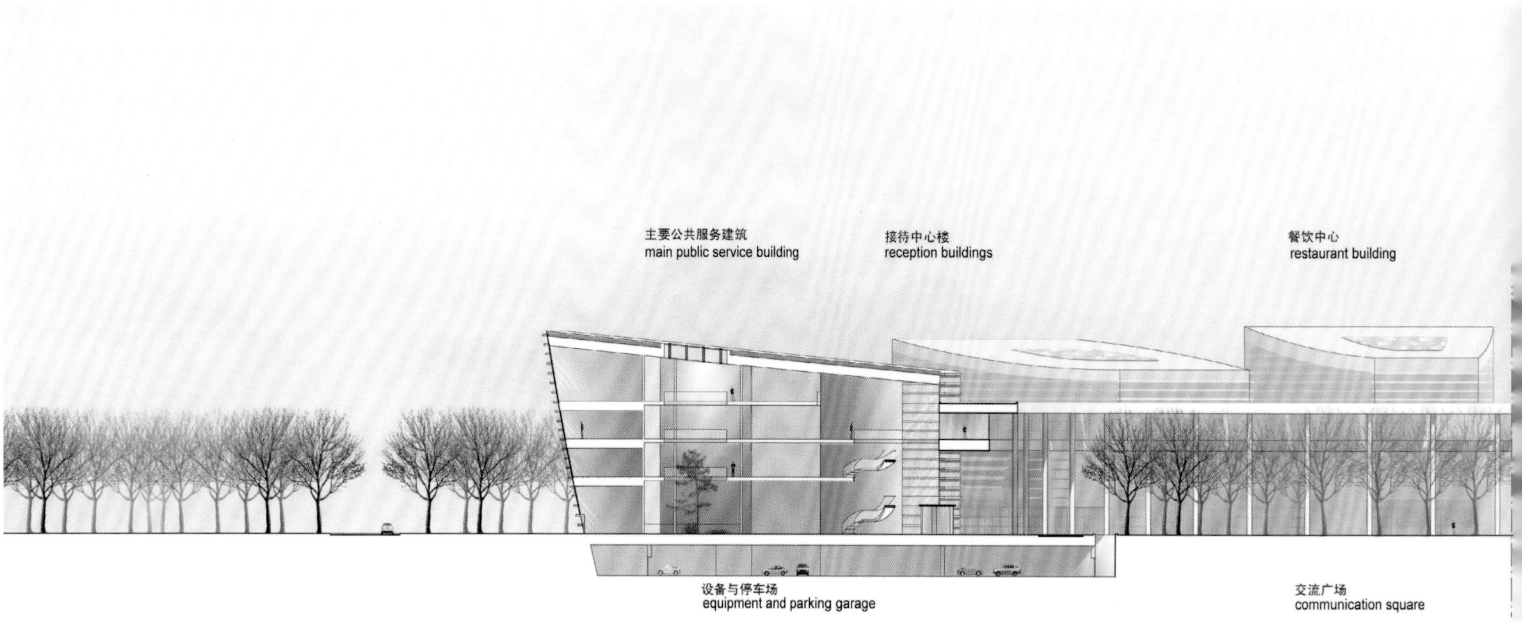

主要公共服务建筑
main public service building
接待中心楼
reception buildings
餐饮中心
restaurant building
设备与停车场
equipment and parking garage
交流广场
communication square

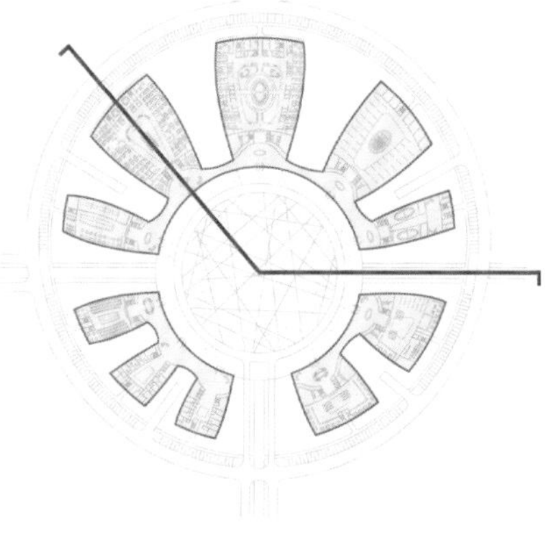

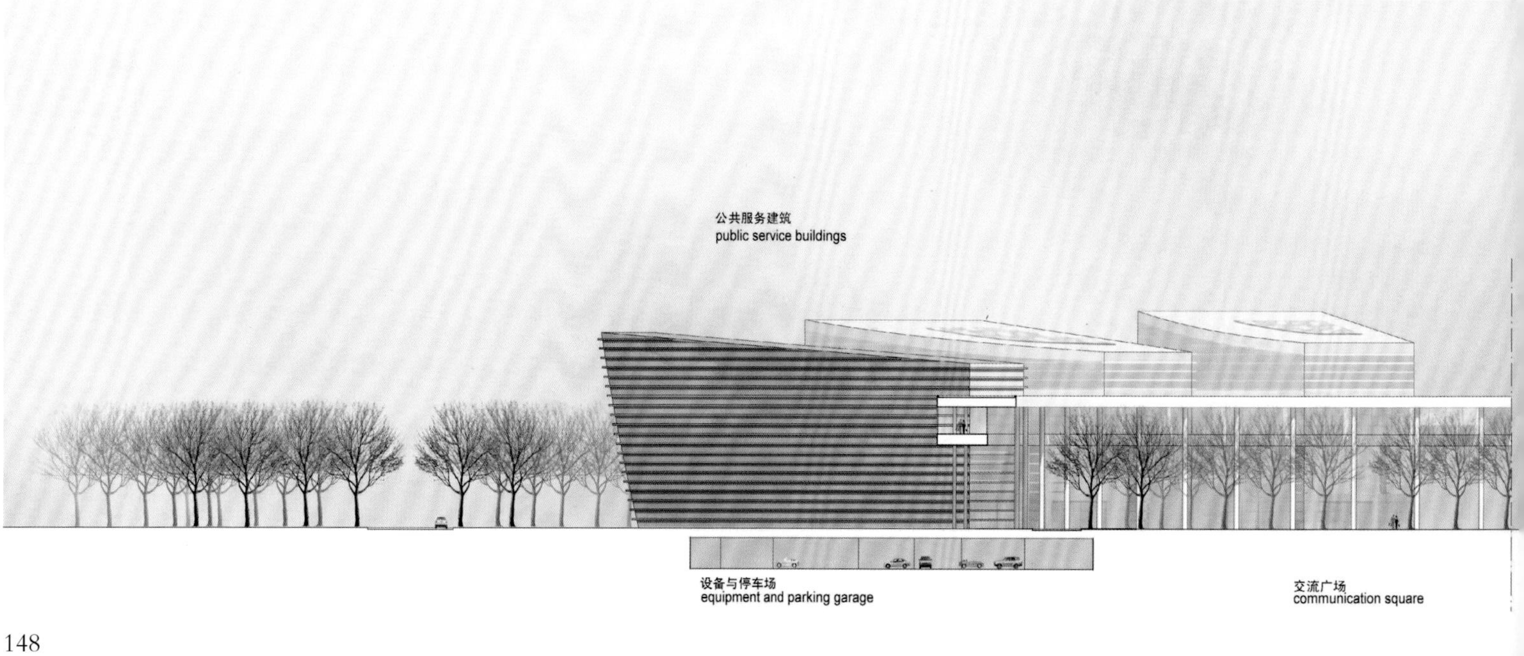

公共服务建筑
public service buildings
设备与停车场
equipment and parking garage
交流广场
communication square

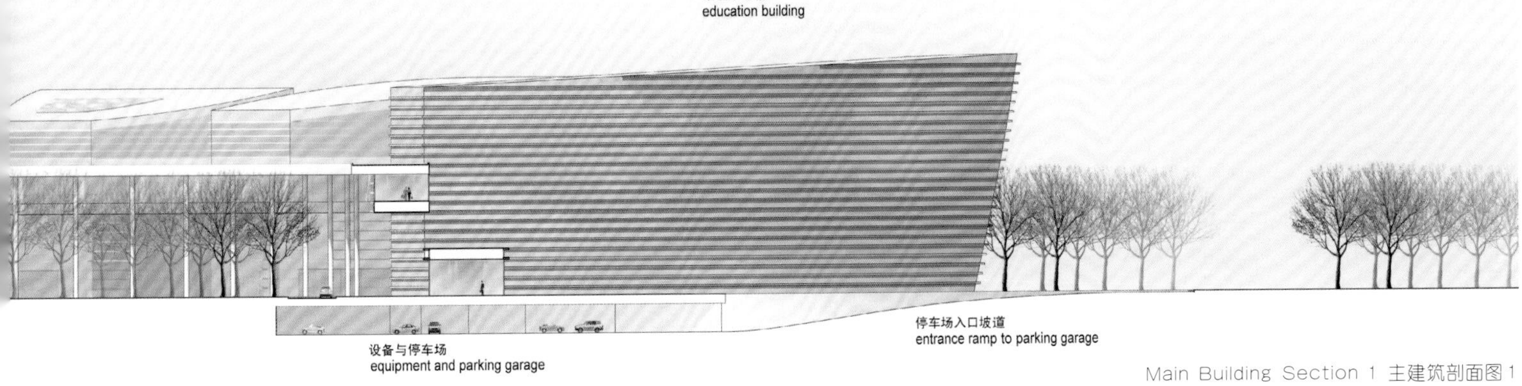

Main Building Section 1 主建筑剖面图 1

Main Building Section 2 主建筑剖面图 2

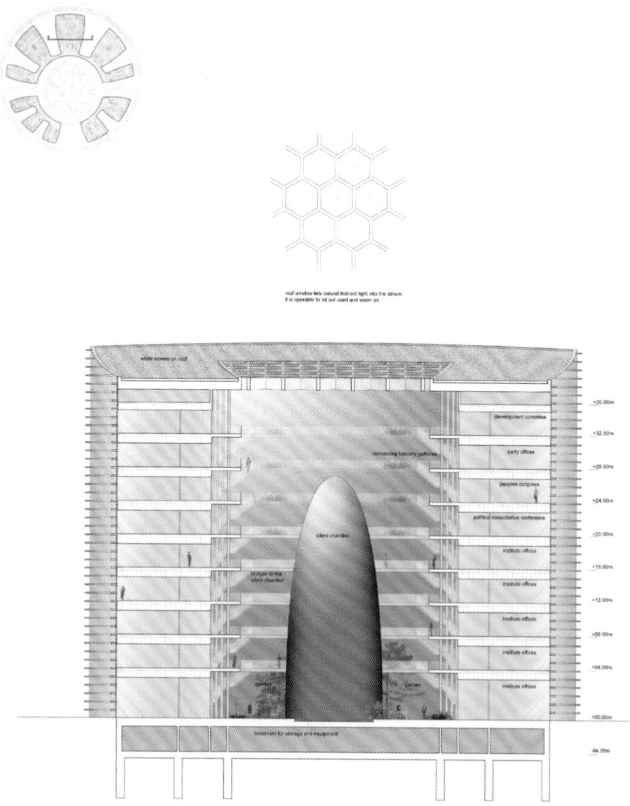

Main Building Section 主建筑剖面图

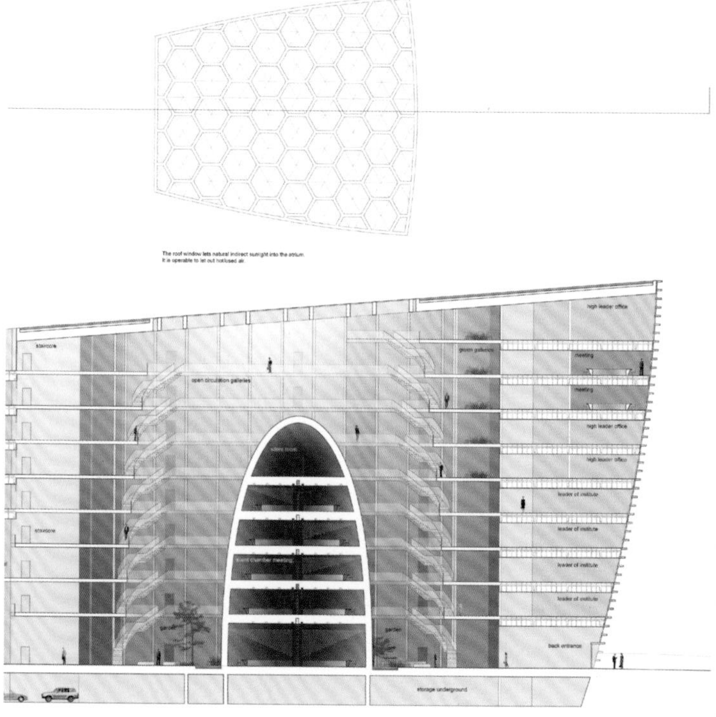

Main Building Section 主建筑剖面图

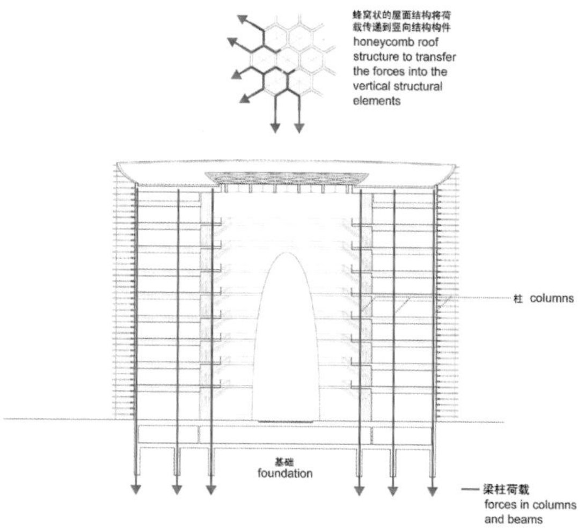

蜂窝状的屋面结构将荷载传递到竖向结构构件
honeycomb roof structure to transfer the forces into the vertical structural elements
柱 columns
基础
foundation
梁柱荷载
forces in columns and beams

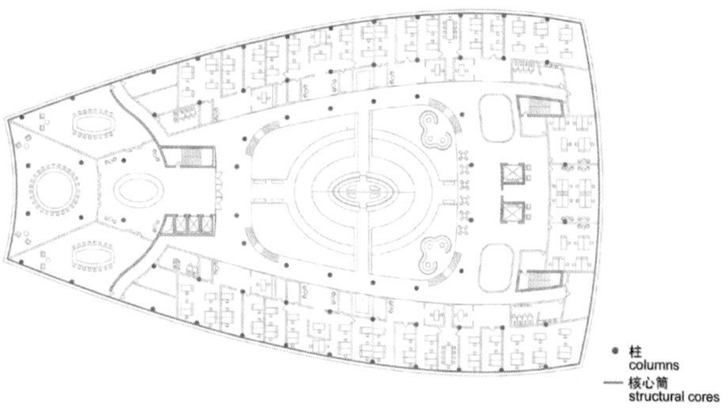

柱
columns
核心筒
structural cores

Structure 结构

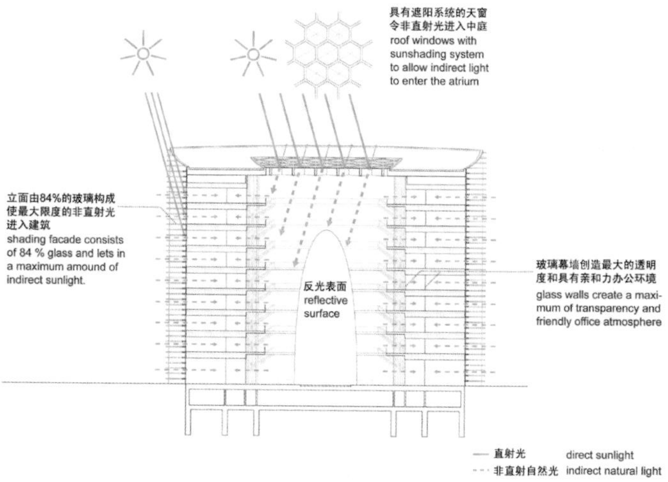

具有遮阳系统的天窗令非直射光进入中庭
roof windows with sunshading system to allow indirect light to enter the atrium
立面由84%的玻璃构成使最大限度的非直射光进入建筑
shading facade consists of 84 % glass and lets in a maximum amound of indirect sunlight.
反光表面
reflective surface
玻璃幕墙创造最大的透明度和具有亲和力办公环境
glass walls create a maximum of transparency and friendly office atmosphere
直射光 direct sunlight
非直射自然光 indirect natural light

Sunlight 光线分析

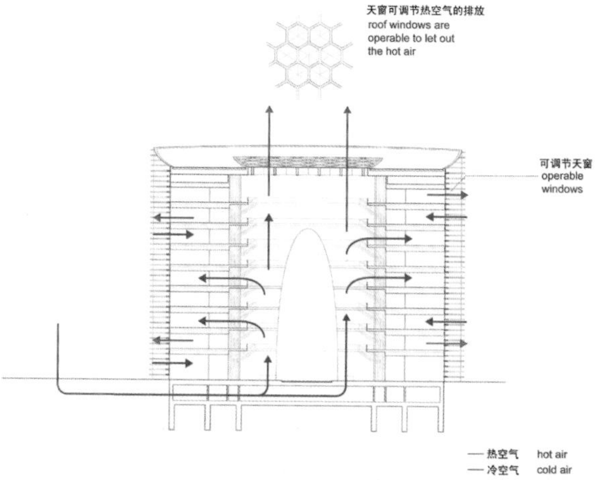

天窗可调节热空气的排放
roof windows are operable to let out the hot air
可调节天窗
operable windows
热空气 hot air
冷空气 cold air

Natural Ventilation 自然通风

Fountain Circulation 喷泉水循环

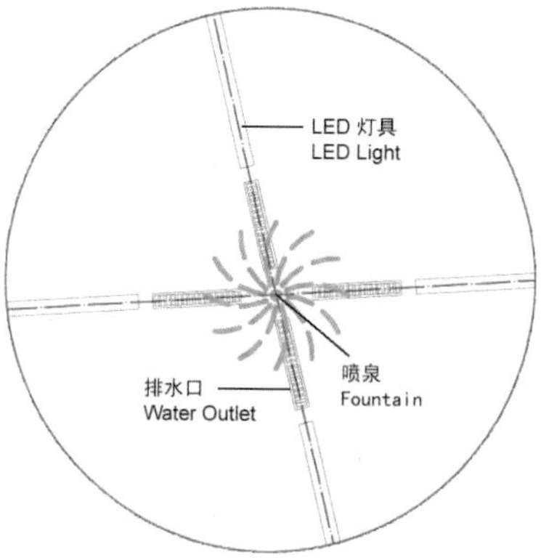

Fountain Detail 喷泉细部

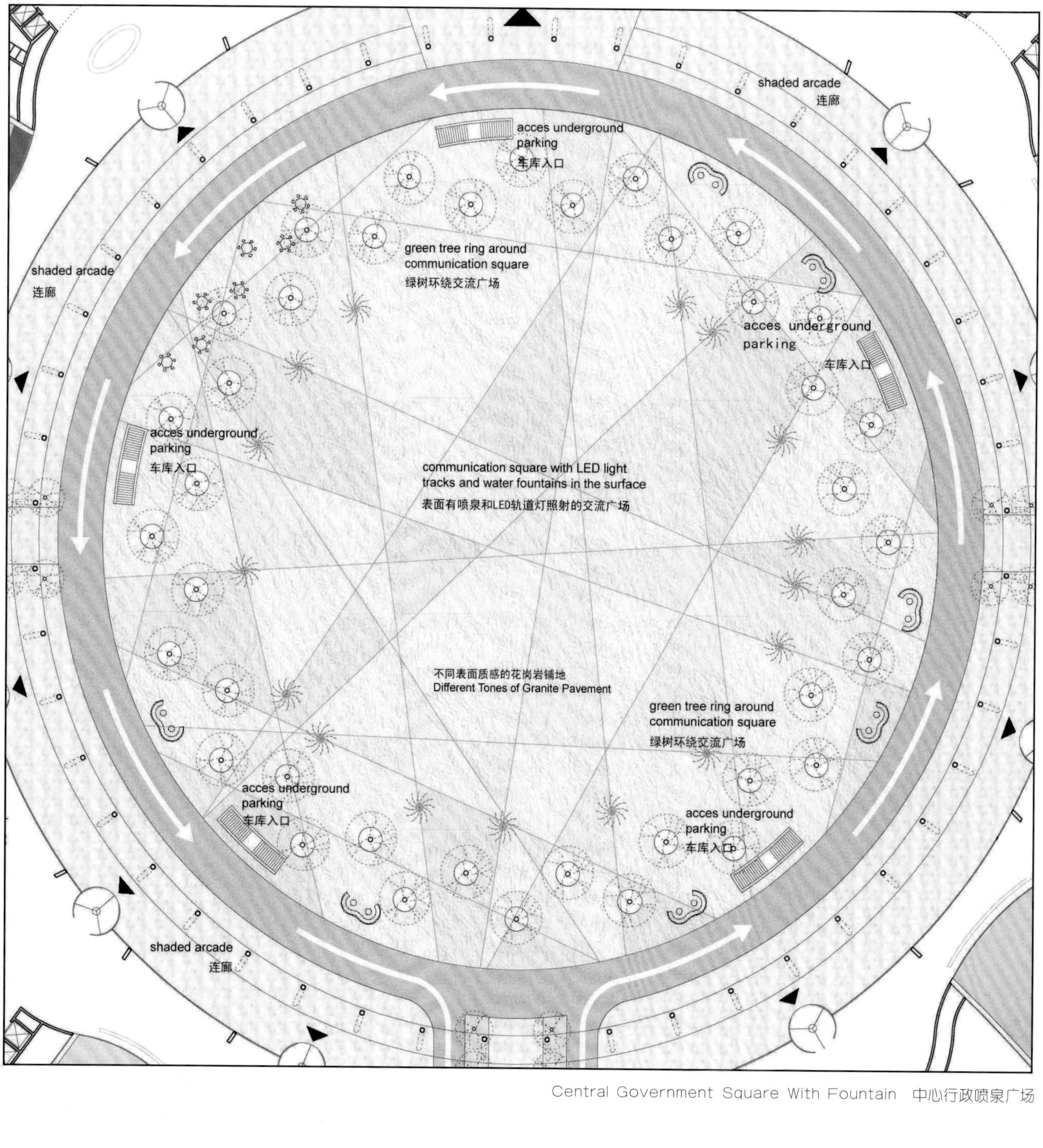

Central Government Square With Fountain 中心行政喷泉广场

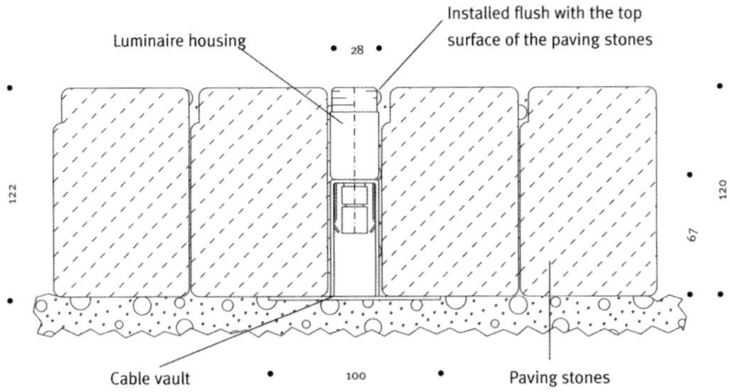

Example of LEL Lights　LEL 灯具应用实例

Zhaoyuan Creative Headquarters, Yantai
烟台招远创意总部

Keywords 关键词

Chinese Style
中式风格

Enclosed Layout
围合式布局

Traditional Art
传统艺术

Location: Yantai, Shandong, China
Architectural Design: Agency C&P Architecture
Total Floor Area: 46,850 m^2
Building Density: 21%
Plot Ratio: 0.61

项目地点：中国山东省烟台市
建筑设计：C&P(喜邦)国际建筑设计公司
总建筑面积：46 850 m^2
建筑密度：21%
容 积 率：0.61

Features 项目亮点

With understanding to traditional culture, the designers combine modern elements with traditional ones, creating elements of traditional charm according to modern aesthetic needs, and letting traditional art play a proper role in modern society.

设计师通过对传统文化的认识，将现代元素和传统元素结合在一起，以现代人的审美需求来打造富有传统韵味的事物，让传统艺术在当今社会得到合适的体现。

■ Overview

Located in Yantai of Shandong Province, the project has a total floor area over 46,850 m^2. It fully integrates the original supporting resources in the site, brings the functions of business reception, commercial activity and business leisure vacation and enriches the function integrity of local tourism and hospitality service as well, which makes positive contribution to attracting tourists, increasing popularity and optimizing regional industrial structure.

■ 项目概况

本项目位于山东省烟台市，总建筑面积为46 850 m^2，项目充分地整合了地块内原有的配套服务资源，同时引入商务接待、商务活动和商务休闲度假等功能，丰富了本区域旅游、度假服务的功能，为留住游客、增加人气、优化区域产业结构都将起到积极作用。

■ Planning

The project adopts modern Chinese architectural style and enclosed courtyard group layout, making the best of the combination of architectural image and volume. It integrates the original supporting resources such as 714 Hotel, Taojin Town and Gold Expo Garden as a whole, combines the points, lines and faces in terms of architectural layout that realizes the organic interaction.

■ 规划布局

招远创意总部基地项目规划采用现代中式建筑风格、围合式院落组团布局，巧妙利用建筑形体和体量的组合，以自然线势和飘积理论为原则，结合宗地地形地貌进行合理布置。通过本项目的建设，使项目所在区域原有的714酒店、淘金小镇、黄金博览苑等配套服务资源形成统一整体，在建筑（群）布局上呈现点、线、面的有机融合，在旅游功能上实现互动。

柒幢

Site Plan 总平面图

■ **Architectural Design**

In terms of design, Chinese style elements are being combined with modern materials ingeniously. A large number of furniture in Ming and Qing Dynasty express the pursuit for elegant, reserved and dignified oriental mental outlook. The Chinese style in this project is not a mere stuff piling-up, but to blend modern with tradition, creating things with traditional charm with modern aesthetic need, letting traditional art play a proper role in modern society.

■ **建筑设计**

在设计上将建筑所蕴含的中式元素与现代材质巧妙兼容，室内大量使用的明清家具、窗棂、布艺床品相互辉映，再现了移步变景的精妙小品。在整体设计上并非完全意义上的复古，而是通过中式风格的特征，表达对清雅含蓄、端庄风华的东方雅韵的追求。项目所应用的中式风格也不是纯粹的元素堆砌，而是通过对传统文化的认识，将现代元素和传统元素结合在一起，以现代人的审美需求来打造富有传统韵味的事物，让传统艺术在当今社会得到合适的体现。

Structure Analysis 结构分析图

Function Analysis 功能分析图

Traffic Analysis Chart 交通分析图

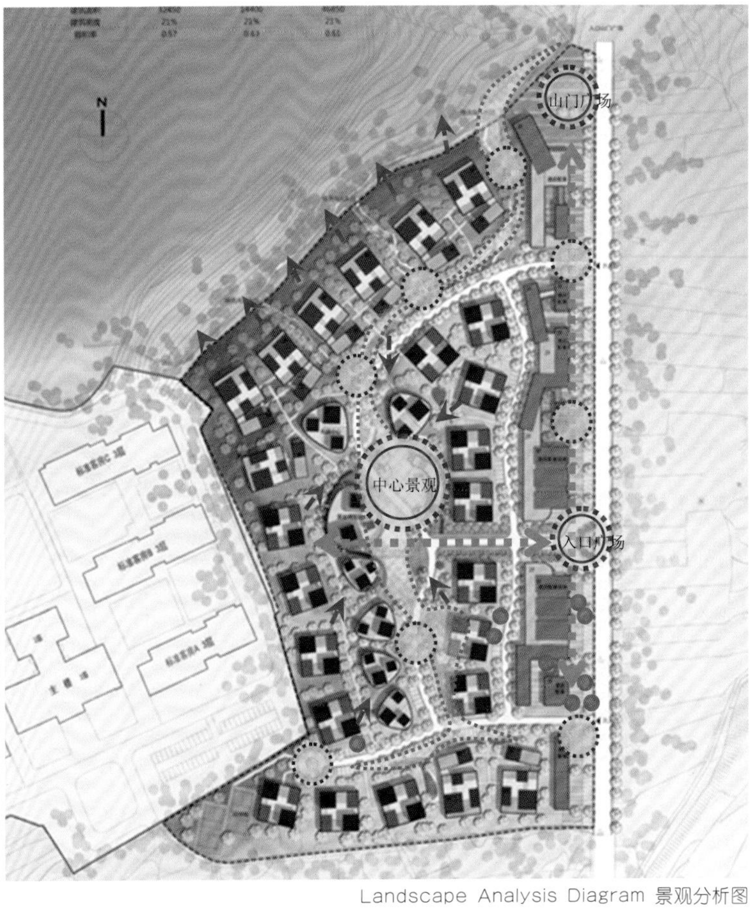

Landscape Analysis Diagram 景观分析图

DAKOTA
FENDI

Lee
RIFLE
RONSON
SPALDING

Taihu Sci-tech Park Administrative & Business Center, Wuxi 无锡太湖科技园行政商务中心

Keywords 关键词

Glass
玻璃

Stone
石材

Landscape Views
观景格局

Features 项目亮点

The facade, made of glass and stone curtain wall, is designed with huge linear blocks interweaving with each other. Characteristic local building materials and facade elements are also adopted.

建筑立面设计采用了直线条为主的大体块穿插设计手法，立面材料以玻璃及石材幕墙为主，局部点缀一些有无锡地方特色的建材与立面元素。

Location: Wuxi, Jiangsu, China
Developer: Wuxi New District Economic Development Group Corporation
Architectural Design: CCI Architecture Design & Consulting Co., Ltd.
Land Area: 48,000 m^2
Total Floor Area: 159,000 m^2

项目地点：中国江苏省无锡市
开 发 商：无锡新区经济发展集团总公司
规划设计：上海新外建工程设计与顾问有限公司
用地面积：48 000 m^2
总建筑面积：159 000 m^2

■ Planning

There is an auxiliary scenic ring road in the plot around the artificial lake. On the west side of the circumferential road, it will be the office area, while the east will be the administrative area. Near the artificial lake, the office buildings will range from 8 stores to 12 stores under 50 m. There are three single buildings standing here. The middle big one is the office building for universal bank, and the other two are office buildings for two banks. There are 7 buildings in the administrative area for Trade and Industry Bureau, Administration, Taxation Bureau, Land Bureau and other national administration authorities.

■ 规划布局

在人工湖周边地块内部增加环形辅助型景观道路，以环形道路为界，位于环形道路西侧的地块规划为金融办公用地，而东侧的区域规划为行政办公用地。金融办公用地离人工湖较近，建筑层数为8~12层，控制高度为50 m。由北至南分为三个地块，布置三栋单体建筑。北侧与南侧的两栋办公楼体量较小，规划为某大型银行的独栋办公楼，而中间的办公楼体量较大，规划为综合银行办公楼。由北至南规划七栋办公塔楼，分别为工商、行政、税务、国土等国家行政机关办公场所。

■ Architectural Design

The facade, which is made of glass and stone curtain wall, is designed with huge linear blocks interweaving with each other. Characteristic local building materials and facade elements are also adopted. To make sure that every office has a good view of the artificial lake and the twin-tower, the buildings range from lower to higher along the lake: buildings nearby the lake are lower, while the office buildings on the east are higher to get a good view.

■ 建筑设计

建筑立面设计采用了直线条为主的大体块穿插设计手法，外立面以玻璃及石材幕墙为主，局部点缀一些有无锡地方特色的建材与立面元素。为了强化基地景观资源，使得行政商务区的每一栋办公建筑都能最大限度欣赏到人工湖和双子塔的优美环境，在建筑的高度布局上采用了东高西低的方式。临湖一侧建筑较低，东侧办公楼的体量则较高，形成了看台式观景格局。

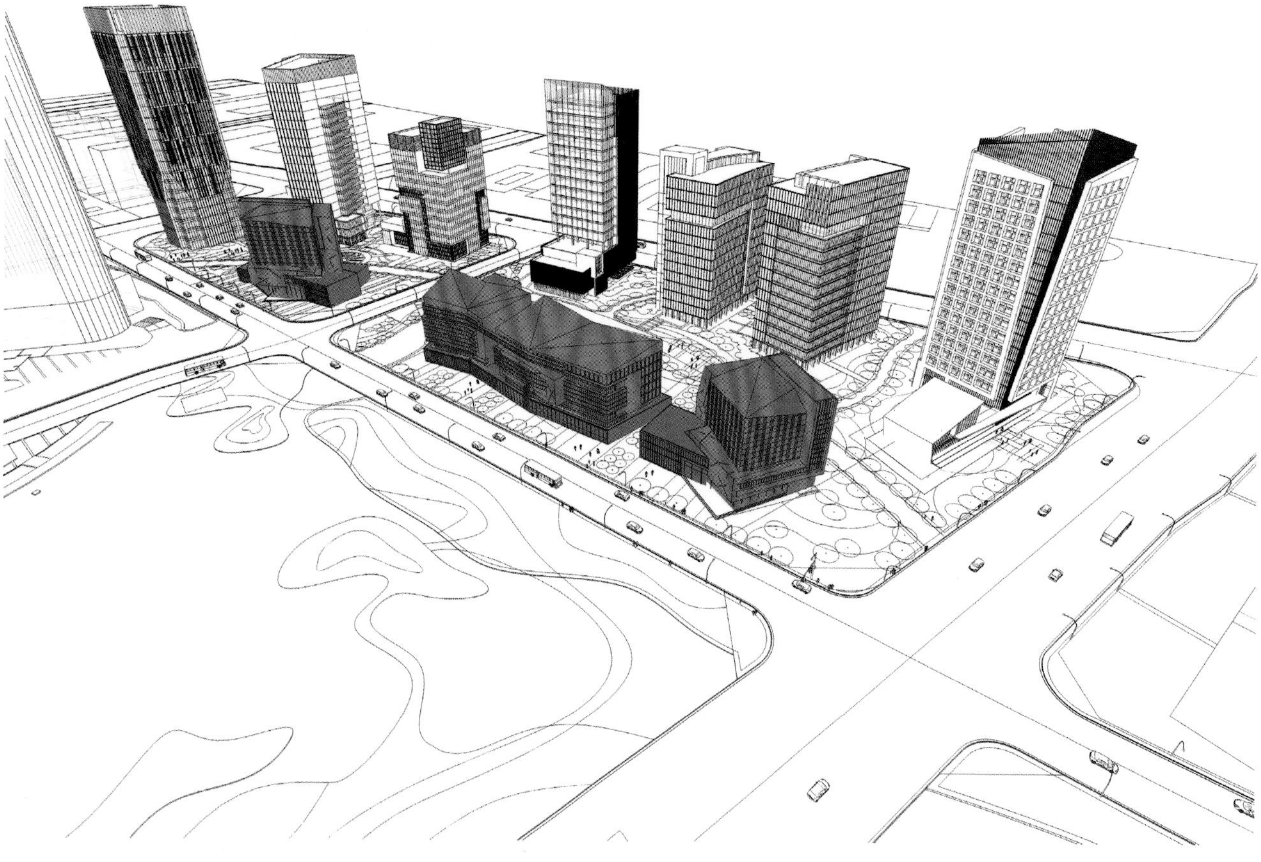

B-09地块技术经济指标：

总建筑面积	C号楼	14000m²	67500m²
	5号楼	12000m²	
	6号楼	19000m²	
	7号楼	22500m²	
地块面积			16000m²
容积率			4.22

B-12地块技术经济指标：

总建筑面积	A号楼	7000m²	89500m²
	B号楼	21000m²	
	1号楼（国土局）	16500m²	
	2号楼（国税局）	17000m²	
	3号楼（地税局）	14500m²	
	4号楼	15500m²	
地块面积			31300m²
容积率			2.92

总体技术经济指标：

总用地面积	47300m²
总建筑面积	159000m²
综合容积率	3.36
原综合容积率	3.31

Site Plan 总平面图

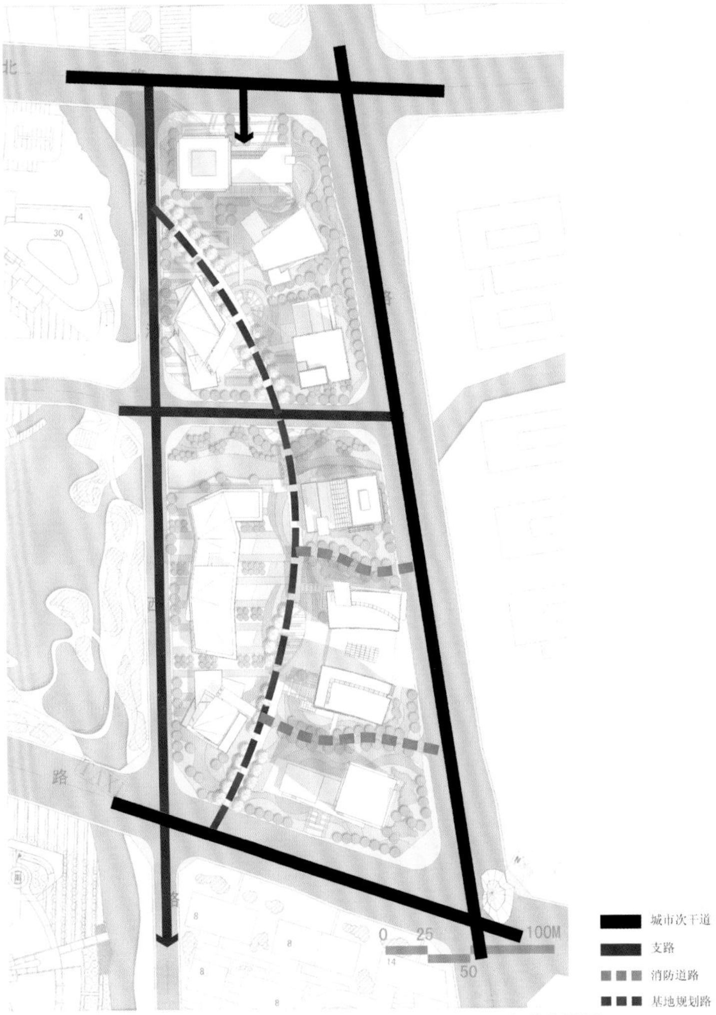

Traffic Drawing 交通分析图

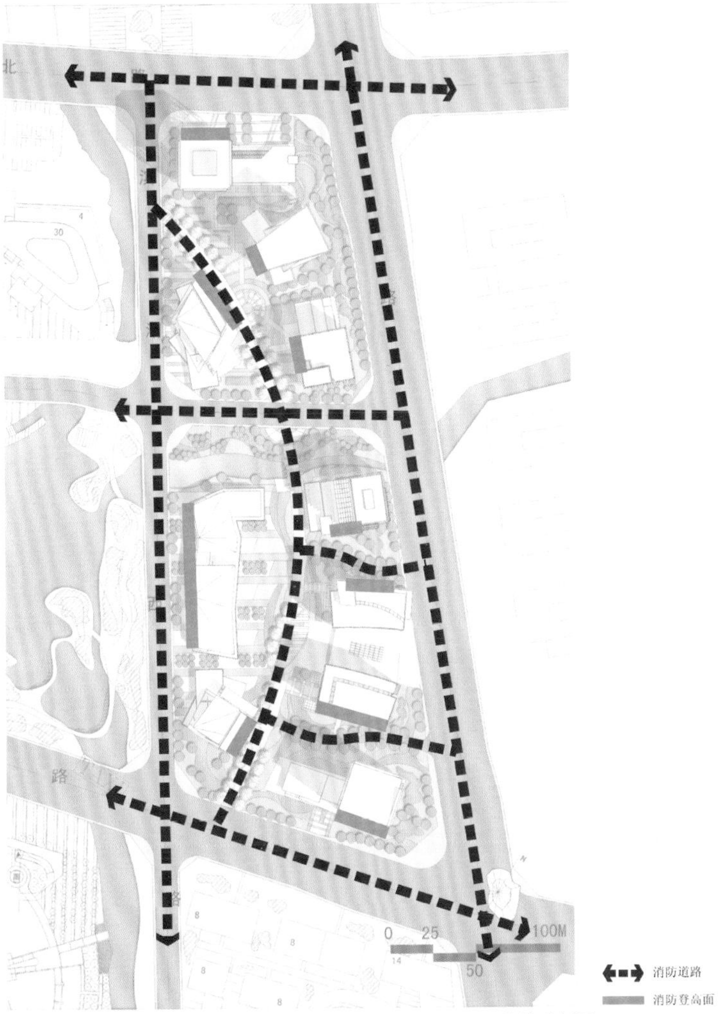

Fire Analysis Diagram 消防分析图

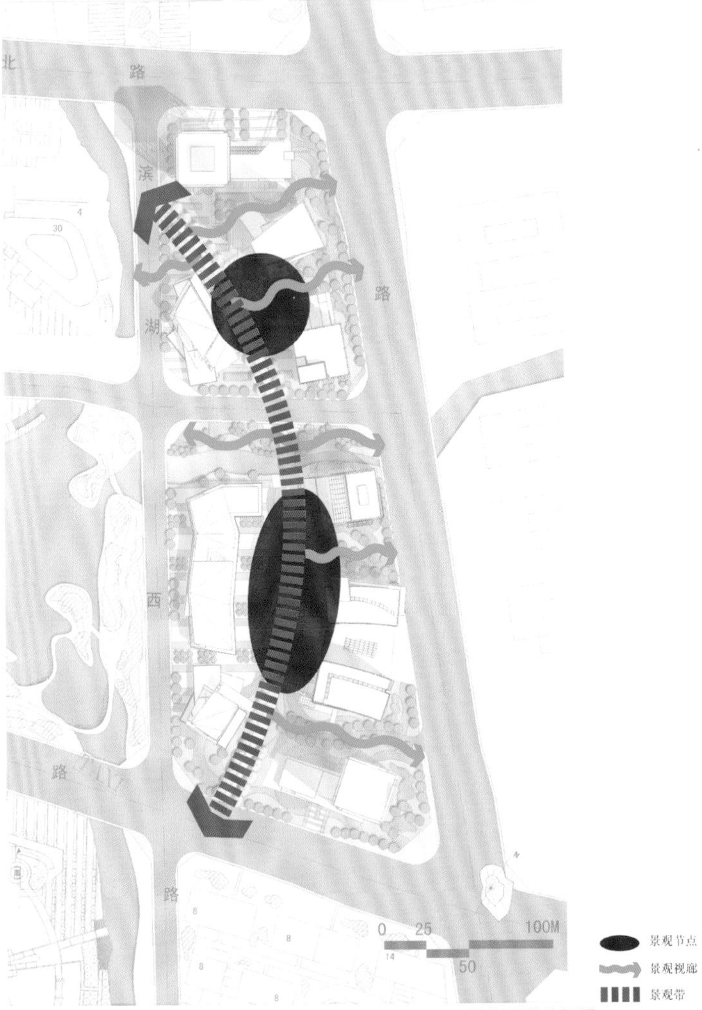

Green Landscape Analysis Drawing 绿化景观分析图

1 两栋高层成为区域最高点，核心代表

2 自然景观系统前后联系，纵向形成一条贯穿的城市景观空间轴线。

3 围绕自然景观和政府行政大楼，由科技文化中心开始，将轨道交通站、办公、住宅和公寓联系起来的一系列2~8层银行商业建筑群，沿带结合周围建筑性质形成独有却又相连的建筑气质。

4 在商业和周围高层建筑之间，以低矮的建筑作为过度，行程向内围合、但总体又开放的城市空间感。

Urban Spatial Analytics 城市空间分析

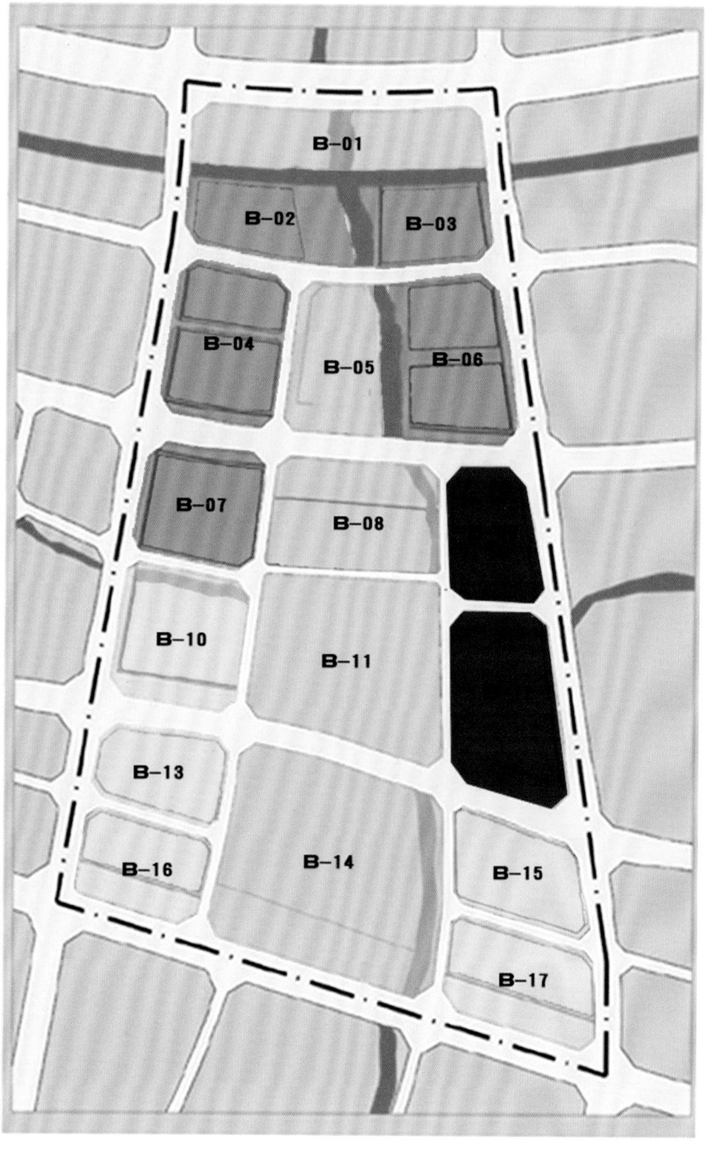

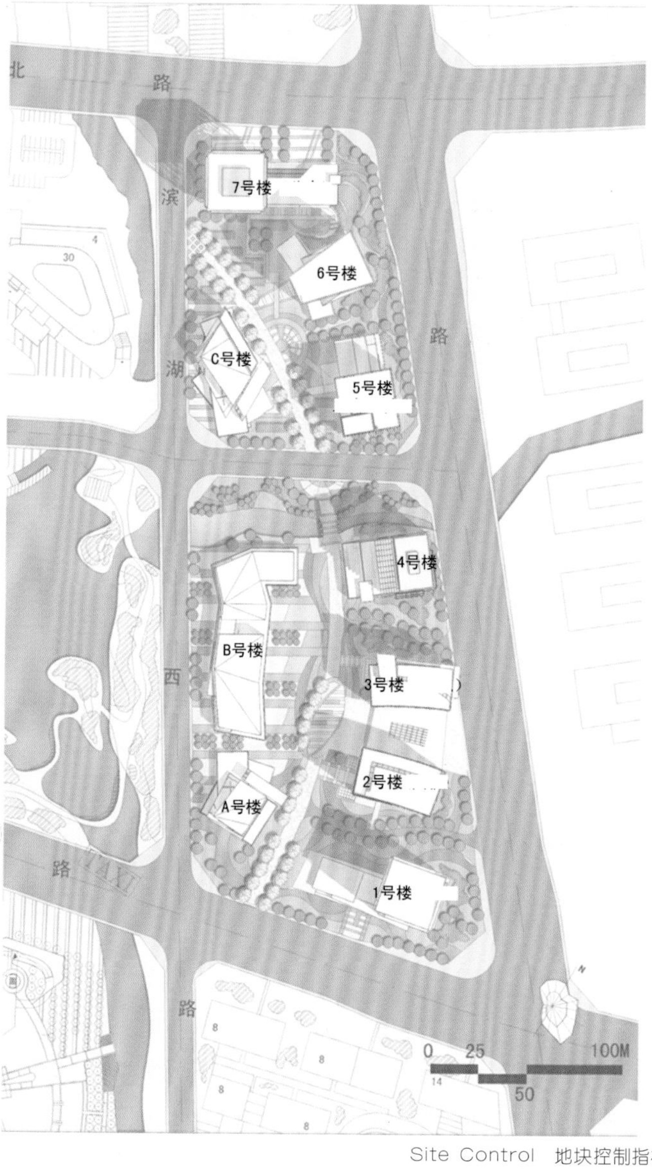

Site Control 地块控制指标

Commercial Building Concept 商业区单体概念

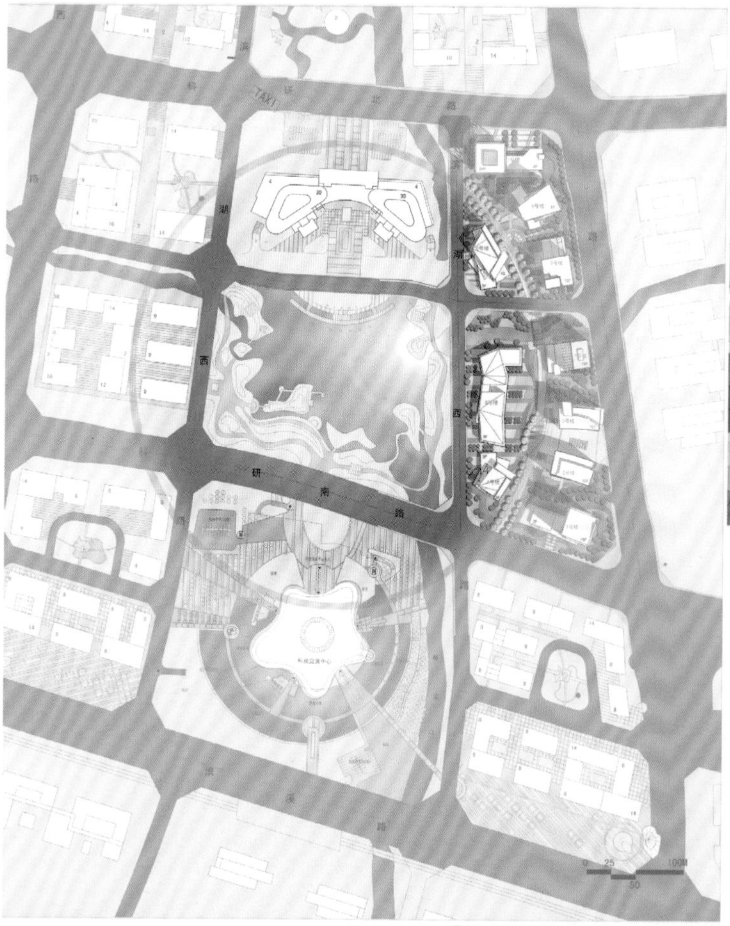

Planning Concept 规划理念

通过规划道路的局部调整创造“聚宝盆”的概念

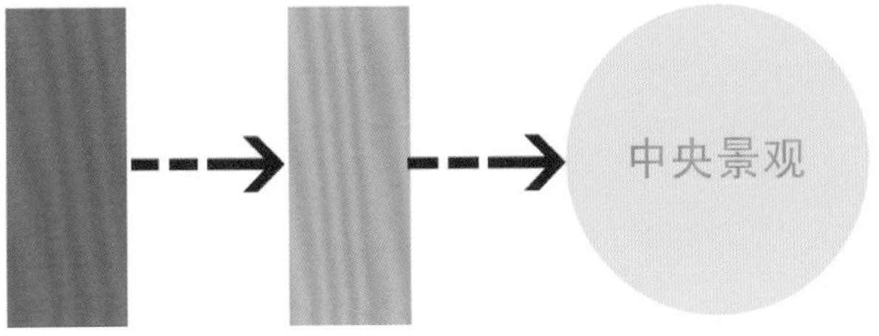

平行布置，景观遮挡

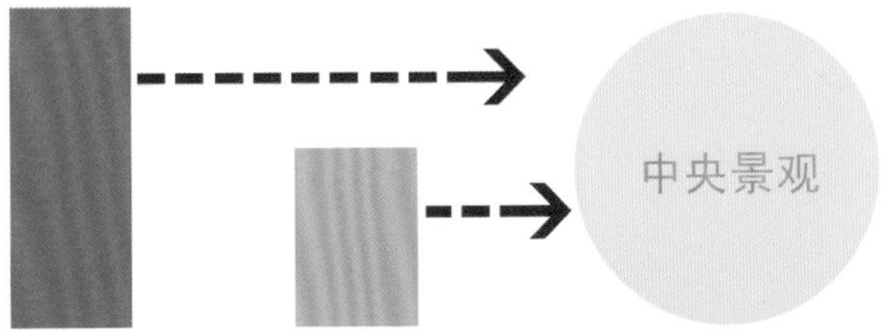

交错布置，景观共享

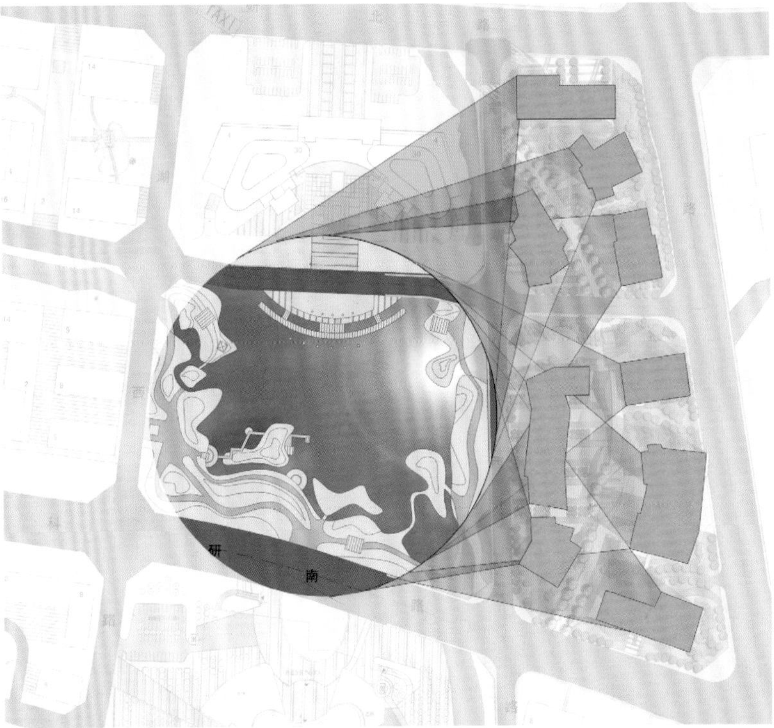

Office Planning Analysis 办公规划分析图

Skyline 1 天际线 1

Skyline 2 天际线 2

Skyline 3 天际线 3

Skyline 4 天际线 4

Office Building of the Land Bureau in New Intelligent City, Taihu Sci-tech Park, Wuxi
无锡太湖科技园新智城国土局办公楼

Keywords 关键词

Vertical Component
竖向构件

Facade Detail
立面细部

Air Garden
空中花园

Features 项目亮点

Since the twin-tower is designed in elegant style with vertical elements, this office building also uses some vertical components to keep a unified style in the central business area.

建筑主体采用竖向构件与双子塔楼形体相呼应，并和其外型相协调，进一步加强了中心商务区形象的整体性。

Location: Wuxi, Jiangsu, China
Architectural Design: CCI Architecture Design & Consulting Co., Ltd.
Floor Area: 16,000 m^2
Plot Ratio: 3.64
Building Density: 29.7%
Green Coverage Ratio: 35.2%

项目地点：中国江苏省无锡市
建筑设计：上海新外建工程设计与顾问有限公司
总建筑面积：16 000 m^2
容 积 率：3.64
建筑密度：29.7%
绿 化 率：35.2%

■ **Overview**

The office building of the Land Bureau is located with the remarkable twin-tower to the north, the central lake view area to the west, the software park to the east, and the sub business center to the south. Situated in the intersection of the main roads, it will be an important node of the New Intelligent City. Therefore the building should be modern and elegant, showing the characteristics of the sci-tech park. What's more, to keep harmonious with the twin-tower, the building is designed to face south with great lake views on the west. All the internal functions are also well organized accordingly.

■ **项目概况**

国土局办公楼北部为中央核心地标双子塔楼，西侧是中心湖面景观区，东侧为软件园区，南侧是次级商务区。本案处于主干道交叉口，是衔接整个新智城项目的重要节点，因此形象上要以现代、简洁为原则，体现科技园区的特点，同时，要与双子塔楼的形象相协调。本案坐北朝南，西侧是优良的中央景观资源区，为此本案的功能布局充分利用西侧景观，并符合南北朝向的特点。

■ **Architectural Design**

Since the twin-tower is designed in elegant style with vertical elements, this office building also uses some vertical components to keep a unified style in the central business area. Offices and meeting rooms are arranged on the south side to have sufficient sunlight. While traffic lines and auxiliary rooms are on the north, and a small air garden is also set on every two floors to get more sunlight and create a garden-like working environment.

■ 建筑设计

双子塔楼形体细部为竖向元素，外形简洁大气，因此本案的建筑主体也采用竖向构件与其呼应，和其外形相协调，进一步加强了中心商务区形象的整体性。塔楼的主要办公、会议用房都布置在南向，享受充分的南向阳光，北侧则布置交通和辅助用房，并且上下每隔两层设置一个小型空中花园。这样，不仅弥补了北向采光欠佳的不足，又能形成花园式的公共办公交流空间。

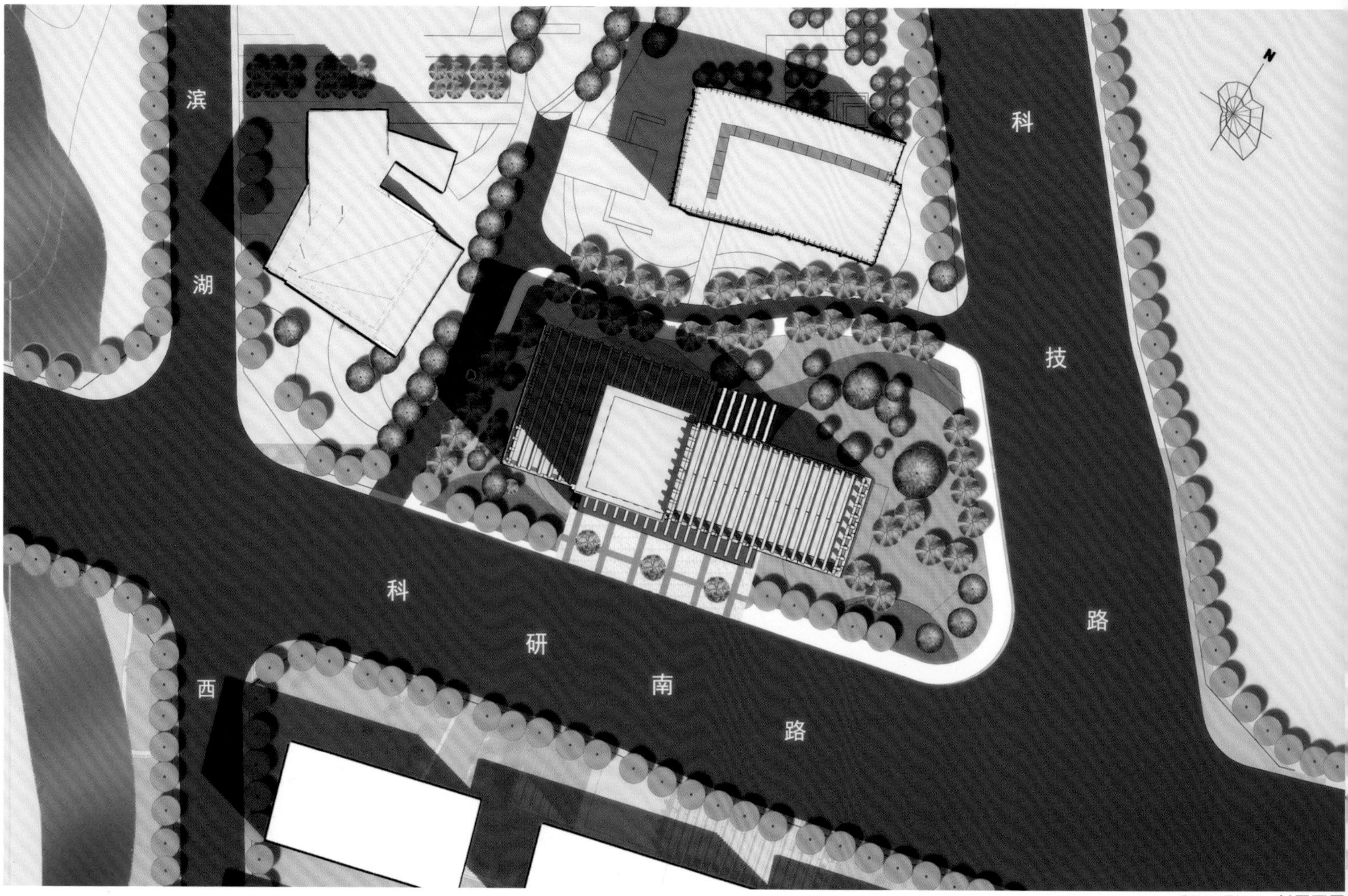

Site Plan 总平面图

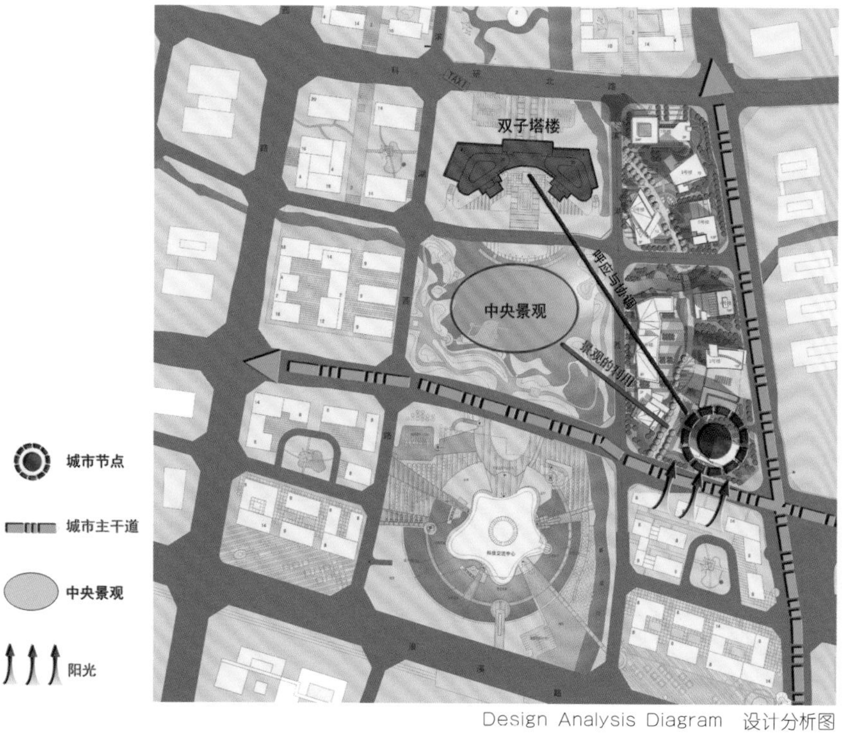

Design Analysis Diagram 设计分析图

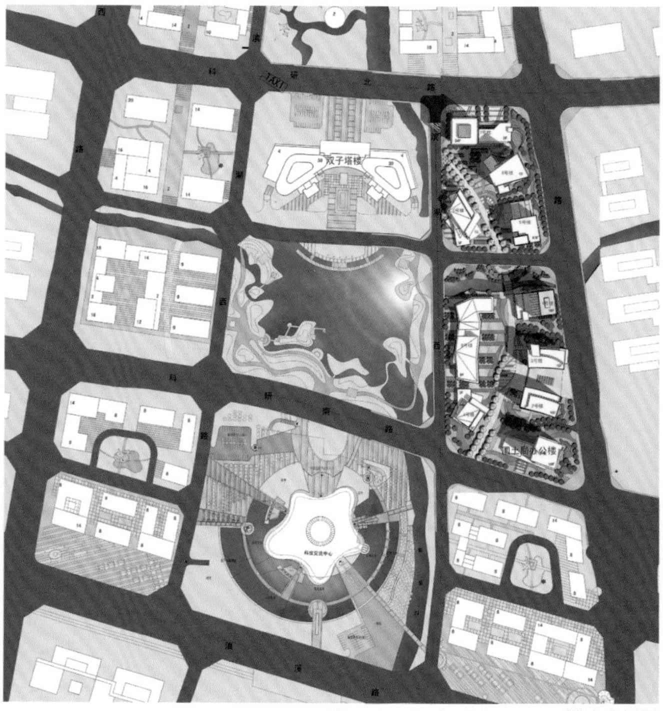

Site Analysis Diagram 基地分析图

Centre Lake Landscape　中心湖景观

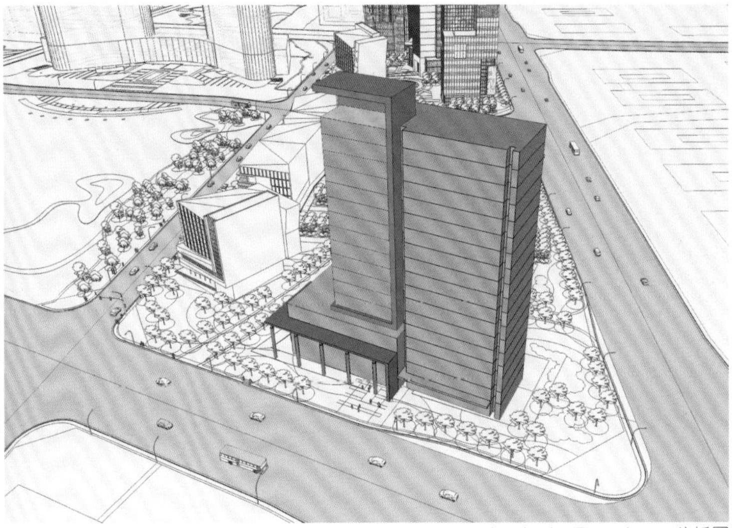

Analysis Drawing　分析图

室外观景平台　　办公入口

空中花园　　　　银行入口

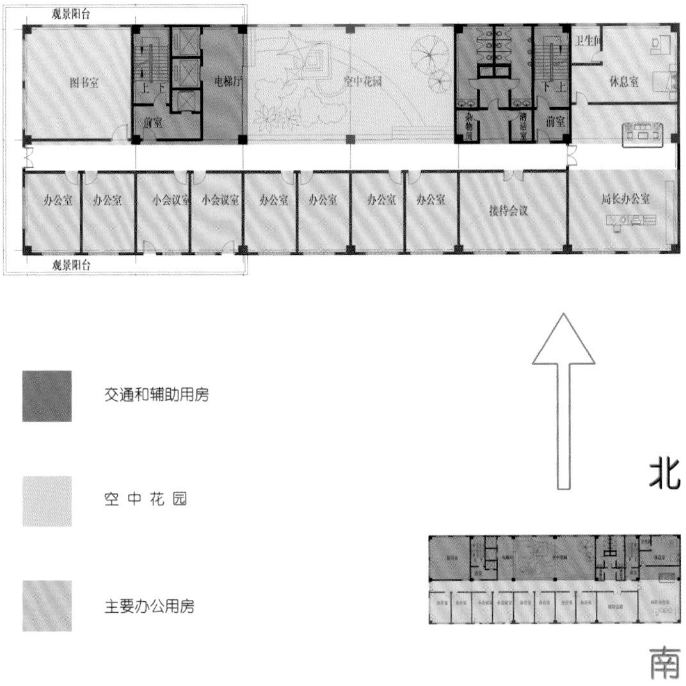

Basement Floor Plan　地下一层平面图

交通和辅助用房

空中花园

主要办公用房

北

南

Functional Layout　功能布局图

First Floor Plan　一层平面图

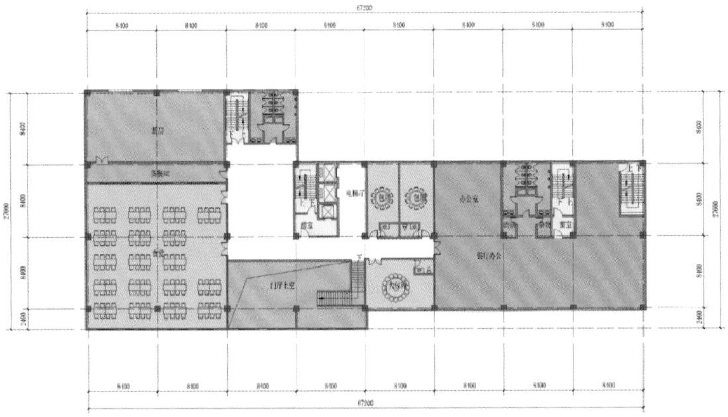

Second Floor Plan　二层平面图

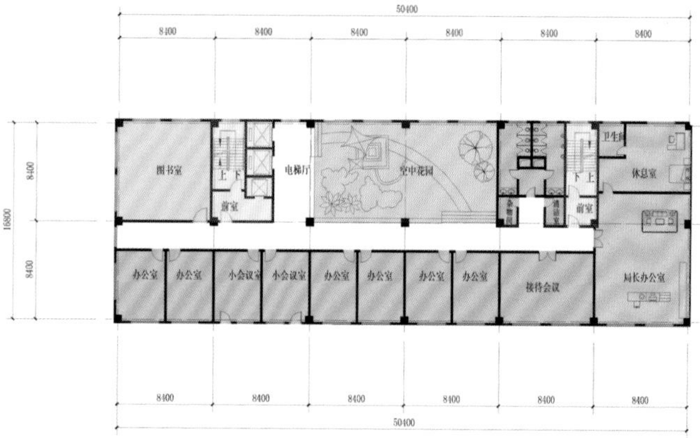

Standard Floor Plan A 标准层平面图 A

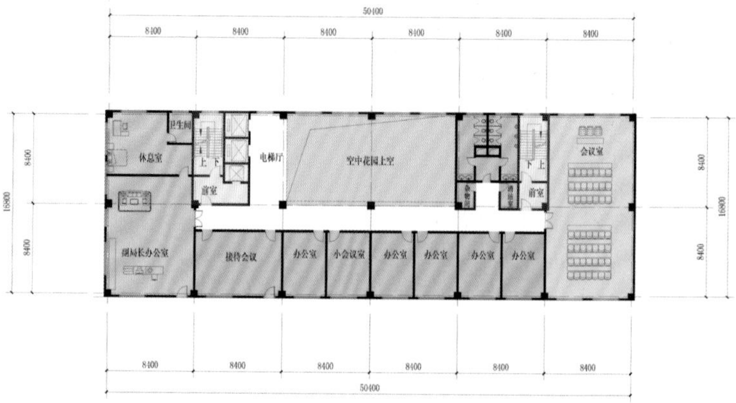

Standard Floor Plan C 标准层平面图 C

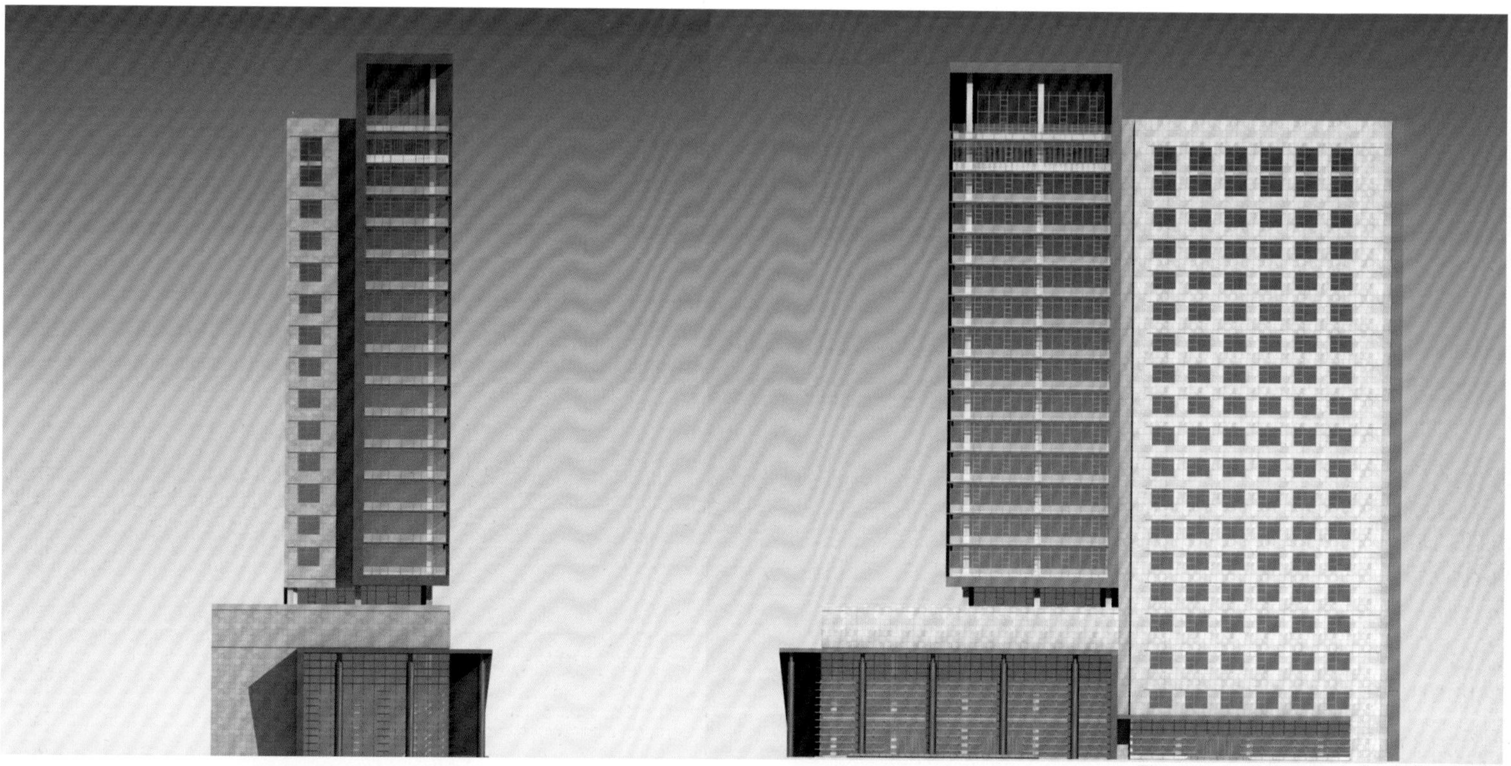
Elevation of Precept 2 方案二立面图

Elevation of Precept 3 方案三立面图

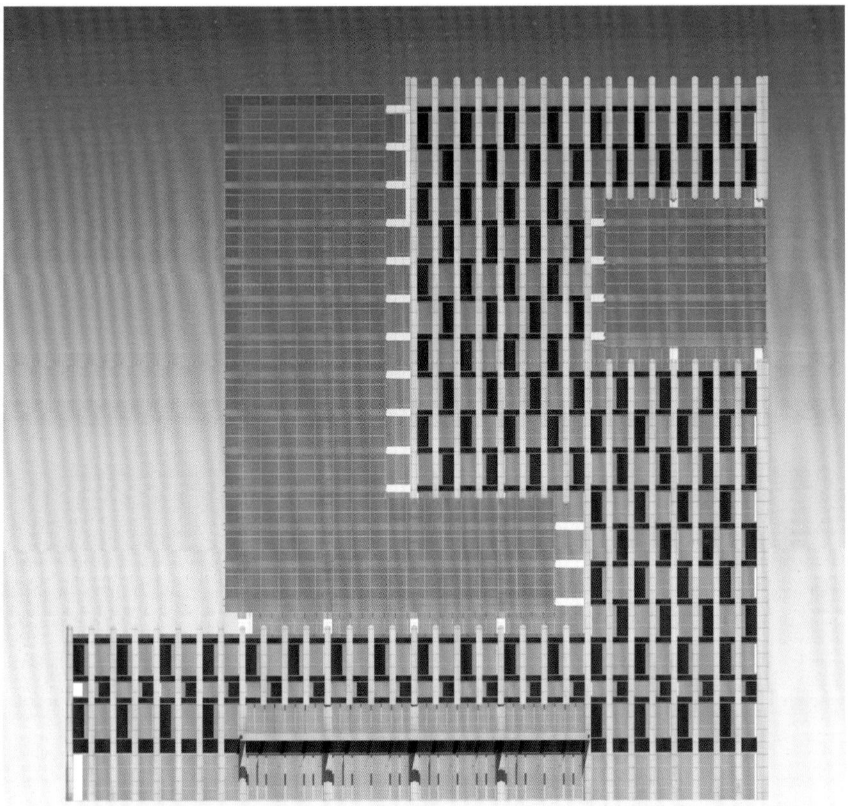

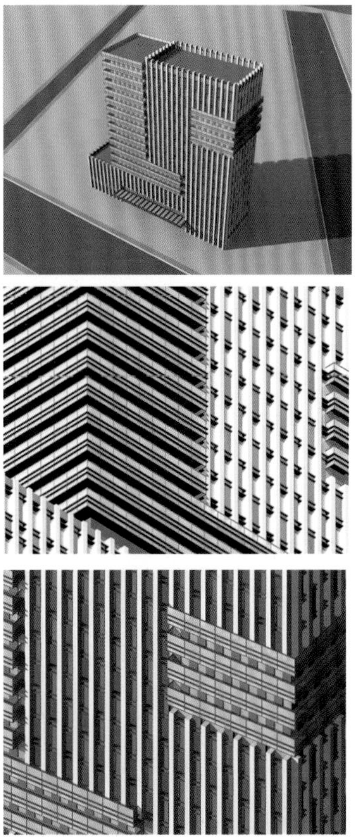

Detailing 细部设计图

Monomer Perspective 单体透视图

Bank of China Office Building in New Intelligent City, Taihu Sci-tech Park, Wuxi
无锡太湖科技园新智城中国银行大楼

Keywords 关键词

Lighting Window
采光窗

Elevated Ground Floor
架空层

Glass Curtain Wall
玻璃幕墙

Features 项目亮点

The facade design makes full consideration of the building functions and the landscape, highlighting the elegance and grace of modern office building with blocks interweaving.

办公楼的立面设计以功能及景观为出发点，以丰富的体块穿插为设计手法，力求体现现代化办公建筑特有的简洁、大气、优雅的形象。

Location: Wuxi, Jiangsu, China
Architecture Design: CCI Architecture Design & Consulting Co., Ltd.
Total Floor Area: 17,408.9 m^2
Building Height: 45 m
Floors: 12

项目地点：中国江苏省无锡市
建筑设计：上海新外建工程设计与顾问有限公司
总建筑面积：17 408.9 m^2
建筑高度：45 m
层　　数：12层

■ Architectural Design

The podium building is divided into several parts. Two-storey building on the north is the business hall of Bank of China which extends horizontally. The facade facing to the road is designed with huge glass, while on the entrance side, it applies solid wall to highlight the entrance and hang the logo of the bank.

The main entrance of the office building is two-storey high, above which there is a two and a half-storey high stone block with slope lighting windows installed randomly. This kind of design makes the entrance more noticeable. In addition with the two-storey hight entrance hall, it looks flexible and dignified.

■ 建筑设计

裙房部分的建筑体量分为几个部分，建筑北侧的两层建筑体量水平向展开，是中国银行营业大厅的位置，面向主要道路的外立面以大玻璃面为主，银行入口部分用局部的实墙面突出建筑出入口，同时也可以用于安装银行的标志。

办公楼主入口部分为两层挑高的建筑架空层，架空层之上为两层半高的石材体块，配置斜向的随机打开的采光窗。这种大气的建筑形象给办公楼的主入口带来了明确的指向性，结合两层高的办公入口门厅，使得整个主入口的感觉庄重中富于变化。

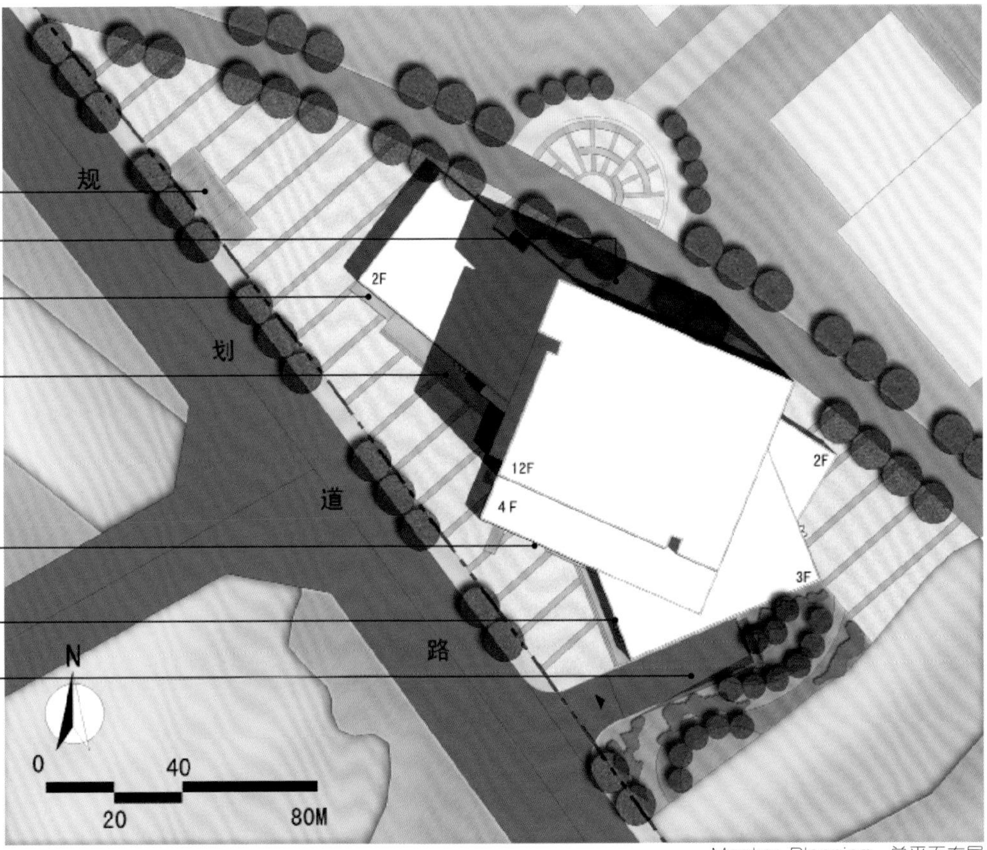

Master Planning 总平面布局

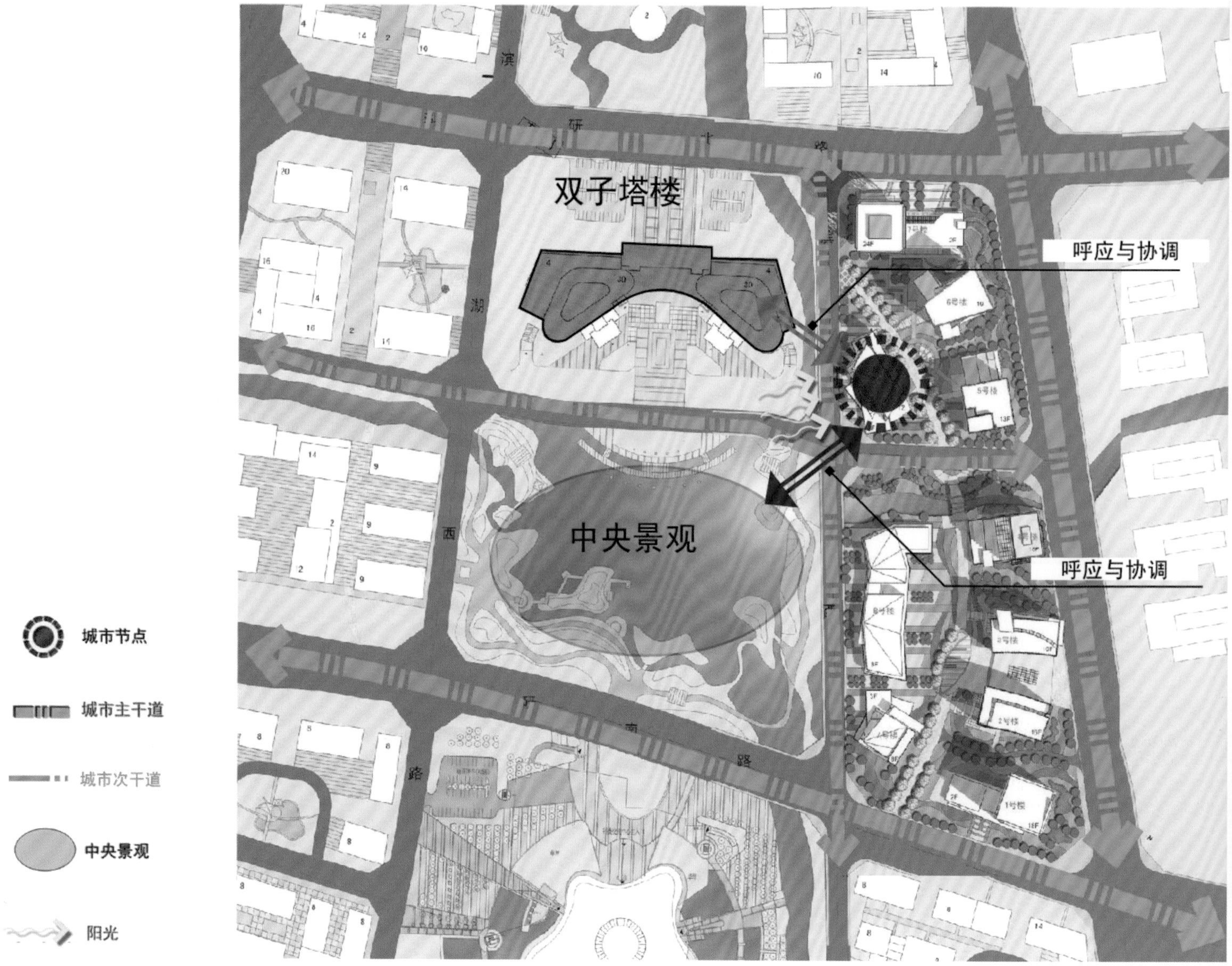

Layout Analysis 布局分析图

First Floor Plan 一层平面图

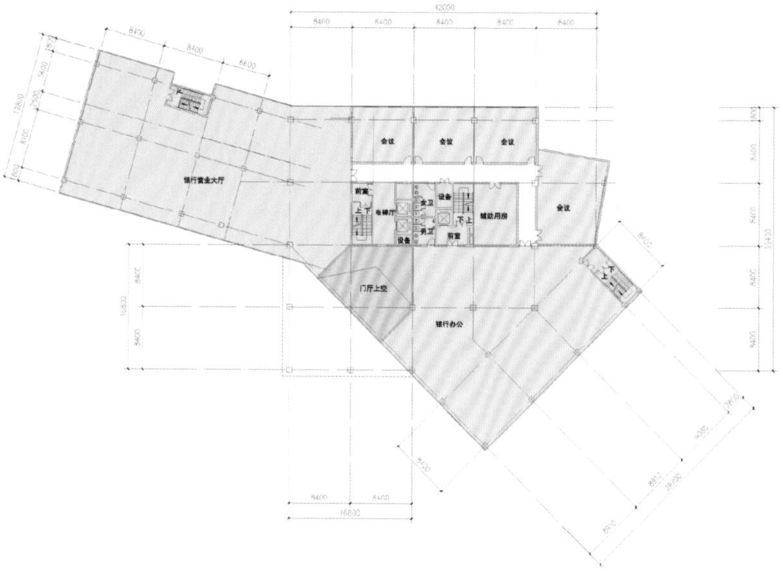

Second Floor Plan 二层平面图

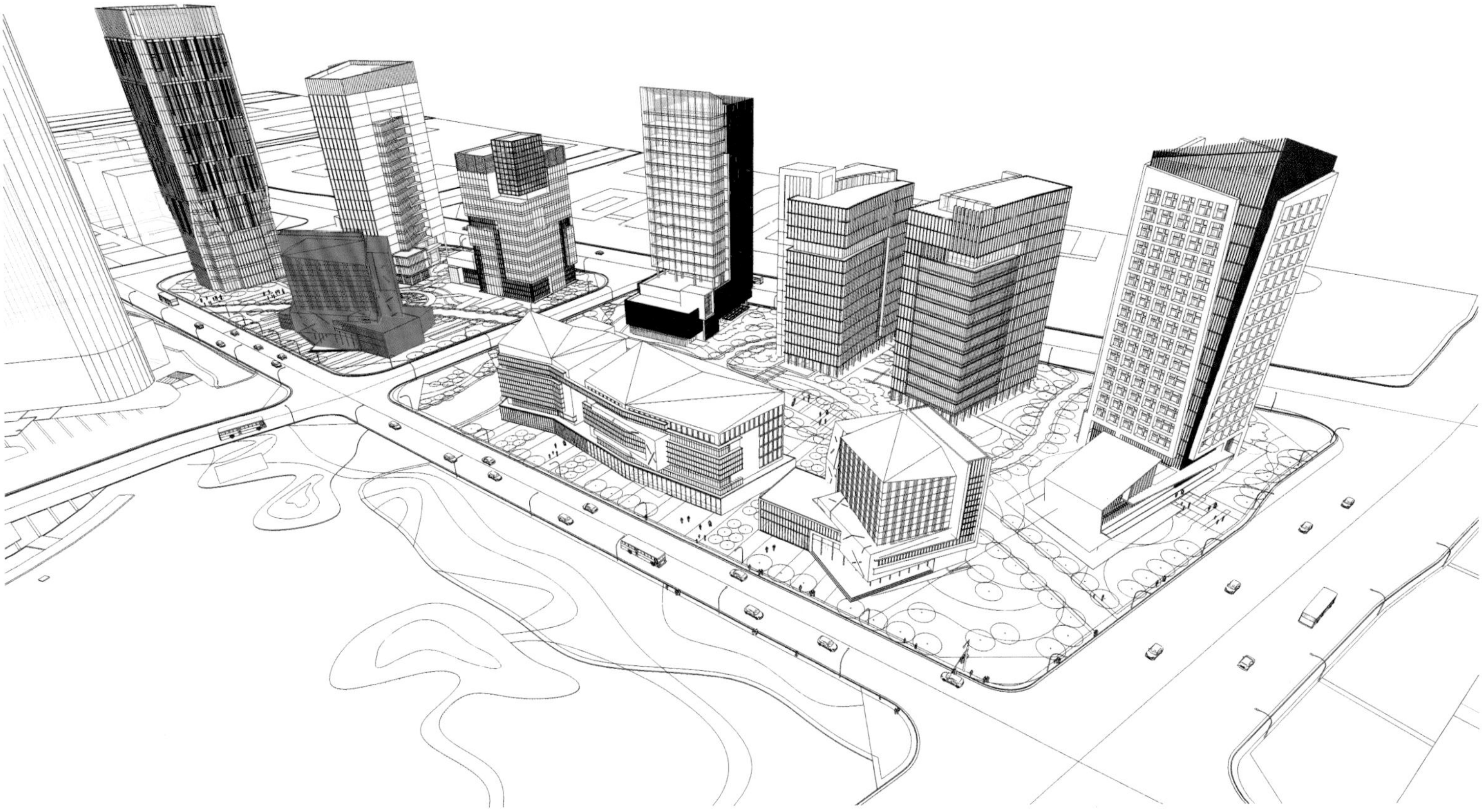

Third Floor Plan 三层平面图

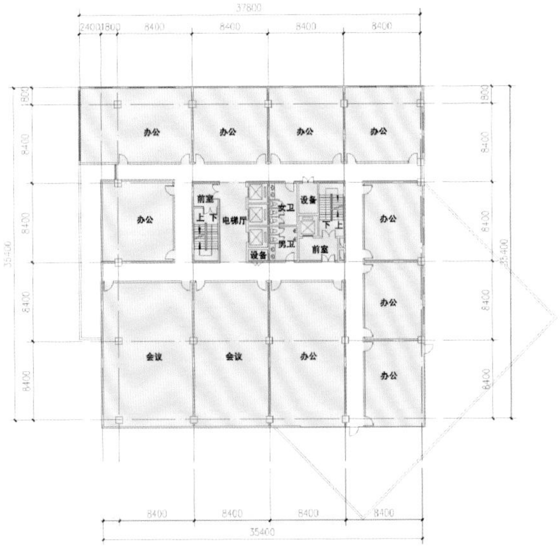

Fourth Floor Plan 四层平面图

West Elevation 西立面图

North Elevation 北立面图

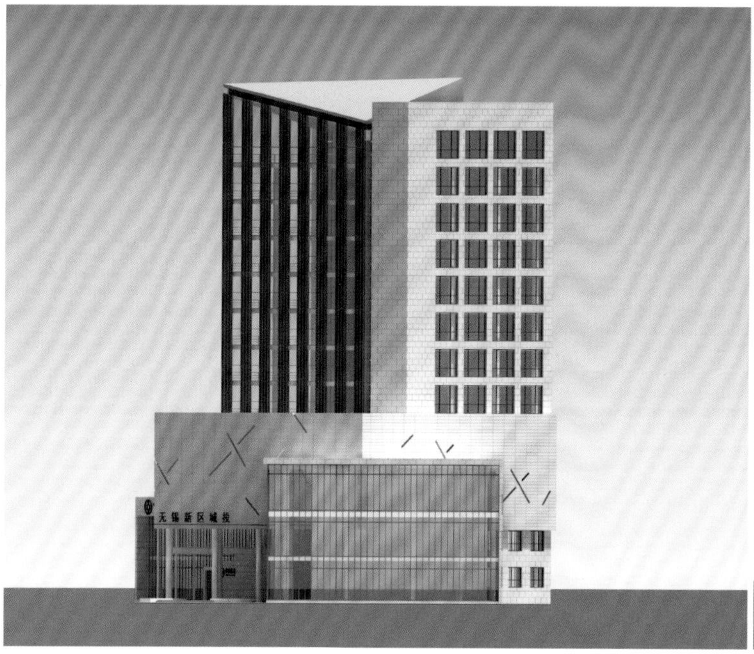

South Elevation 南立面图

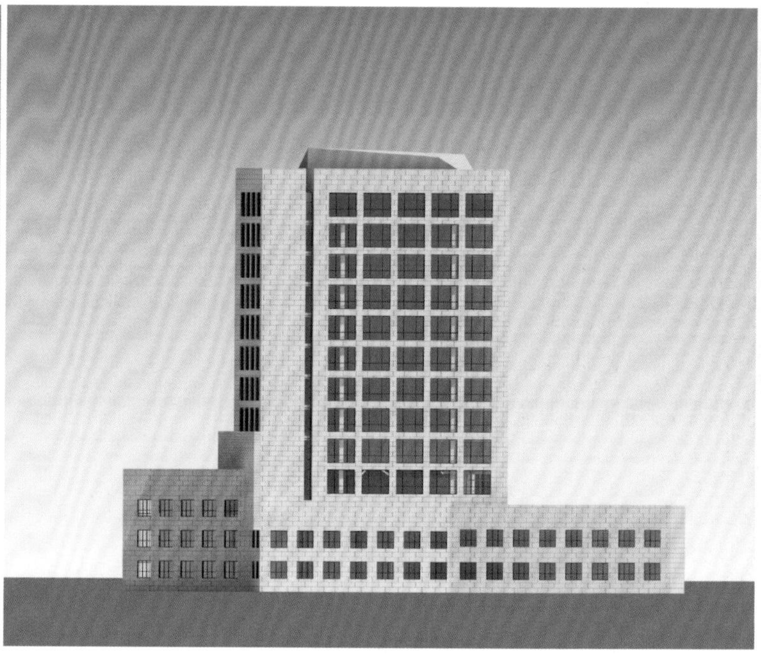

East Elevation 东立面图

■ Facade Design

The facade design makes full consideration of the building functions and the landscape, highlighting the elegance and grace of modern office building with blocks interweaving. The facade of the tower facing to the lake is designed with large-area glass curtain wall to get the most landscape views. And the reverse side is made of elegant stone curtain wall and well-arranged windows to create an integrated facade and greatly reduce energy consumption.

■ 立面设计

办公楼的立面设计以功能及景观为出发点，以丰富的体块穿插为设计手法，力求体现现代化办公建筑特有的简洁、大气、优雅的形象。塔楼部分面向人工湖方向的立面采用了大面积玻璃幕墙的做法，最大限度地将景观资源引入办公空间，背向湖面的立面采用了简洁的石材墙面和规整的开窗形式，保持了完整协调的立面感觉，同时也有利于提升建筑整体的节能保温性能。

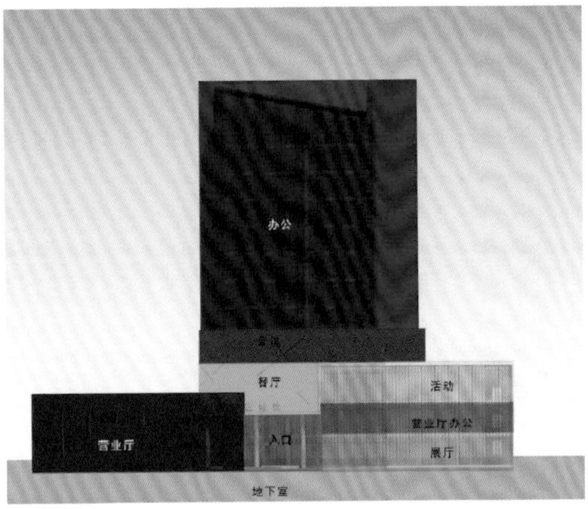
办公
会议
餐厅
活动
营业厅办公
入口
营业厅
展厅
地下室

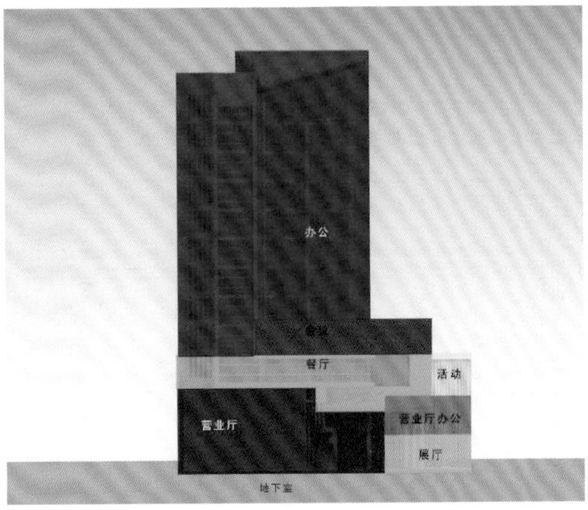
办公
会议
餐厅
活动
营业厅
营业厅办公
展厅
地下室

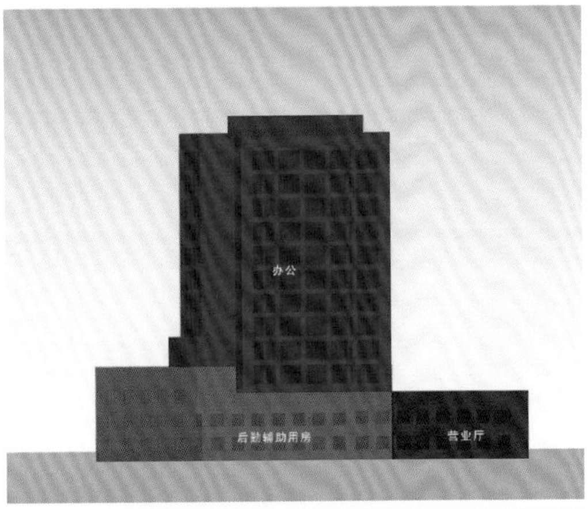
办公
后勤辅助用房
营业厅

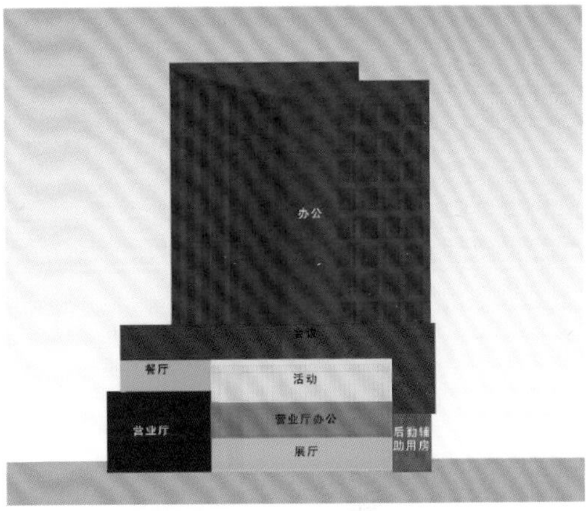
办公
会议
餐厅
活动
营业厅办公
营业厅
展厅
后勤辅助用房

中国银行

Headquarters of Huainan Animation Technology Industrial Park
淮南志高动漫文化科技产业园指挥部

Keywords 关键词

Architectural Texture
建筑肌理

Abundant Facade Form
立面丰富

Metal Curtain Wall
金属幕墙

Location: Huainan, Anhui, China
Developer: Huainan Zhigao Group
Architectural Design: Sunlay Design
Designers: Zhang Hua, He Wei, Liu Weiwei, Stefanie, Chen Na, Peng Xuefeng
Floor Area: 12,306 m^2
Plot Ratio: 0.67

项目地点：中国安徽省淮南市
开 发 商：淮南志高实业有限公司
建筑设计：三磊建筑设计有限公司
设计人员：张华　何威　刘维维　Stefanie　陈娜　彭雪峰
建筑面积：12 306 m^2
容 积 率：0.67

Features 项目亮点

The idea of the design comes from "melting iceberg". Irregular building texture expresses the concept well and forms the abundant facade.

整个建筑的概念设计源自“融化的冰块”。不规则的建筑肌理将“融化的冰块”的构思很好地表现了出来，形成了建筑丰富的外立面。

■ Overview

Located at Huainan, Anhui, Headquarters of Huainan Animation Technology Industrial Park is an office and exhibition building including five floors aboveground and one floor underground. It is a supersize world-class animation industrial park that integrates sightseeing, entertainment, recreation, exhibition, popular science education and industry development. It is comprised of eight theme areas such as Fangzhou Square, Qiqi's Paradise, Huainan Kingdom, Fiery-red Wood, Titanic & Iceberg, the 9th Milky Way and Time Island.

■ 项目概况

本项目坐落于安徽省淮南市，是淮南志高动漫文化科技产业园的指挥部，是一栋地上五层、地下一层的办公及展览建筑。淮南志高动漫文化科技产业园是一座融观赏、娱乐、休闲、博览、科普教育和产业开发于一体的超大型世界顶级动漫产业园。包括方舟广场、奇奇乐园、淮南王国、火红迷林、泰坦冰川、达芬奇小镇、第九星河、时光岛八个主题区组成。

■ Planning and Layout

In terms of planning layout, core function of the project is taken into account. Underground floor is for garage, worker's restaurant and related matching rooms; 1st floor is for exhibition & marketing hall, the 2nd to 4th floors are for offices and their matching rooms; 5th floor is for reception rooms.

■ 规划布局

在规划布局上面，考虑到本案作为整个动漫产业园的指挥部的核心作用，将整个建筑的地下一层设置为汽车库、职工餐厅及相关设备配套用房，一层为展览及营销大厅，二层至四层为办公用房及办公配套用房，五层为办公接待用客房。着力打造产业园区内文化创意产业基地和科普教育的指挥中枢。

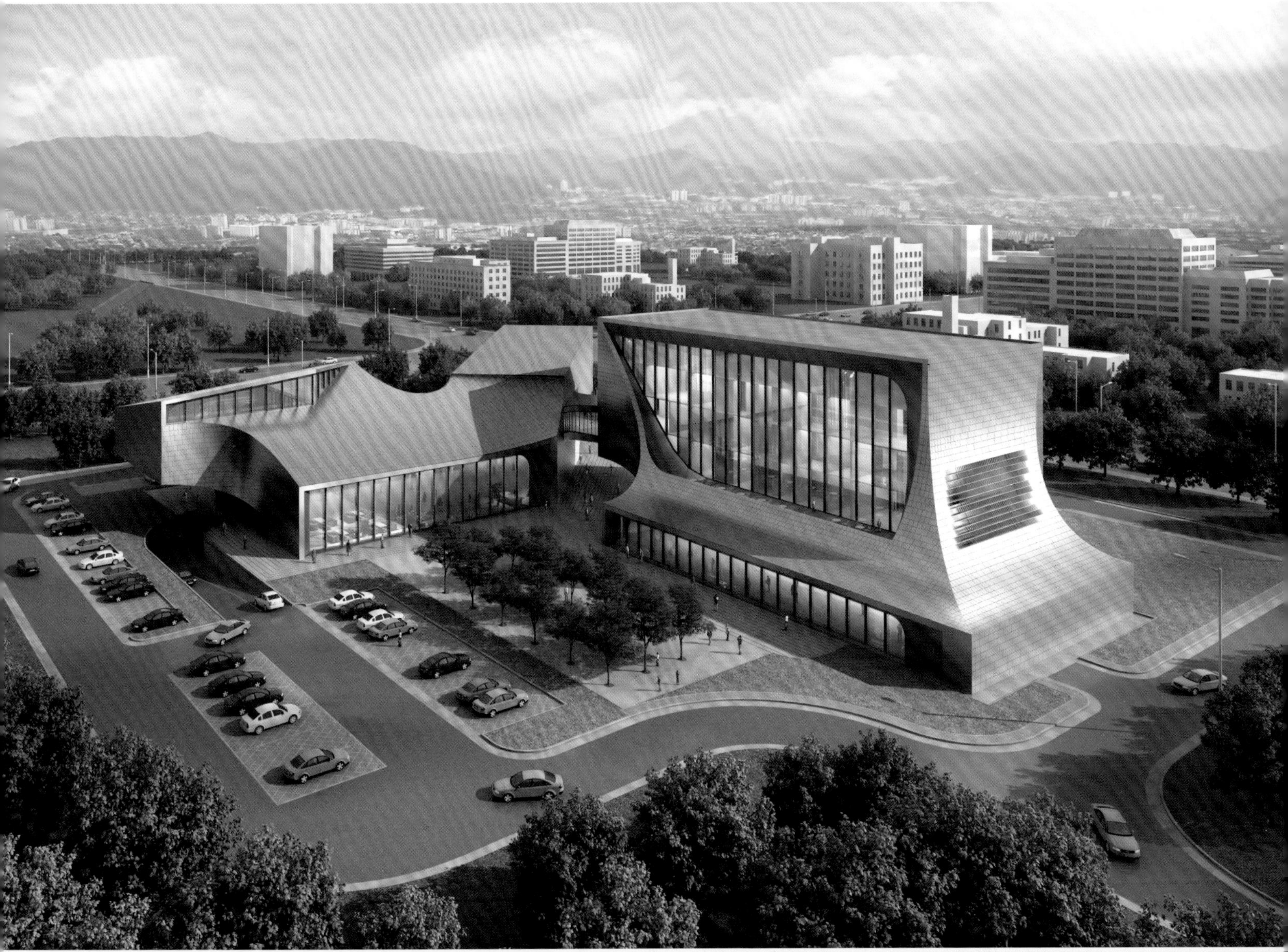

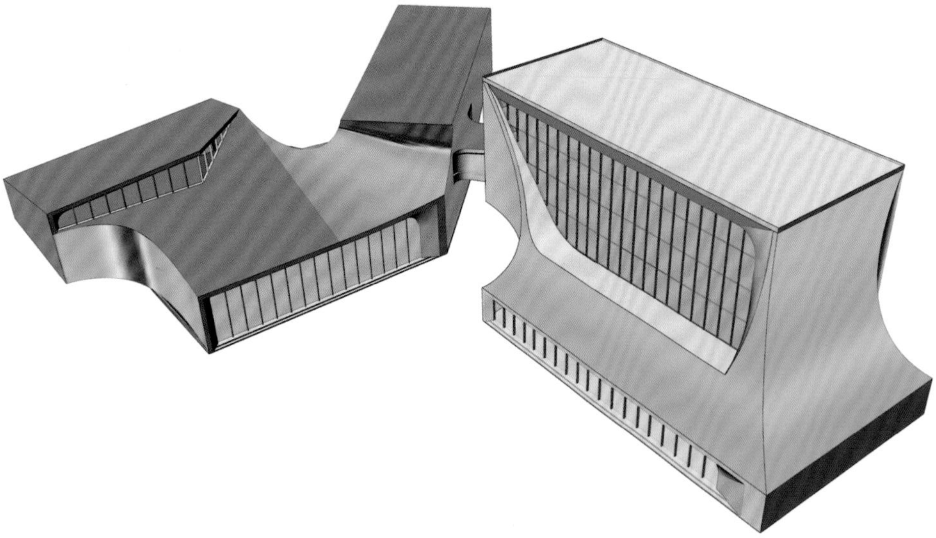

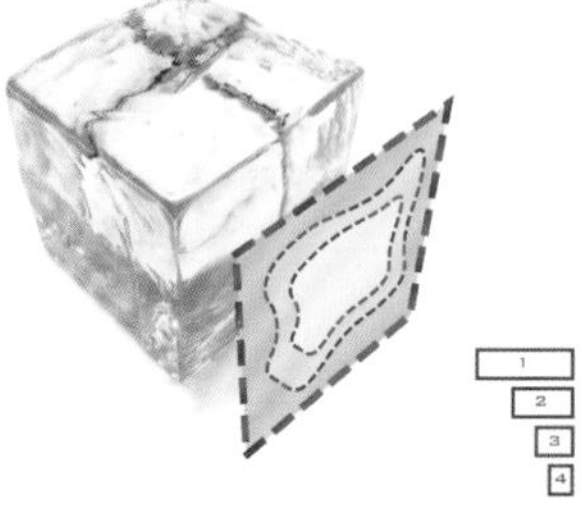

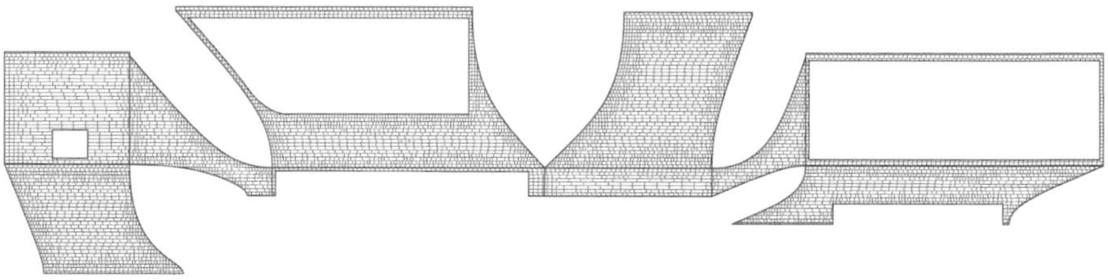

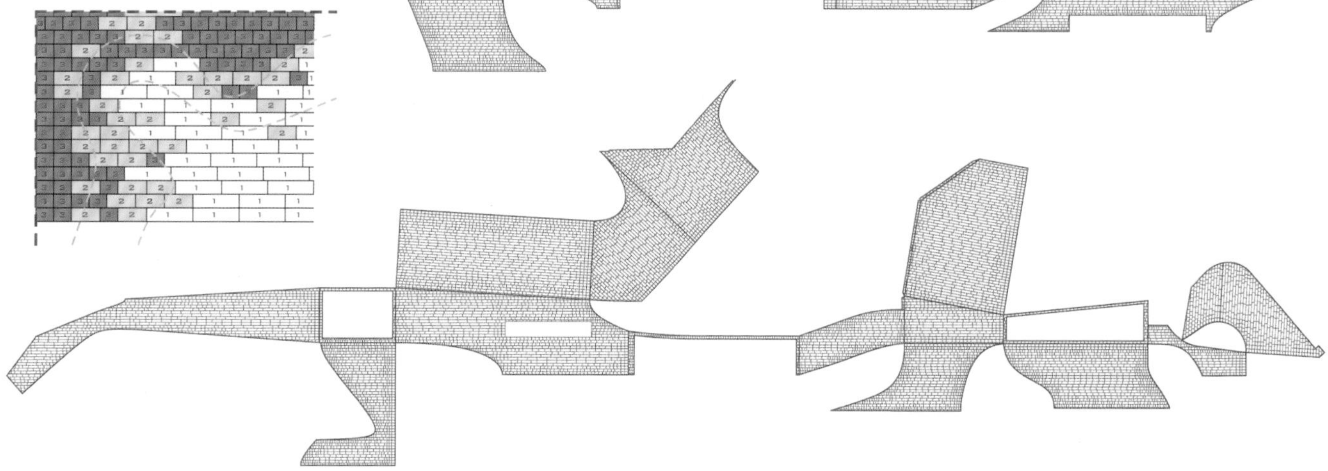

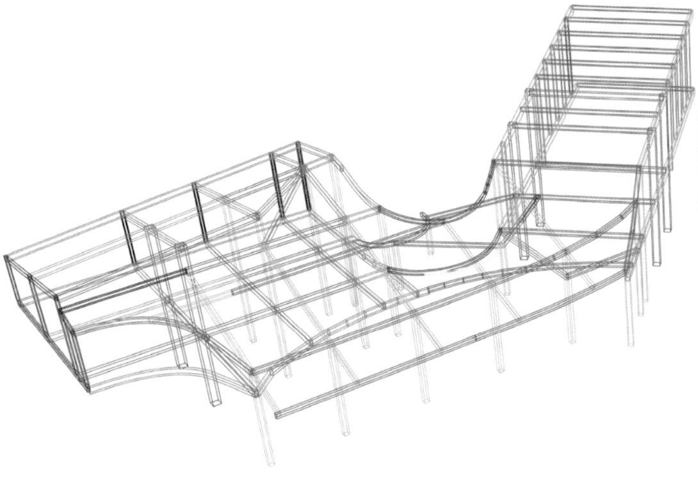

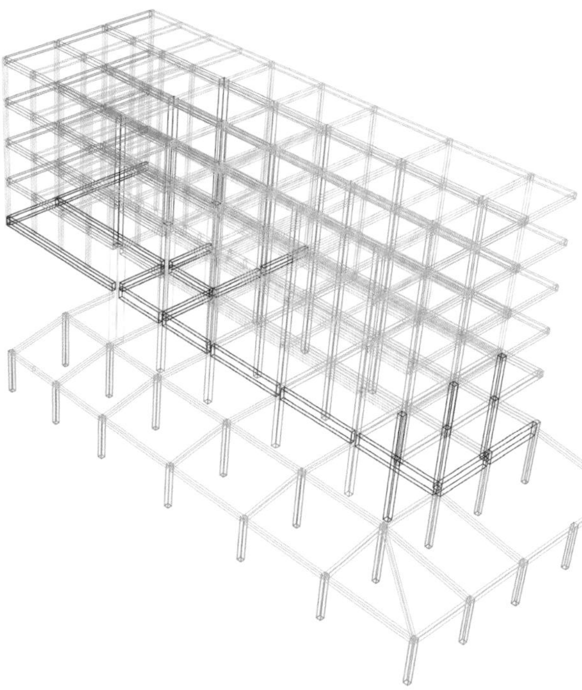

■ Architectural Design

The idea of the design comes from "melting iceberg". Irregular building texture expressed the concept well and formed the abundant facade. An overhead corridor connects the two buildings properly and a semi-enclosed structure is shaped in the interlaced network. In terms of selecting suitable materials, metal curtain wall, glass curtain wall and standing seam roofing system are used in the exterior; while building blocks, infilled wall, light gauge steel joist and glass partition are in the interior; in addition.

■ 建筑设计

整个建筑的概念设计源自“融化的冰块”。不规则的建筑肌理将“融化的冰块”的构思很好地表现了出来，形成了建筑丰富的外立面。两栋主体建筑通过一个空中的过道恰到好处地连接了起来，整体在横竖交错中形成了一种半围合的结构。在建筑的表面材料的选用上，幕墙部分采用金属幕墙、玻璃幕墙及直立锁边屋面系统。内装部分采用砌块填充墙、轻钢龙骨轻质板材及玻璃隔断。

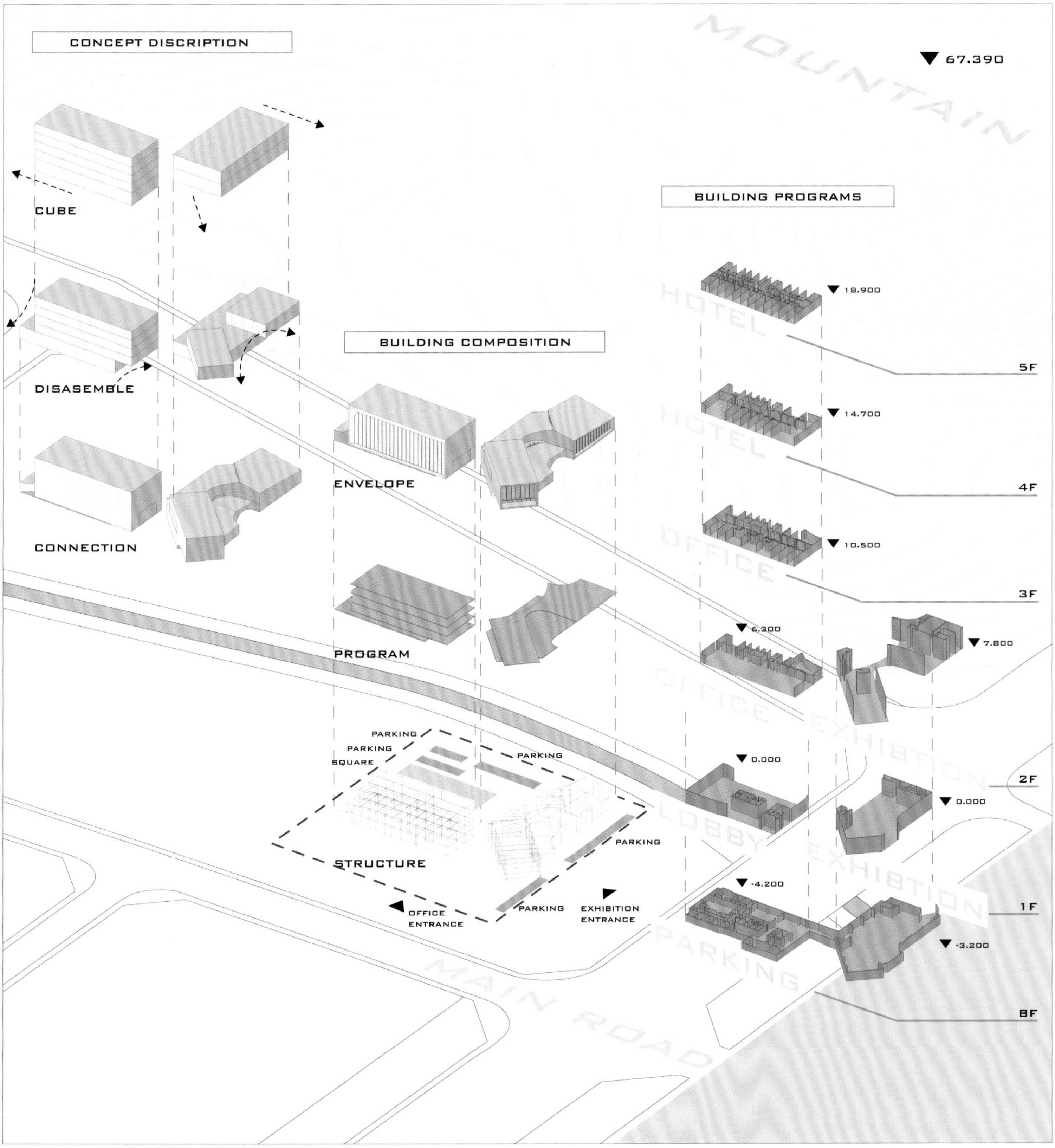

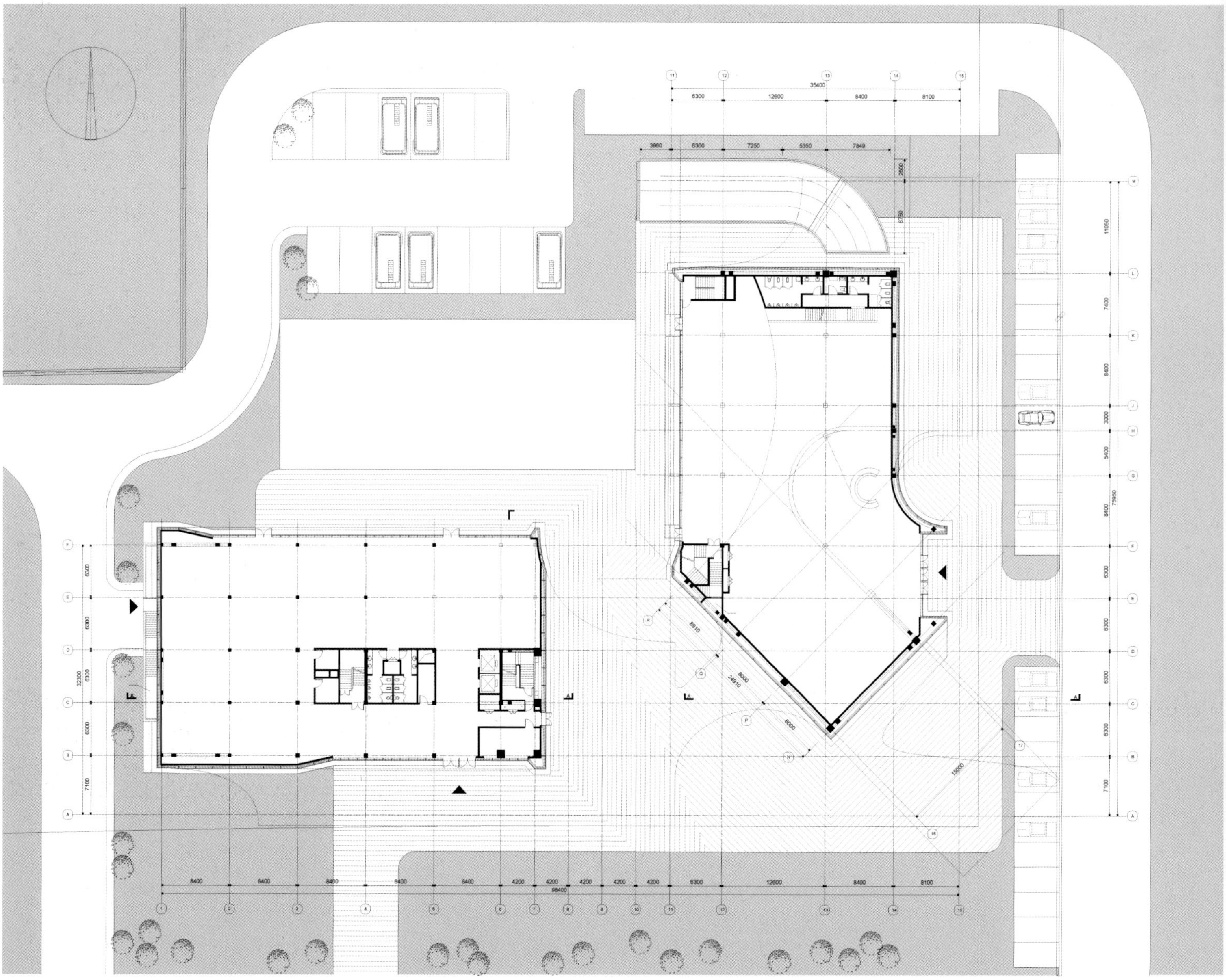

Floor Plan 平面图

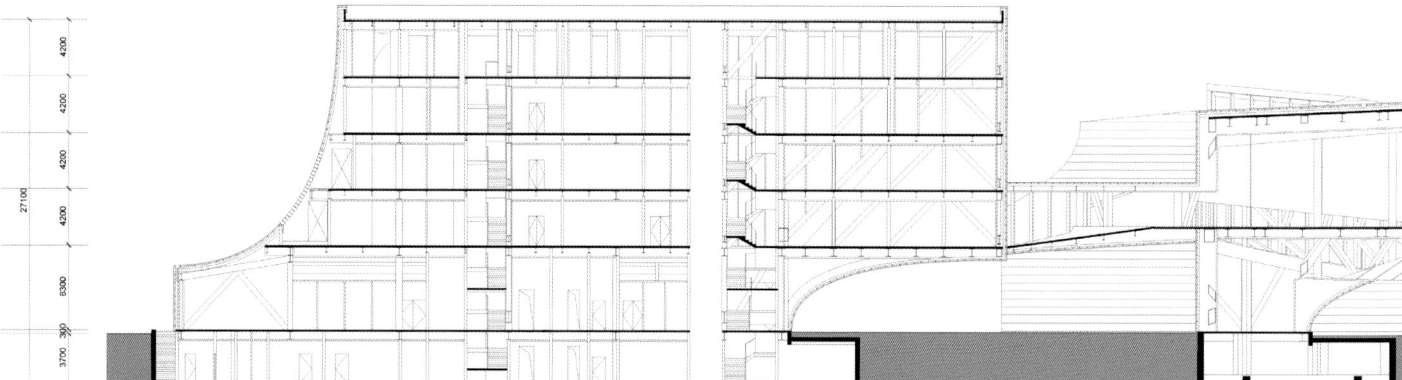

Sectional Drawing 1 剖面图 1

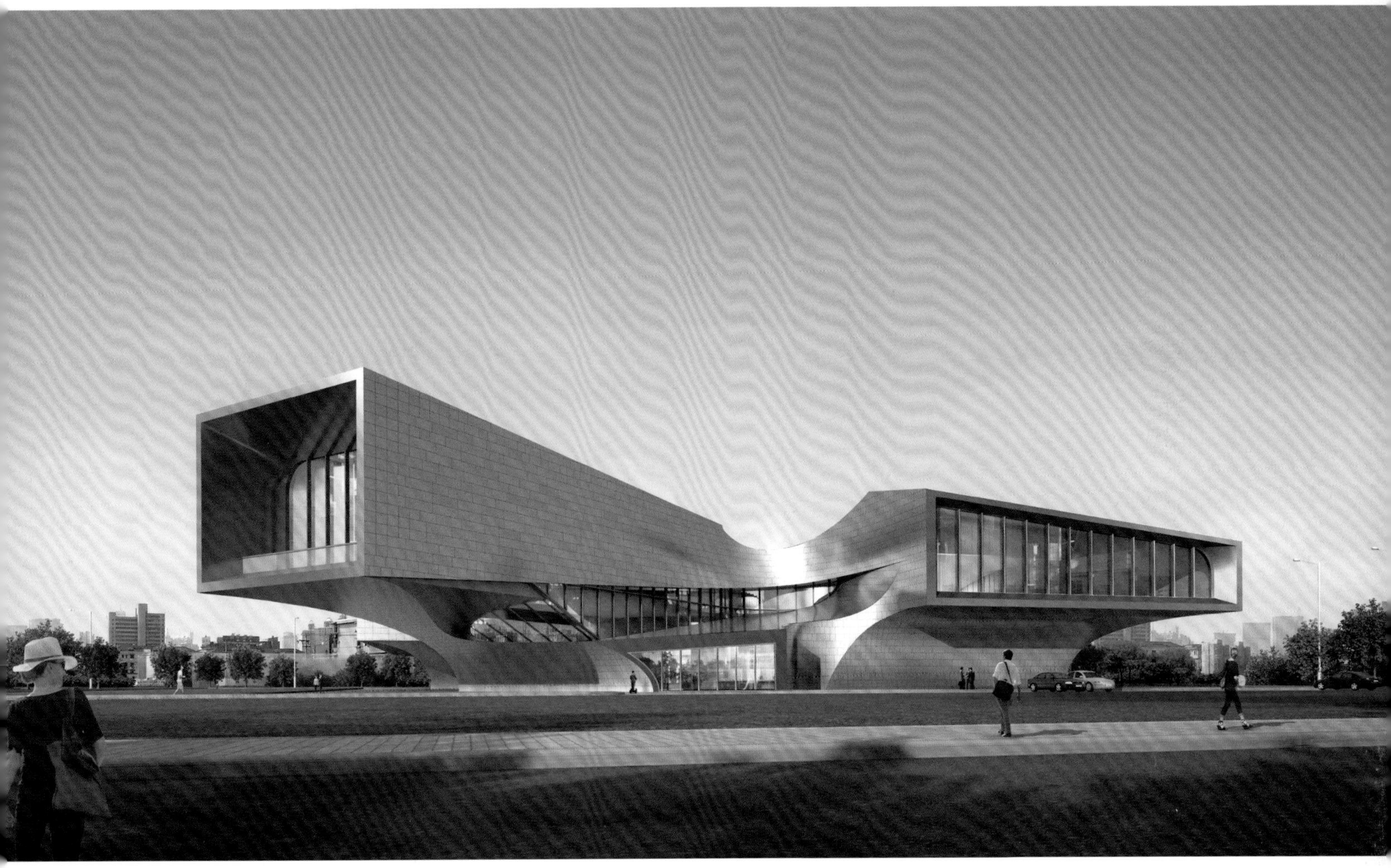

Sectional Drawing 2　剖面图 2

Experimental & Testing Center of Avic Aero-polytechnology Establishment

中国航空综合体技术研究所实验与检测中心

Keywords 关键词

Reasonable Layout
布局合理

Green Ecology
绿色生态

Distinctive Characteristics
个性鲜明

Location: Huairou District, Beijing, China
Architectural Design: Agency C & P Architecture
Designers: Lan Jian, Fan Hongtao
Land Area: 117,429.00 m^2
Floor Area: 128,418.00 m^2
Plot Ratio: 1.30
Green Coverage Ratio: 38%

项目地点：中国北京市怀柔区
建筑设计：C&P(喜邦)国际建筑设计公司
设计人员：兰剑　范洪涛
用地面积：117 429.00 m^2
建筑面积：128 418.00 m^2
容 积 率：1.30
绿 化 率：38%

Features 项目亮点

With the help of abundant ecological resources in Yanxi Development Zone, the project creates interior green environment in which the industrial park is not a factory as cold as a wagon tire any more but an interesting research space brimming over with vigor and vitality.

项目借助雁栖开发区的丰富生态资源，打造园区内部绿色环境，建筑对绿地敞开，人被绿色环绕，产业园不再是冷冰冰的厂房，而是一个生机盎然、充满趣味的科研空间。

■ Overview

Located at the Huairou, Beijing, the project covers an area of 117,429 m^2 and the floor area is 128,418 m^2 with a building density of 29.80% and 699 parking lots. Design of the whole project reflects the idea of "green, humanity and technology".

■ 项目概况

中国航空综合体技术研究所实验与检测中心位于北京市怀柔区，项目用地面积为117 429 m^2，建筑面积为128 418 m^2，建筑密度为29.80%，设置有699个停车位。整个项目的设计体现了"绿色、人文、科技"的理念。

■ Design Concept

With the help of abundant ecological resources in Yanxi Development Zone, the project creates interior green environment in which the industrial park is not a factory as cold as a wagon tire any more but an interesting research space brimming over with vigor and vitality.

In accordance with people-oriented principles, it fully takes staffs' need of work and life, distributes the functions in proper status. Construction, central landscape and leisure landscape footpath combined with each other organically, forming a characteristic office space.

It brings in sustainable design techniques, fully respects the natural condition of the site and composites properly in terms of overall plan. While in terms of architectural design, it introduces technologies such as solar energy, sunshading board and water recycling system to create a world-class modern park with low pollution, low noise and low-carbon green.

■ 设计理念

绿色：项目借助雁栖开发区的丰富生态资源，打造园区内部绿色环境，建筑对绿地敞开，人被绿色环绕，产业园不再是冷冰冰的厂房，而是一个生机盎然、充满趣味的科研空间。

人文：坚持以人为本，赋予最细致的人文关怀，充分考虑人员的工作和生活需求，将办公、研发、培训、服务等功能进行最合理分布，地表建筑和中心景观绿地、休闲景观步行道有机结合，打造具有鲜明个性化办公空间。

科技：引入可持续发展的设计手法，充分尊重基地的自然条件，在总体规划上合理布局，在建筑设计中引入太阳能利用、遮阳板、水循环使用等技术手段，打造低污染、低噪音、低碳环保的世界一流现代化园区。

技术经济指标

名称				数值	单位	备注
总用地面积				117429.31	m^2	
总建设用地面积				80134.11	m^2	
总建筑面积				128418	m^2	
其中	地上总建筑面积			104133	m^2	
	其中	科研大楼		24020	m^2	
		科研试验大楼		20922	m^2	建筑单元面积1000-1500平方米 柱网8M×8M，七个单元
		培训中心		16283	m^2	
		科研交流中心		6048	m^2	
		独立科研楼		24000	m^2	共5栋
		配套服务用房		3750	m^2	
		其中	服务人员生活用房	3000	m^2	
			动力站	750	m^2	
	地下总建筑面积			24285	m^2	包括地下车库及配套用房
容积率				1.30		
建筑密度				28.80%		
绿地率				38%		
停车位				699	辆	
其中	地上			6	辆	大型客车停车位
	地下			693	辆	

Site Plan 总平面图

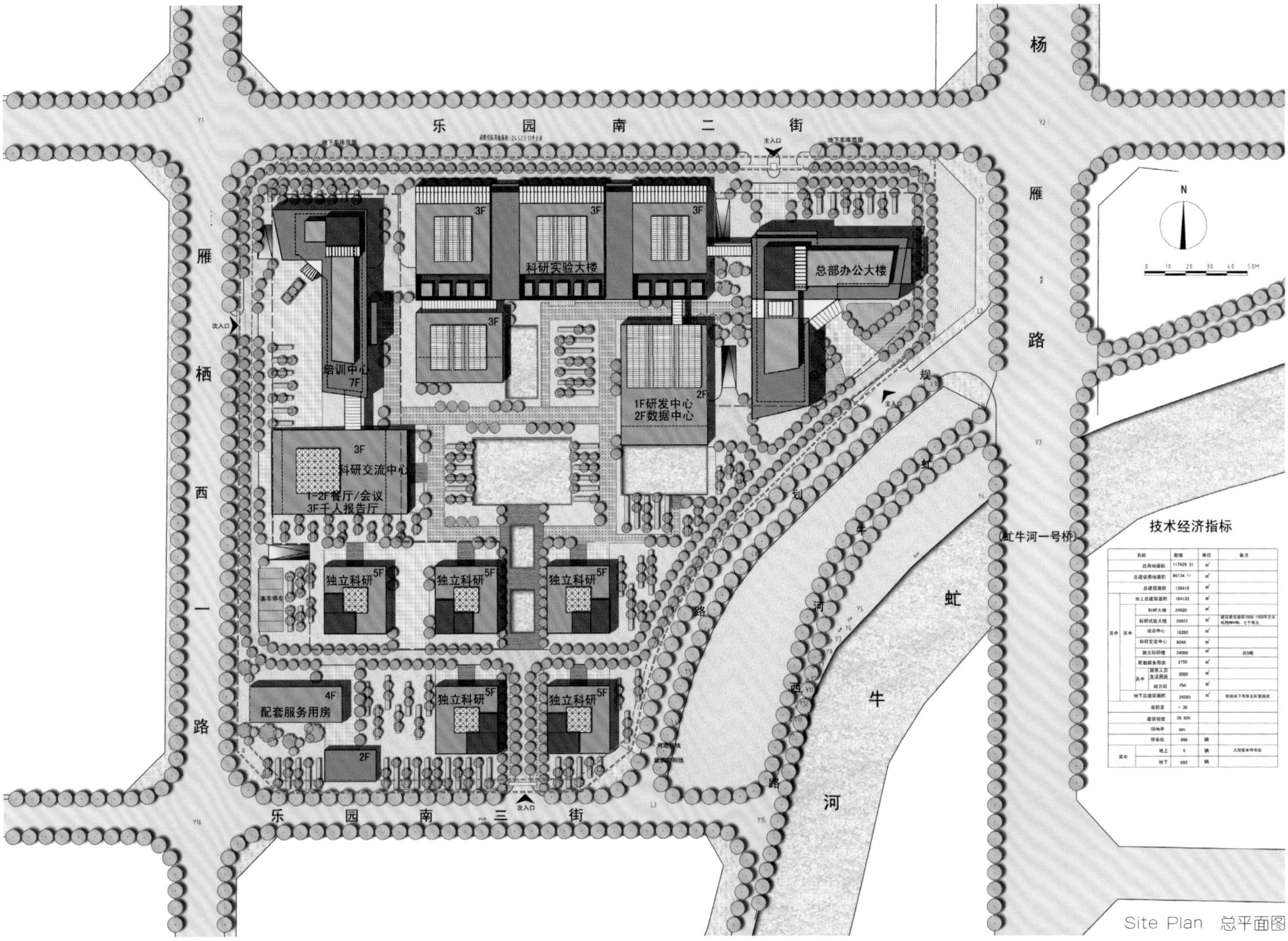

Headquarters Office Standard Layer plan 总部办公标准层平面

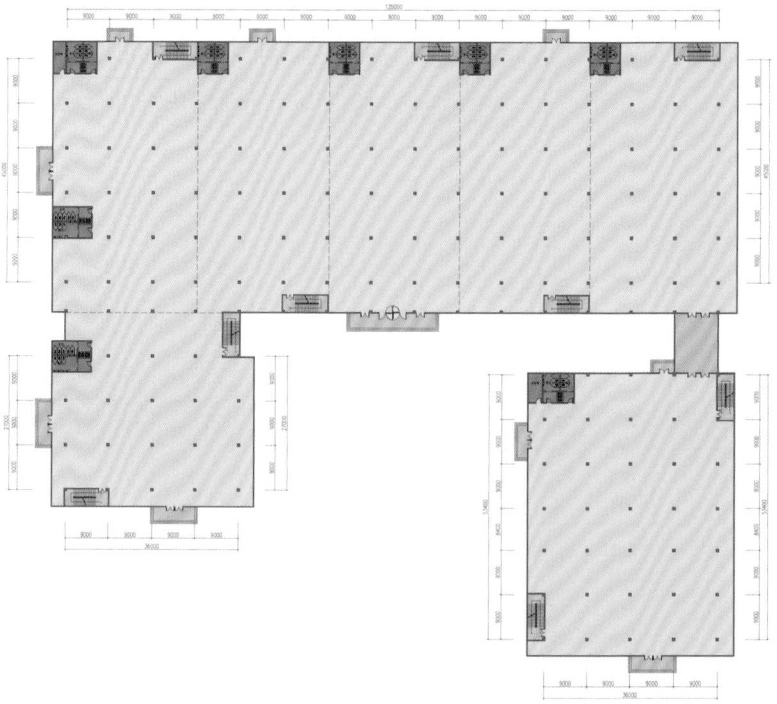

Research and Test Building Standard Layer Plan 科研试验楼标准层平面

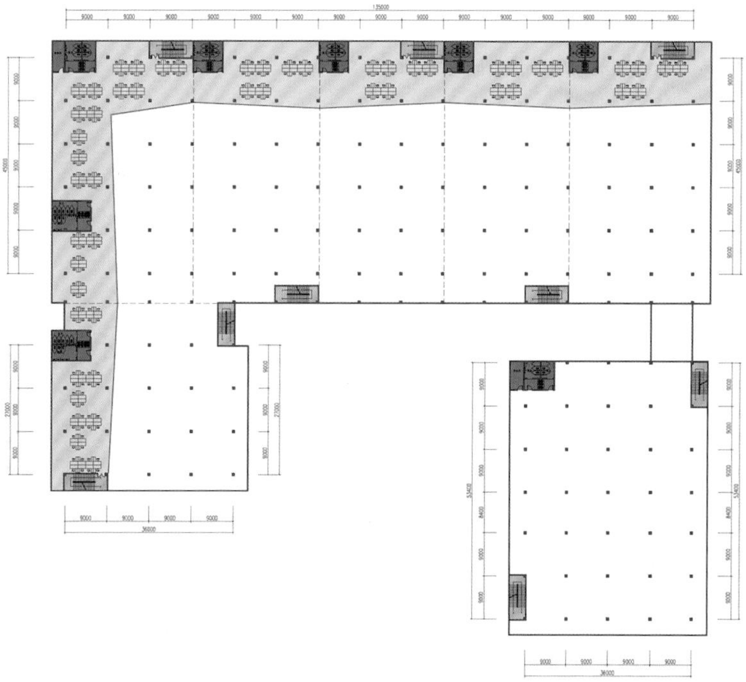

Research and Test Building Mezzanine Plan 科研试验楼夹层平面

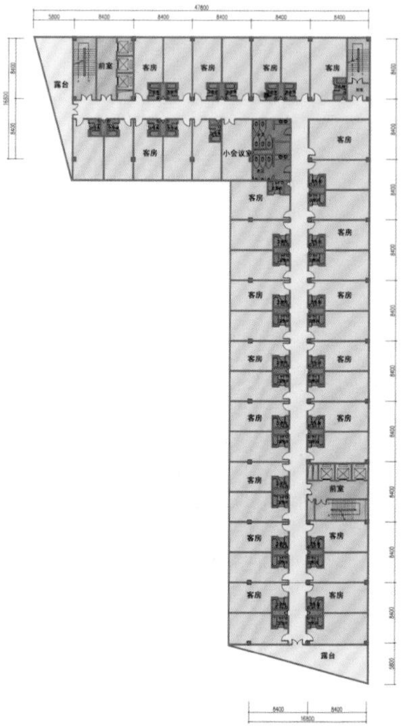

Training Center Standard Layer Plan 培训中心标准层平面

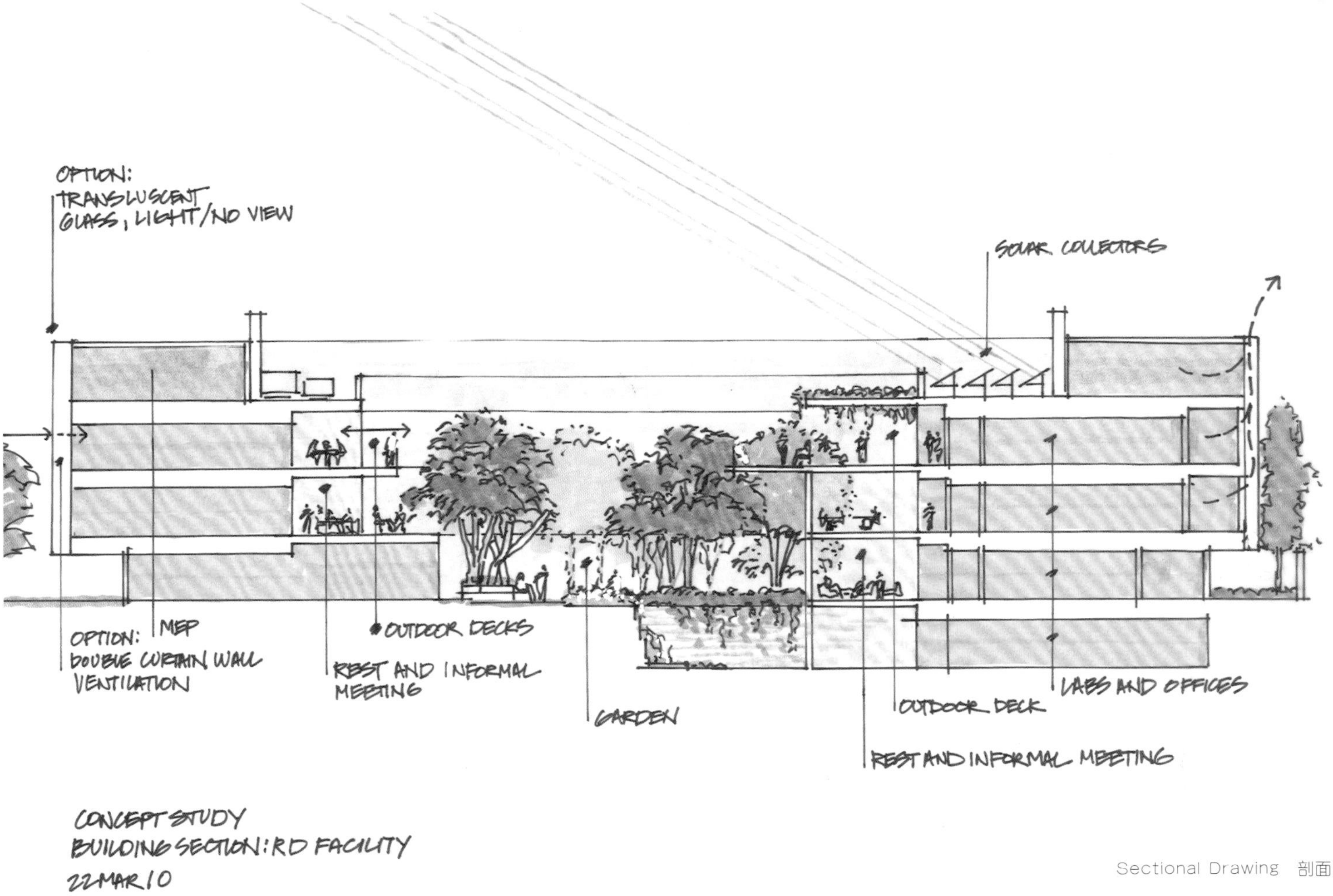
OPTION:
TRANSLUSCENT GLASS, LIGHT/NO VIEW
SOLAR COLLECTORS
OPTION: DOUBLE CURTAIN WALL VENTILATION
MEP
OUTDOOR DECKS
REST AND INFORMAL MEETING
GARDEN
OUTDOOR DECK
LABS AND OFFICES
REST AND INFORMAL MEETING
CONCEPT STUDY
BUILDING SECTION: RD FACILITY
22 MAR 10

Sectional Drawing 剖面图

Financial Street of Kunshan Economic & Technical Development Zone, Phase I
昆山开发区金融街一期

Keywords 关键词

Glass Curtain Wall
玻璃幕墙

Hanging Garden
空中花园

Partly Elevated Floor
局部架空

Location: Suzhou, Jiangsu, China
Architectural Design: China Architecture Design & Research Group Shanghai Johnson
Land Area: 55,000 m^2
Floor Area: 280,000 m^2

项目地点：中国江苏省苏州市
建筑设计：中国建筑设计研究院上海中森
用地面积：55 000 m^2
建筑面积：280 000 m^2

Features 项目亮点

The five main buildings interlock from the bottom up, forming a rhombic tangent plane. Enclosed by the transparent Low-E glass, the towers stand in the city like five sparkling crystals.

五栋主楼从下往上相互错动，形成菱形切面，再以透明的Low-E玻璃将其包裹，使塔楼宛如五块晶莹剔透的水晶立于城市之中。

■ **Overview**

The project is located to the south of Qianjin Road, to the north of Jingwang Road, to the east of Xiadong Street, to the west of No. 1 of Shimao East. The project functions with bank operation and office, forming the open and modern office group.

■ **项目概况**

项目所处基地位于前进路南侧、景王路北侧、夏东街东侧、世茂东一号西侧。项目以银行营业和办公功能为主，形成开敞现代的办公组群。

■ **Planning and Layout**

Starting from the surrounding environment, the plan fully considers the relation between the project and the residential houses around as well as the landscape corridor of Xiajia River. The site is divided into three plots as three separate functional zones. Since the site is of strip shape, the buildings are arranged in linear distribution. The main street area of the linear financial plaza and the financial city is set on the west. The three plots are mainly used for the operation and office spaces of the four state-owned banks. The podiums of the three plots are linked on the fourth floor, forming a unified roof garden. The four towers have two shifted floors on the roof floor of the podium, creating the penetrating landscape vision.

■ **规划布局**

项目规划以周边环境为主要设计切入点，充分考虑本项目与基地周围住宅以及夏驾河的景观走廊的关系。规划将基地划分为三个地块，形成三组独立使用的功能区。由于基地呈带状分布，因此建筑沿线性分布，并在西侧布置线性的金融广场与金融城的主街。三个地块主要考虑设置四大国有银行的营业及办公空间。三个地块的裙房在第四层相连，形成统一的屋顶花园。四座塔楼在裙房屋顶层架空两层，创造出通透的景观视野。

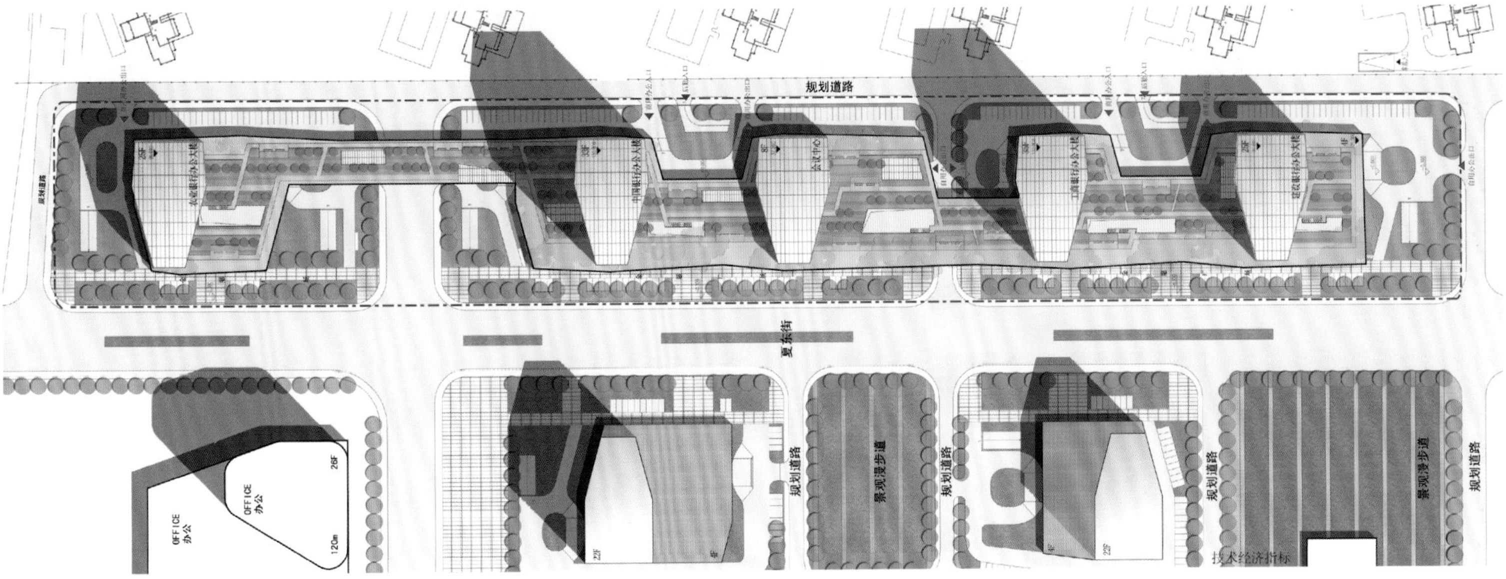

Site Plan 总平面图

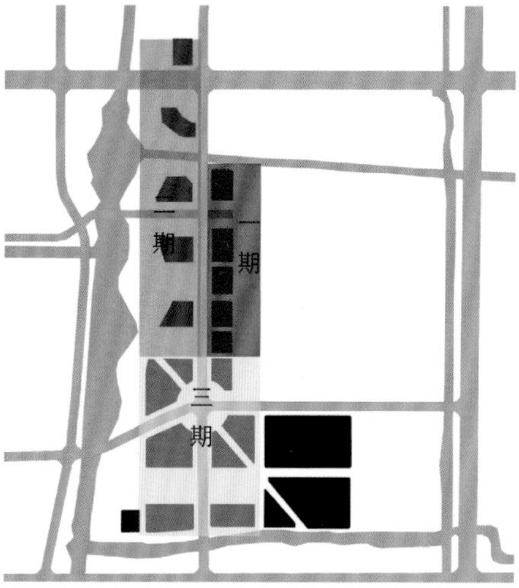

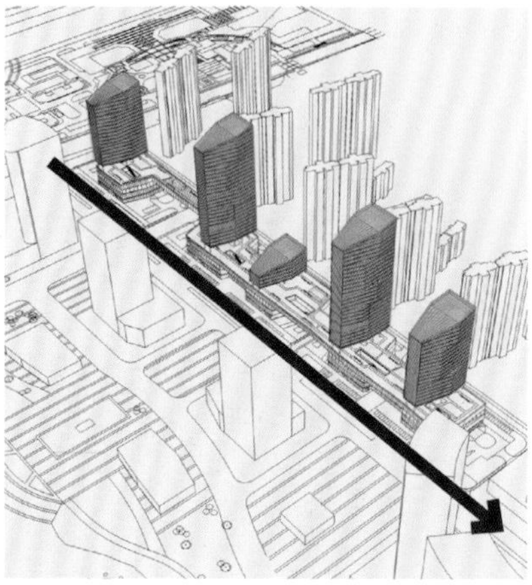

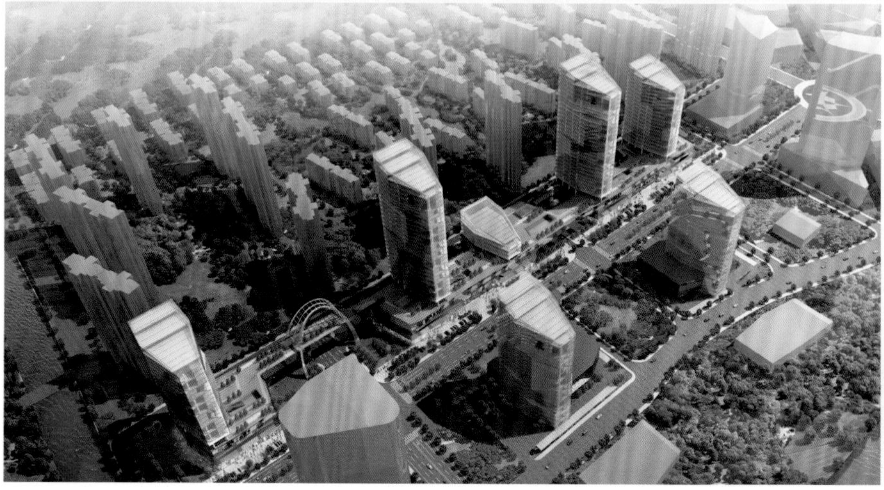

Financial City 金融城

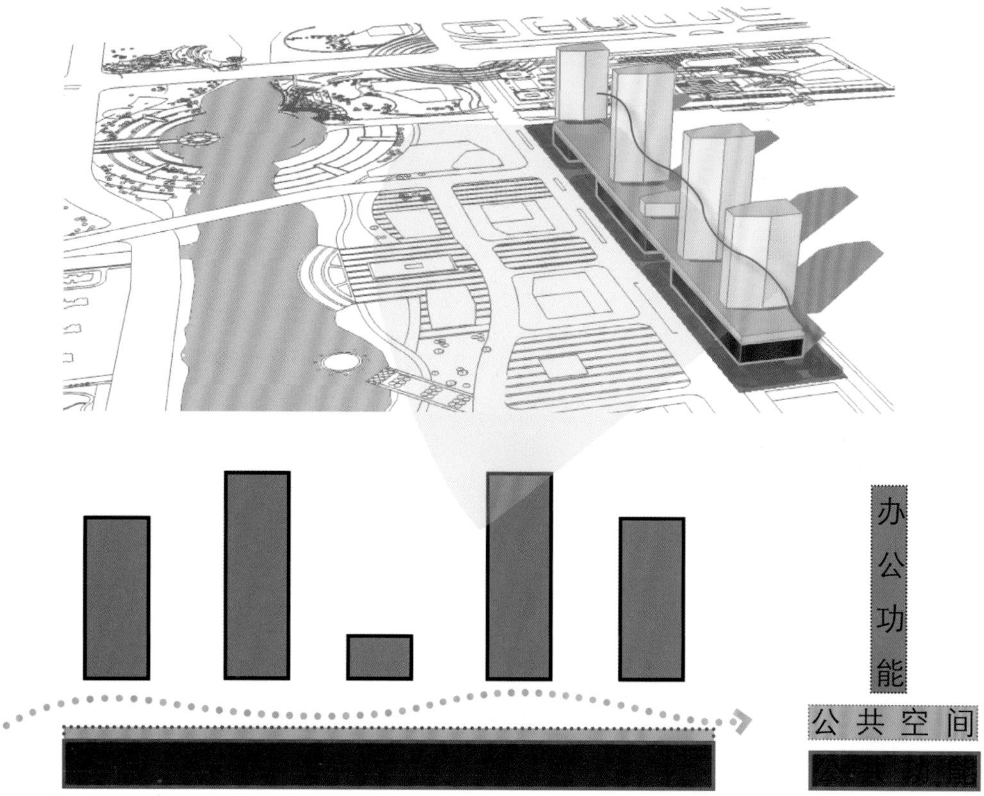

Hanging Gardens Effect Schematic Diagram 空中花园效果示意图

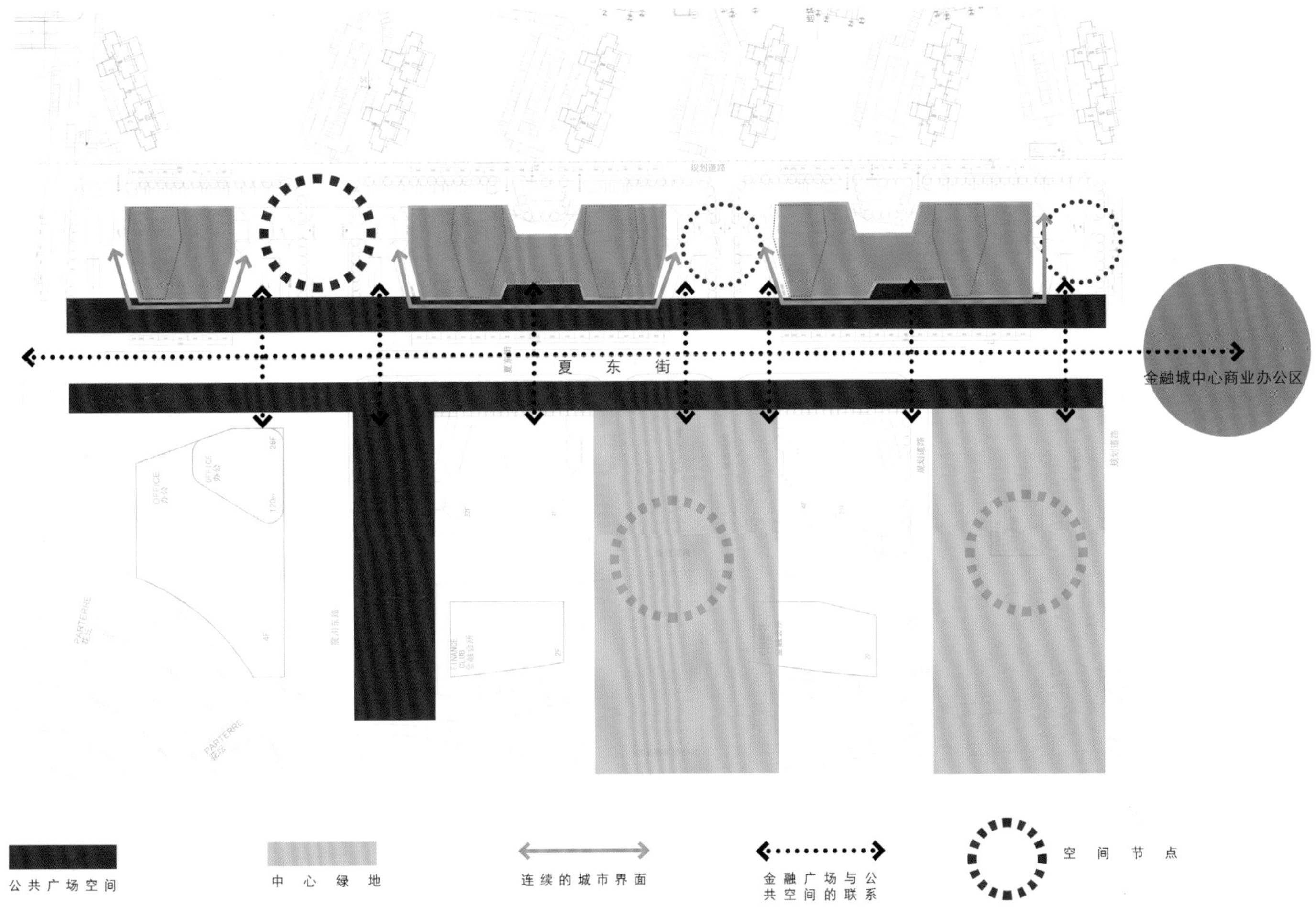

Layout Analysis Diagram 布局分析图

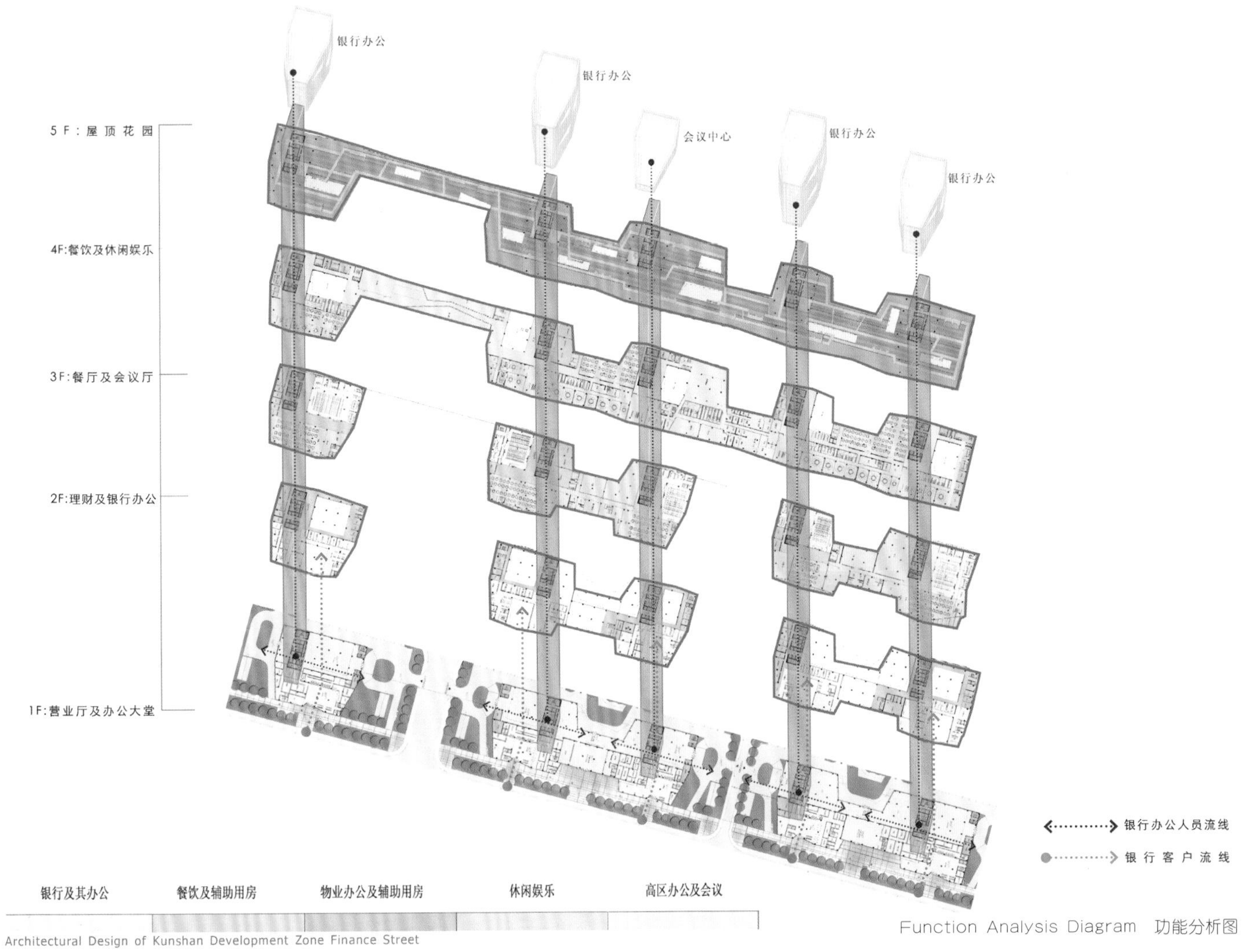

Function Analysis Diagram 功能分析图

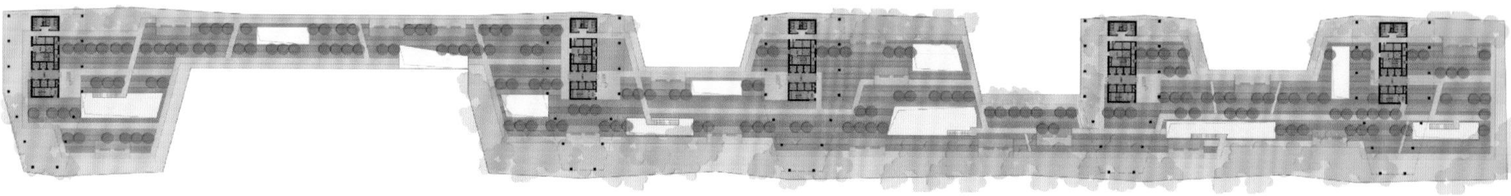

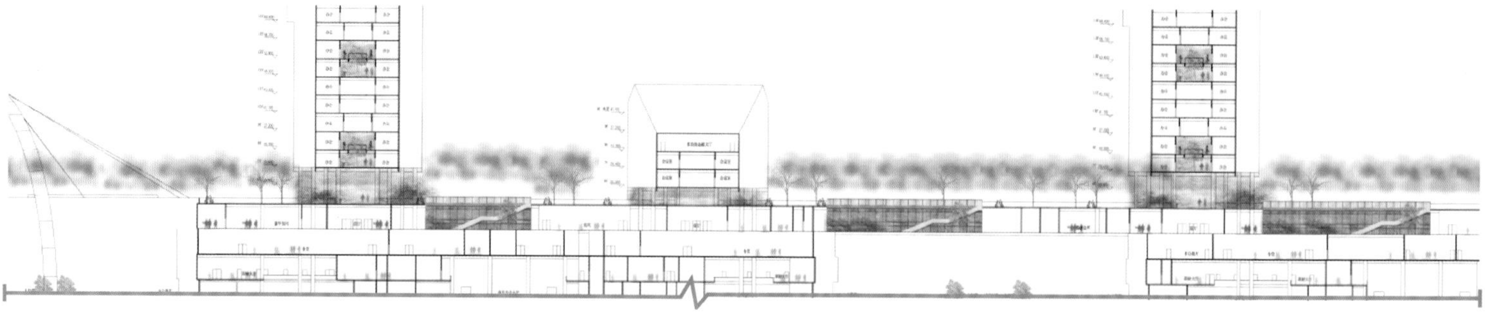

线 性 绿 地 示 意

木板、草地与家具示意

花园夜景示意

Landscape Analysis Diagram 景观分析图

形体演生示意图

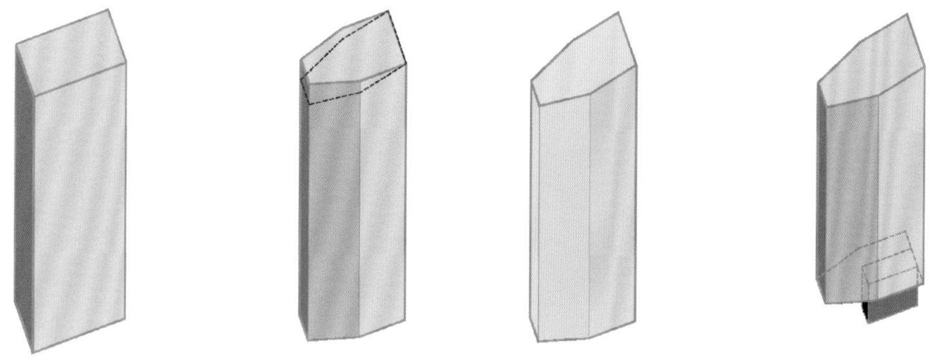

常见矩形塔楼形体

形体斜向切割

增加视野面

塔楼底层架空
增加屋顶花园通透性

Body Genesis Schematic Diagram 形体演生示意图

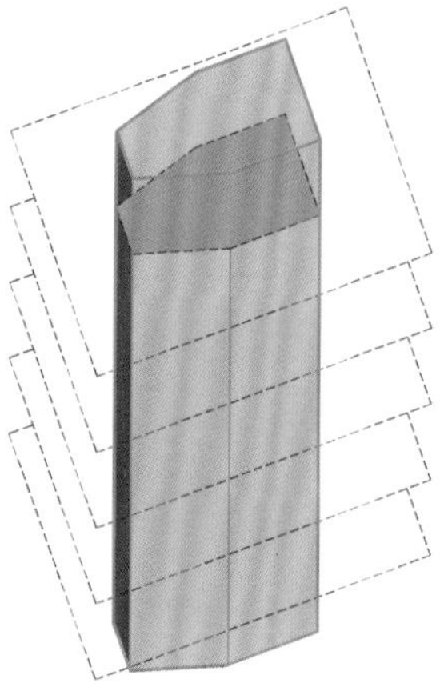

将塔楼分成五段

体量的不断转化变形

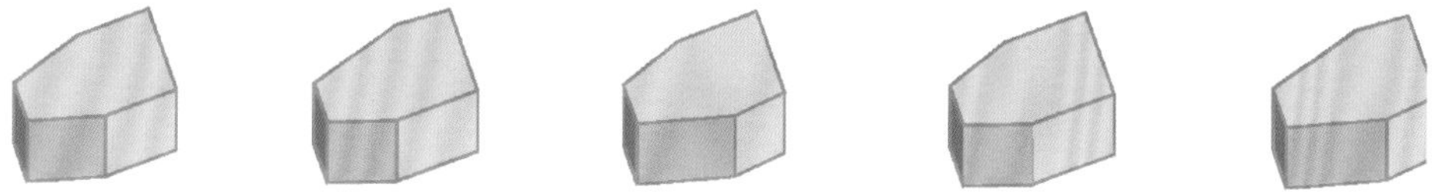

不同体量的叠加

Tower Morphological Analysis Diagram
塔楼形态分析图

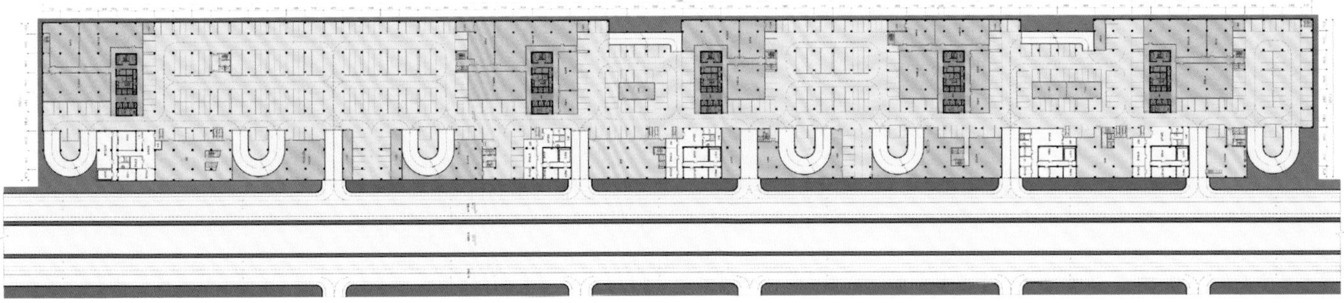

Plan of Basement Floor 地下一层平面图

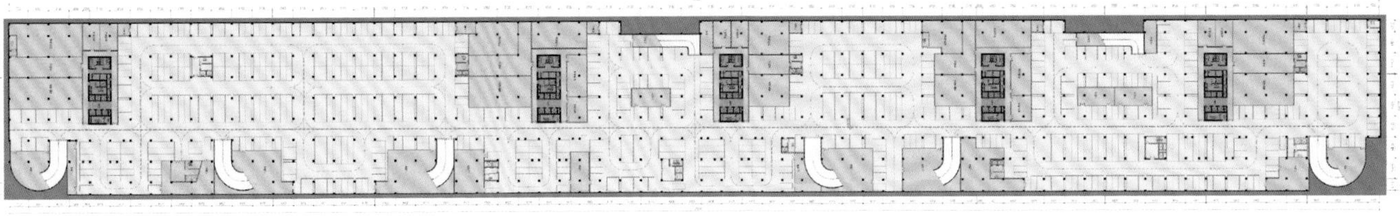

Plan of Basement Two Floor 地下二层平面图

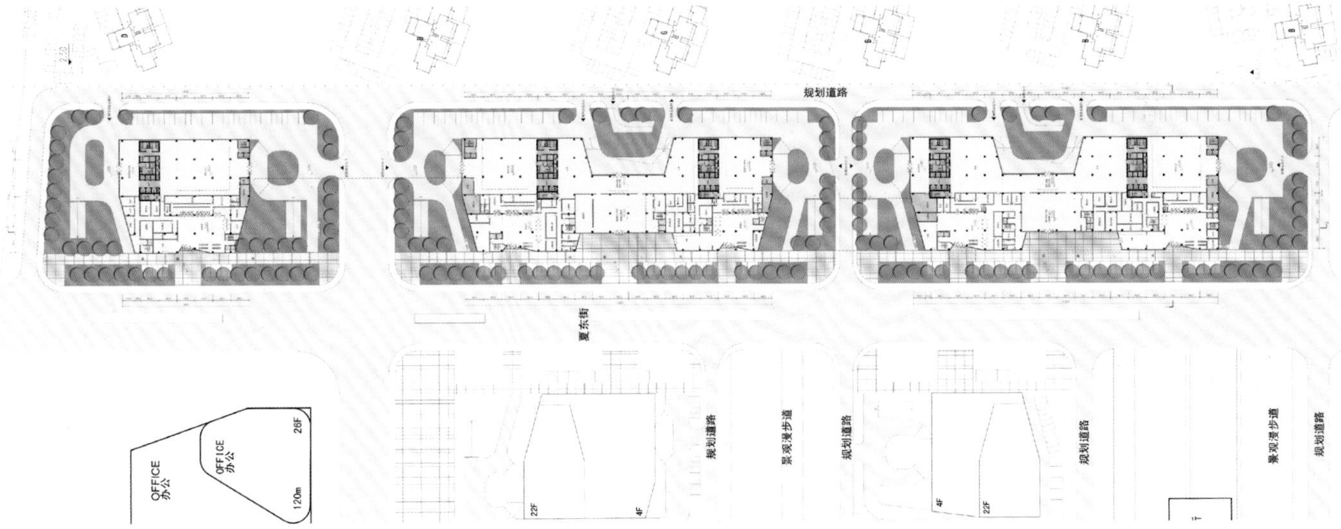

First Floor Plan 层平面图

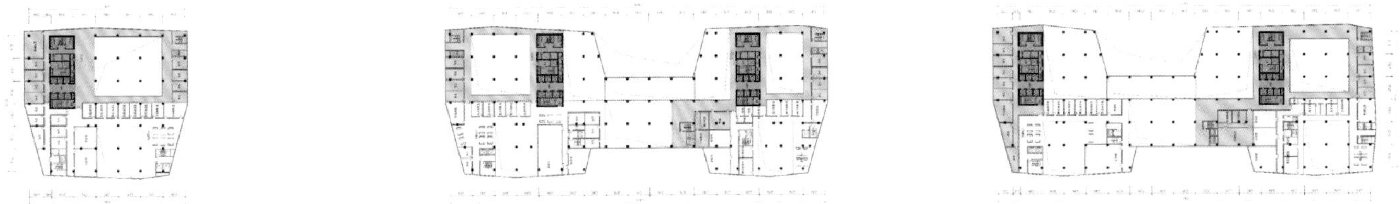

Second Floor Plan 二层平面图

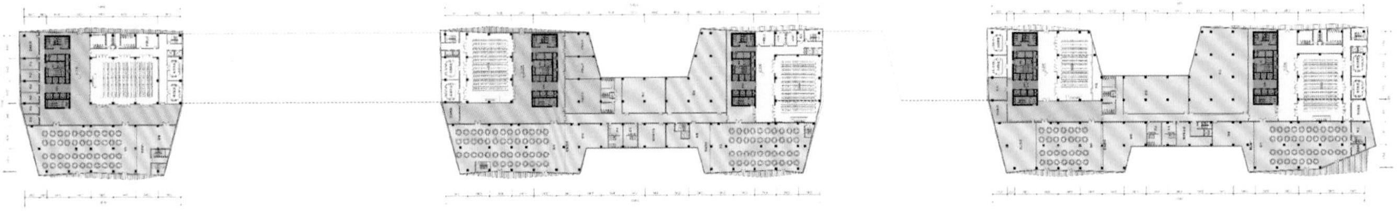

Third Floor Plan 三层平面图

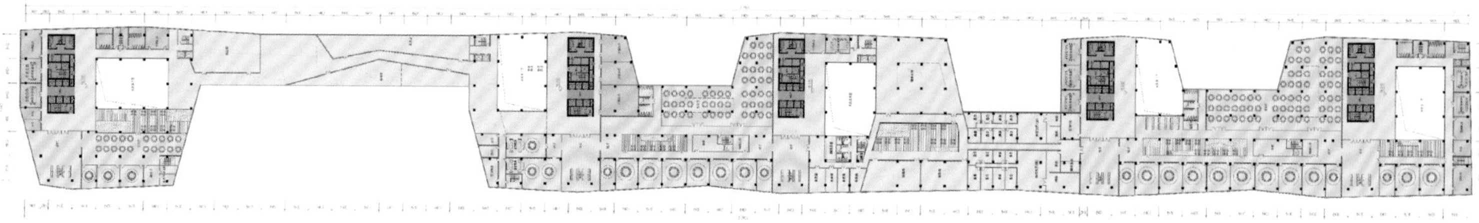

Fourth Floor Plan 四层平面图

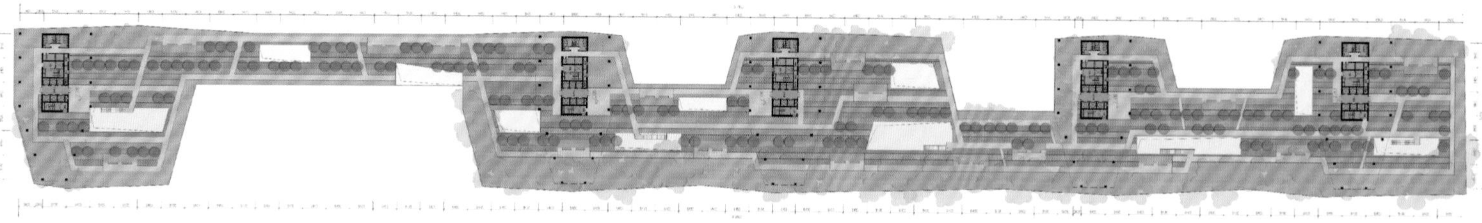

Fifth Floor Plan 五层平面图

银行及其办公	餐饮及辅助用房	物业办公及辅助用房	休闲娱乐	高区办公及会议

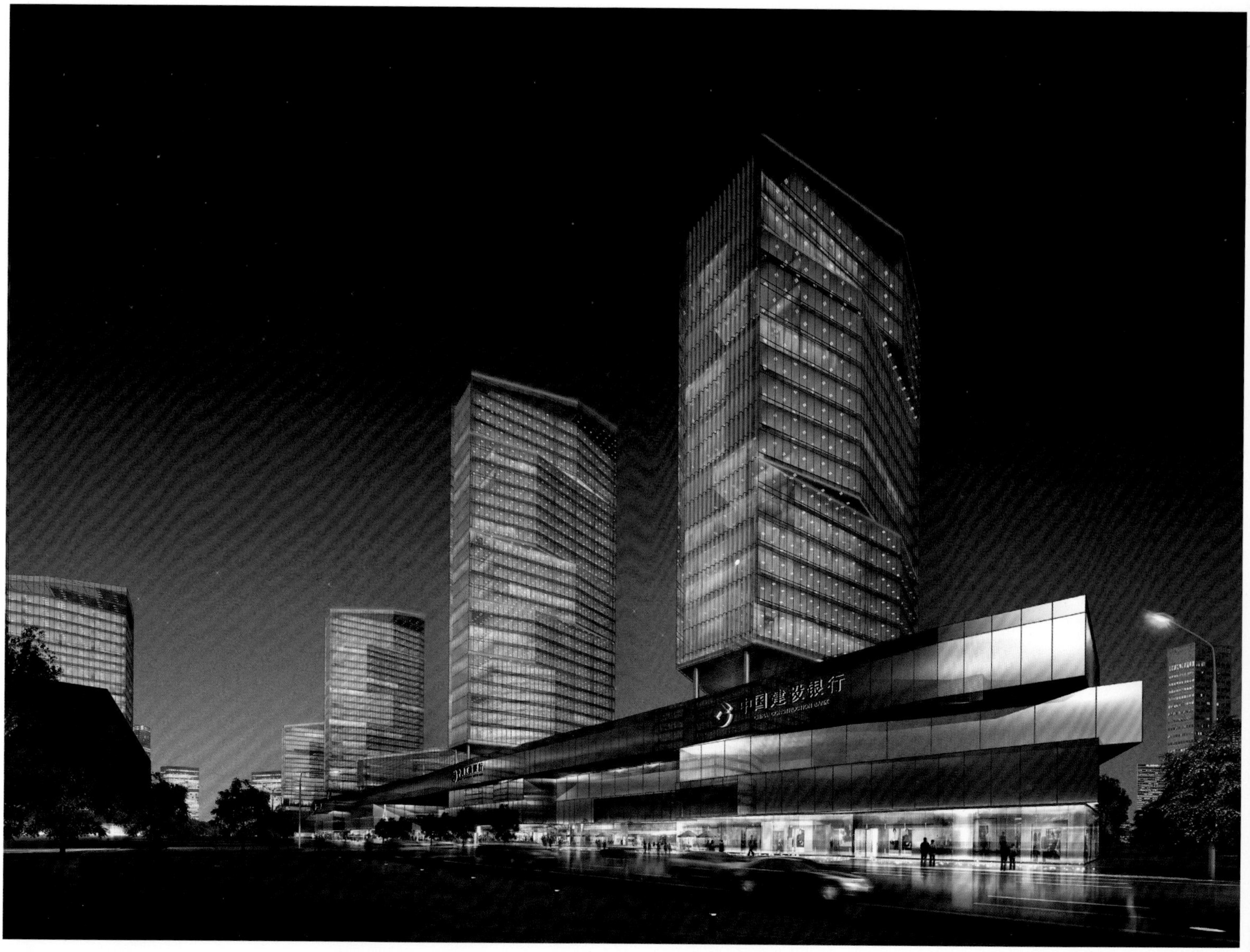

■ Architectural Design

The entire building exerts to highlight the steady and grand image of the bank office with sense of modern architecture.

Tower: The five main buildings interlock from the bottom up, forming a rhombic tangent plane. Enclosed by the transparent Low-E glass, the towers stand in the city like five sparkling crystals. What's more, the high-tech double glass imitating waves is adopted for the front of the towers, expressing the bright and modern sense of the building.

Podium: Glass curtain wall is utilized for the first floor, ensuring the vision penetration of indoor and outdoor and creating a good office environment. Using dark coated glass, from Floor 2 to Floor 4 are coherent from south to north, outlining the steady and grand image of the bank office building.

Ecological Design: The podium adapts roof greening. Between the tower and the podium, shifted floor is partly built to make the roof greening penetrate thoroughly, forming the natural ecological oxygen bar. The air courtyards are added to make the natural light reach every indoor space, maximizing the natural lighting.

■ 建筑设计

整体建筑力求突出银行办公建筑稳重、大气的建筑形象，却又不失建筑时代感。

塔楼部分：五栋主楼从下往上相互错动，形成菱形切面，再以透明的Low-E玻璃将其包裹，使塔楼宛如五块晶莹剔透的水晶立于城市之中。另外，塔楼沿街面采用了模拟水波纹的高科技双层玻璃，突出了建筑轻快现代的气质。

裙房部分：首层采用了玻璃幕墙，保证室内外视线的通透性，营造良好的对外办公环境。2～4层采用了深色的镀膜玻璃，由南向北连贯统一，突出了银行办公建筑稳重大气的建筑形象。

生态节能设计：裙房采用屋顶绿化，塔楼与裙房之间局部架空，让屋顶绿化渗透贯通，形成自然生态氧吧。增加空中庭院使自然光能到达各个室内空间，最大限度地利用自然采光。

■ Landscape Design

The square is higher than the sidewalk and divided by greening belts, forming the street space with rich layers. The urban furniture is set to outline the humanity of the square. For the roof Hanging Garden the continuous roof garden is created by using the shifted-floor method in which the lawns, the hedge and the wooden pavement are arranged in linear, while the natural colors and shapes and the landscape of the Financial Street bring the best of each other.

■ 景观设计

广场高于人行道，间以绿化带分隔，形成层次丰富的街道空间，局部设置城市家具小品，突出广场的人性关怀。屋顶空中花园：塔楼采用局部架空的方式，创造出一处连续的屋顶花园。屋顶花园以线性分布的草地、绿篱和木板铺地组成，自然的色彩形态与金融街的景观相映成趣。

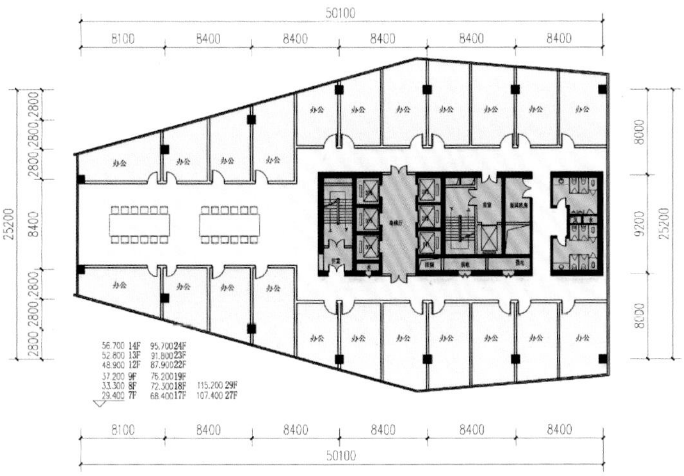

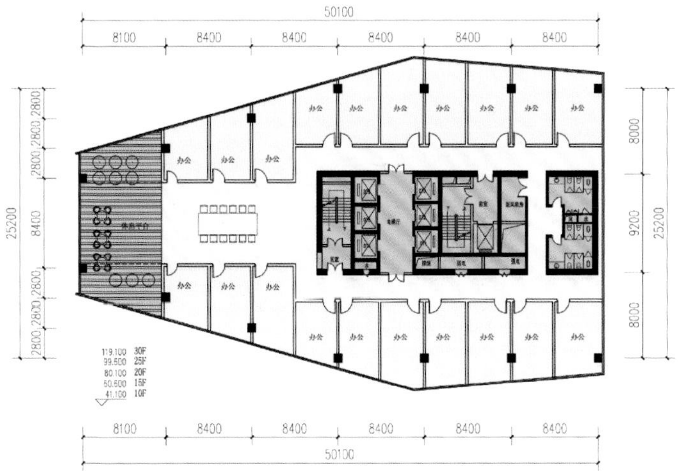

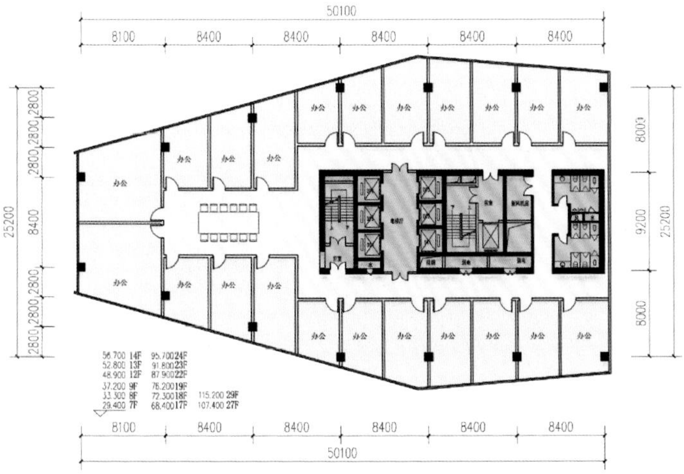

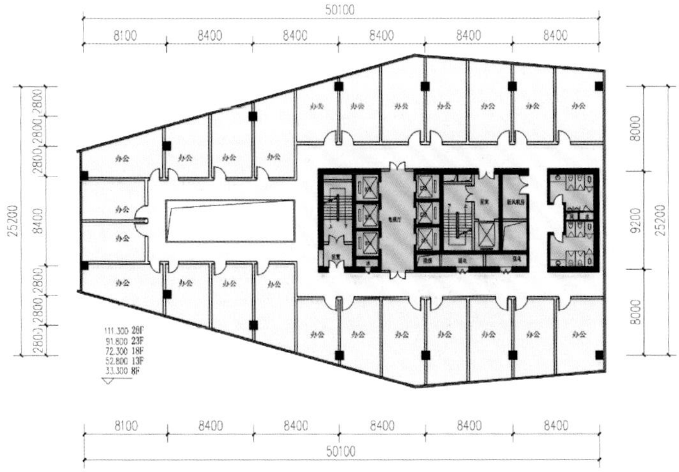

Floor Plans 平面图

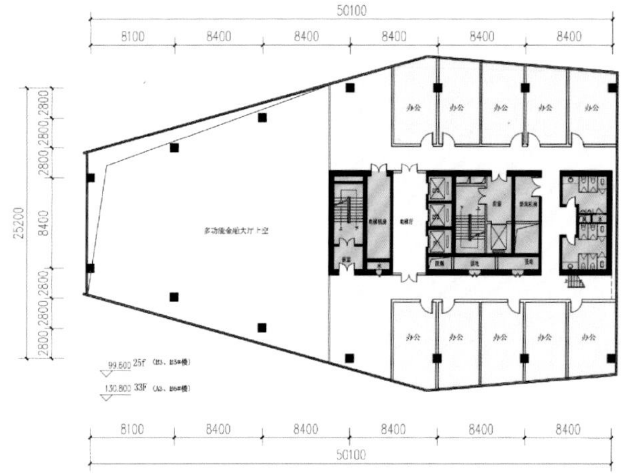

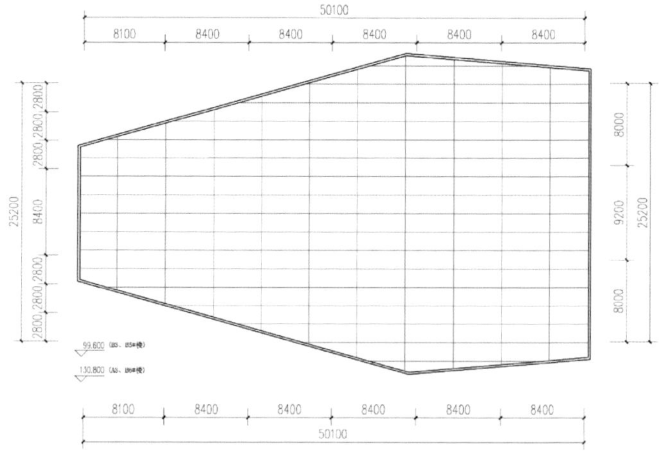

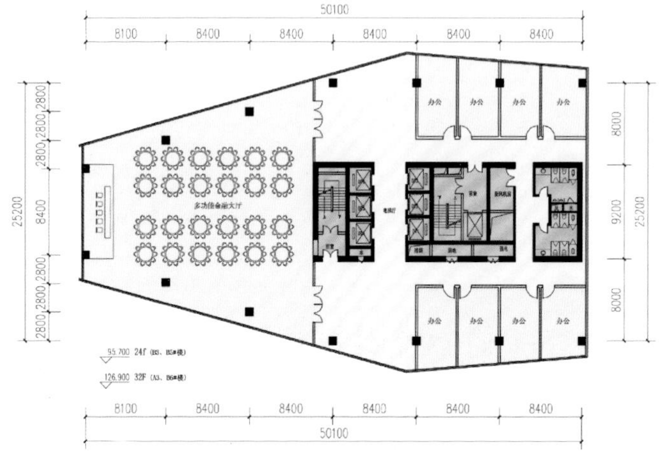

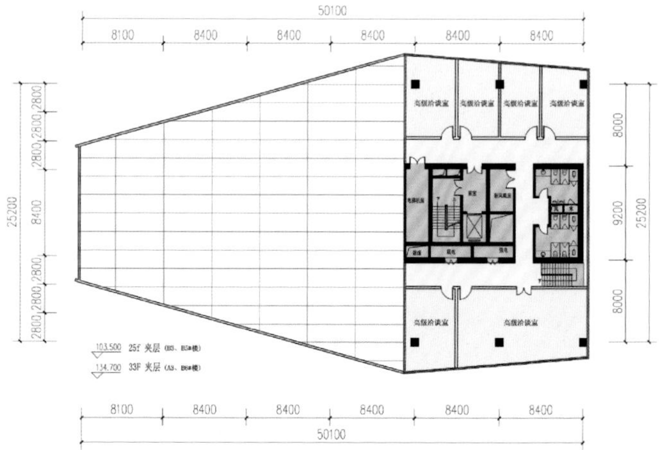

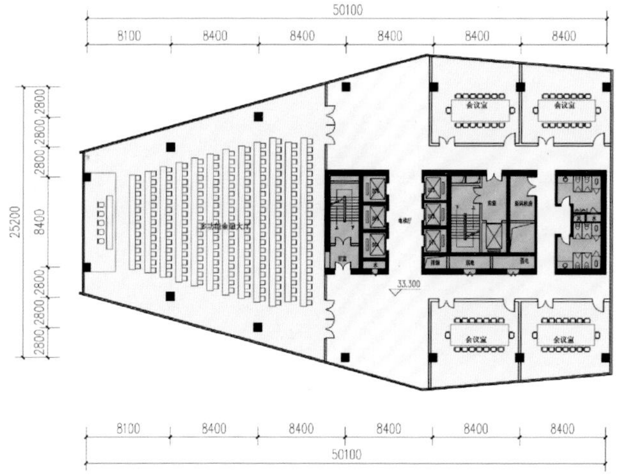

8层平面

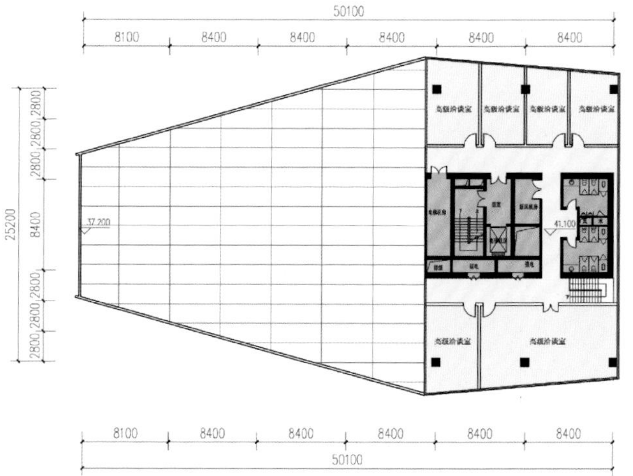

9层夹层平面

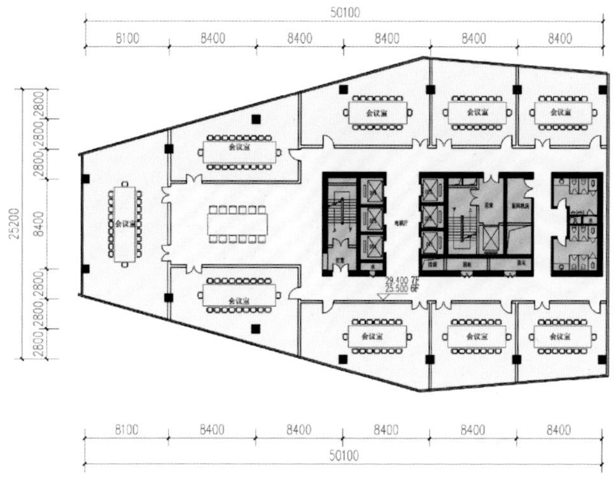

6、7层平面

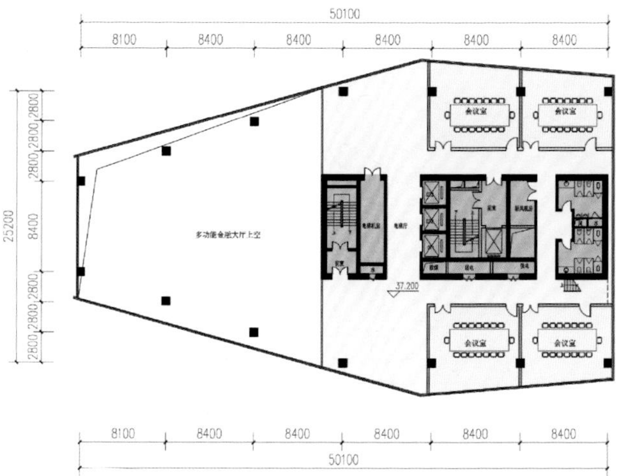

9层平面

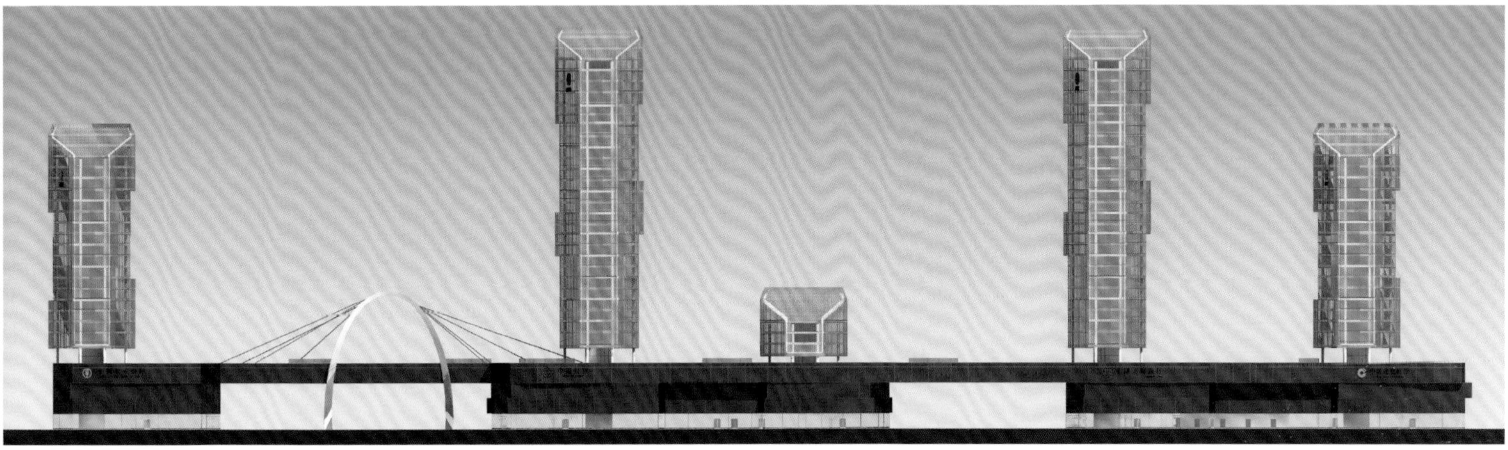

China Micro-Nano Industrial Park Phase Ⅲ, Wuxi 无锡中国微纳产业园三期

Keywords 关键词

Brevity and Brightness
简约明快

Unified Color
色彩统一

Elevated Ground Floor
底层架空

Location: Wuxi, Jiangsu, China
Architectural Design: China Architecture Design & Research Group Shanghai Johnson
Land Area: 400,000 m^2

项目地点：中国江苏省无锡市
建筑设计：中国建筑设计研究院上海中森
用地面积：400 000 m^2

Features 项目亮点

The materials and colors of the building are inspired by the local feature and unified with Phase I and II based on black, white and gray colors of Jiangnan and interweaved by greening of the wooden suspended ceiling and the mid-air courtyard, creating the fresh and elegant style.

建筑的材质和色彩源于地域特征并且和一、二期协调统一，以江南的黑白灰作为建筑的主基调，穿插以木吊顶和空中庭院的绿化，创造出清新优雅的风格。

■ Overview

Covering total land area of 400,000 m^2, the project is developed in several phases. The goal is to attract the creative projects from both home and abroad and the talented person to the park within 3 to 5 years, then make the park the world's leading national micro-nano industry base and provide a long-tern green, harmonious and humane research environment for the micro-nano international enterprises as the one and only sustainable research office area within the country.

■ 项目概况

中国微纳国际创新园总占地约400 000 m^2，分多期建设，力争通过3至5年的努力，吸纳海内外创新项目、创新人才来园区发展，进而把园区建设成为世界领先的国家级微纳产业基地，为高端微纳国际创新企业提供一个长期的绿色、和谐、人性化的办公研发环境，并将成为国内独一无二的可持续性研发办公园区。

■ Design Concept

1. One of the main materials of micro-nano — mini-crystal is used as the main part for the facade of the individual office while the hi-tech PCB is used for the outside part, forming the strong contrast of substance and void.

2. Through the high-rise and multi-level office enclosure, the continuous courtyards are built, corresponding to the planning structure of the entire area and staying unified with Phase I on space and shape. Every courtyard is connected to each other through the bottom shifted floor, forming a complete leisure footway system.

3. The high rise is located at the marginal side of the site. On one hand, it forms the sense of sequence with Phase I and II on Wuyue Road; on the other hand, the superior landscape of the site is left to the individual office of the enterprises, blocking the noise from the city's expressway. Through the bottom shifted floors, mid-air courtyards and roof gardens, the greening is introduced into the inner courtyard and indoor space, creating the green ecology office environment.

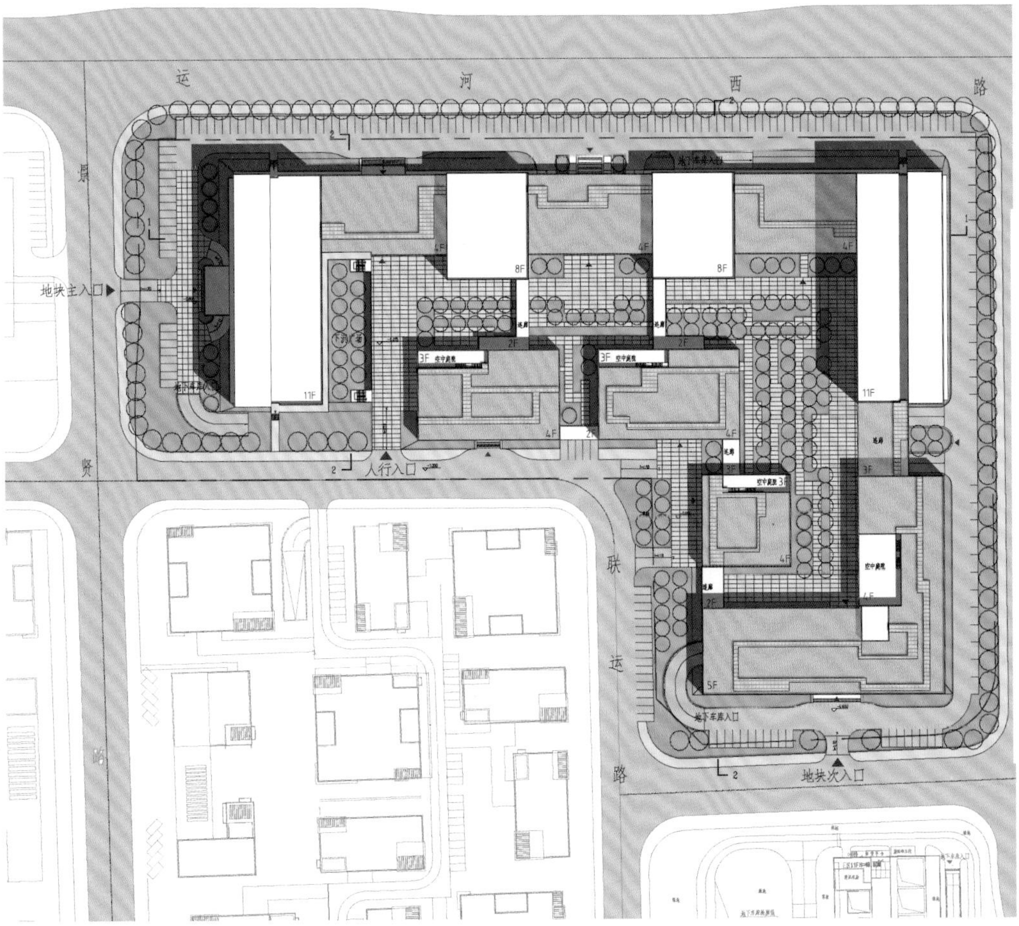

Site Plan 总平面图

■ 设计理念

1. 以微纳产业的主要材料之一——微晶体作为独栋办公楼立面的设计灵感，外围以高科技线路板为设计灵感的建筑和独栋办公楼形成鲜明有致的对比。

2. 建筑通过高层和多层办公楼的围合创造出连续的庭院，与微纳全区的规划结构一致，和一期、二期达到空间形态上的统一。每个庭院空间通过底层架空层联系在一起，形成一条完整的休闲步行道。

3. 高层位于地块的最外侧，一方面在吴越路上和一期、二期建筑形成高层的序列感；另一方面把地块内部的优越景观留给独栋企业办公楼，屏蔽城市快速路噪音。通过底层架空、空中庭院与屋顶花园等方式，将茵茵绿意引入内院与室内空间，营造出绿意盎然的生态办公环境。

■ Functional Layout

There are two kinds of office functions. The high rise and multi-level part with unit office are located at the marginal side adjacent to the city roads, while the multi-level part with enterprise offices is located at the center of the site. The facilities that provide service for the whole area are set on the north side of the site with pedestrian lanes along Yunhe West Road, which is convenient for the future operation.

■ 功能布局

方案设计的办公功能分为两种，以单元式办公室为主的高层和多层部分主要位于紧邻城市道路的外侧，而以企业办公场所为主的多层部分位于地块的中心位置。服务于全区的配套功能放在地块北侧并且沿运河西路配置人行道，方便日后的运营。

■ Traffic Lines Design

The vehicle entrances are set on Jingxian Road to the north and Langxin Road to the southwest. In combination with the underground parking lot entrances, the traffic roads are set around the site and the discharge area is set on Yunhe West Road. Along the outer ring road, every high-rise office, enterprise office and facility function are set with individual entrance. The pedestrian entrances are set together with landscapes in the pedestrian squares and inner courtyards in order to separate the pedestrian and the traffic.

■ 交通流线设计

在北向的景贤路和西南向的浪新路开设机动车出入口，在出入口附近设置地下车库出入口，车行路在地块最外围环通，在运河西路设置地面卸货区。沿着这条外环路每个高层办公楼、企业办公楼以及配套功能建筑都有独立的出入口。在人行广场与内院结合景观设置便捷到达各个门厅的人行入口，达到人、车的彻底分流。

■ Materials and Colors

The materials and colors of the building are inspired by the local feature and unified with Phase I and II based on black, white and gray colors of Jiangnan and interweaved by greening of the wooden suspended ceiling and the mid-air courtyard, creating the fresh and elegant style. The facade is made of aluminum plates and glass, presenting the concise and bright style of modern architecture.

■ 材质及色彩

建筑的材质和色彩源于地域特征并且和一、二期协调统一，以江南的黑白灰作为建筑的主基调，穿插以木吊顶和空中庭院的绿化，创造出清新优雅的风格。立面材料主要使用铝板和玻璃，体现出现代建筑简约明快的风格。

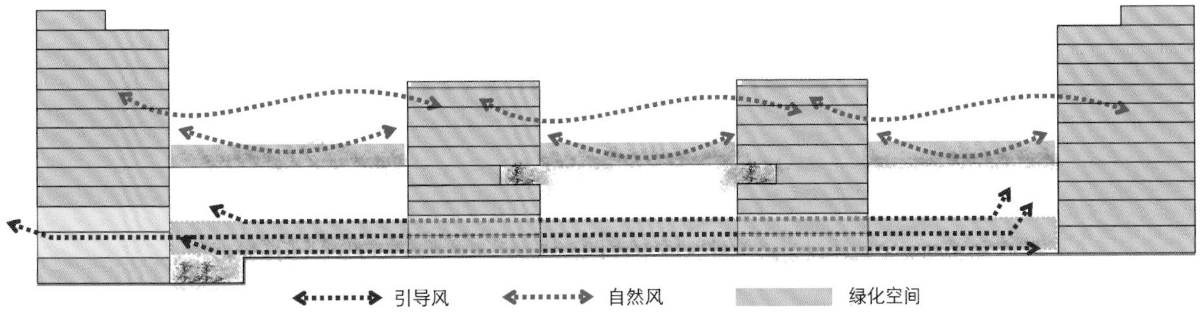

Eco Landscape Idea 生态景观理念

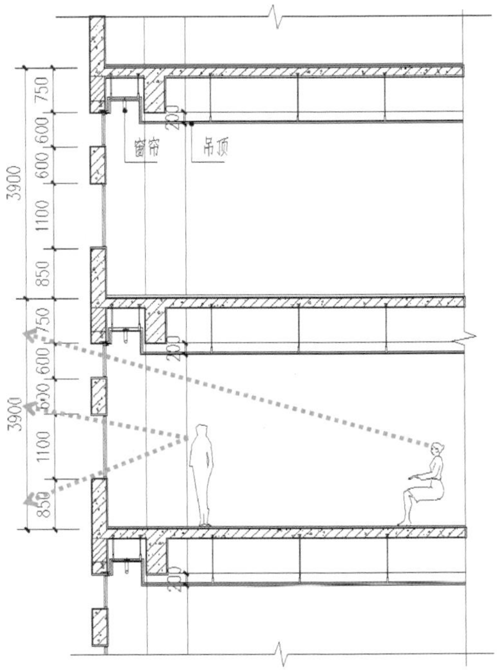

高窗满足最大采光要求

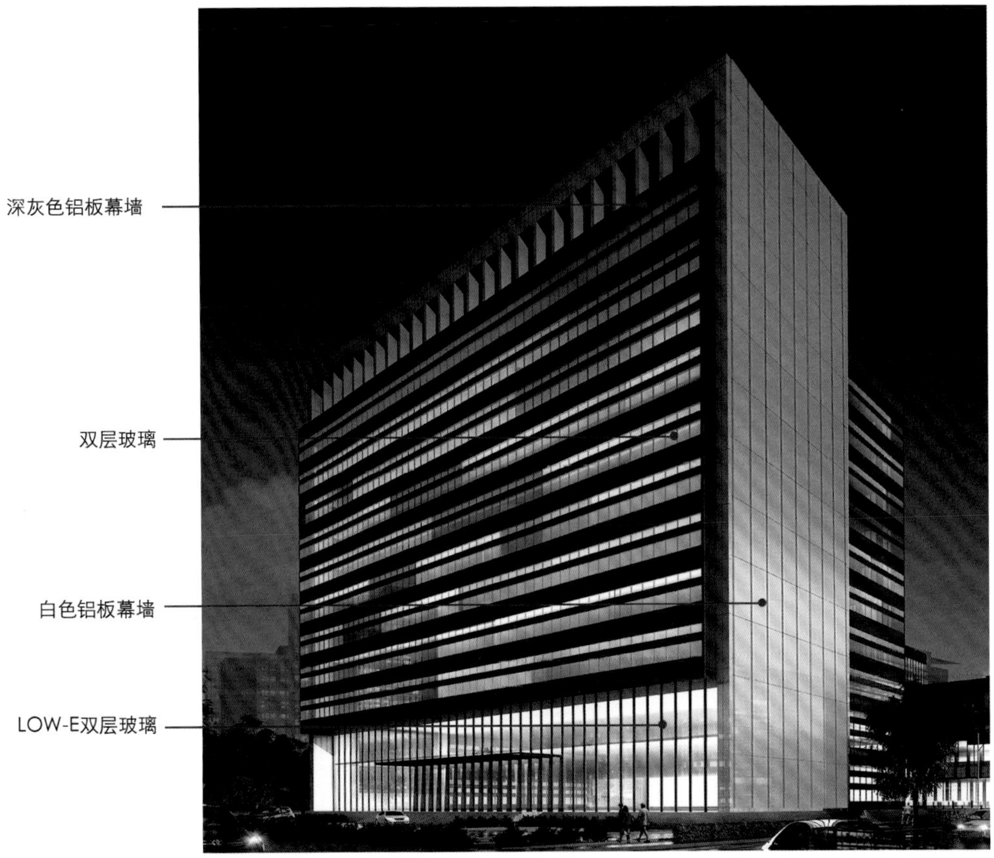

Facade Material 立面材料

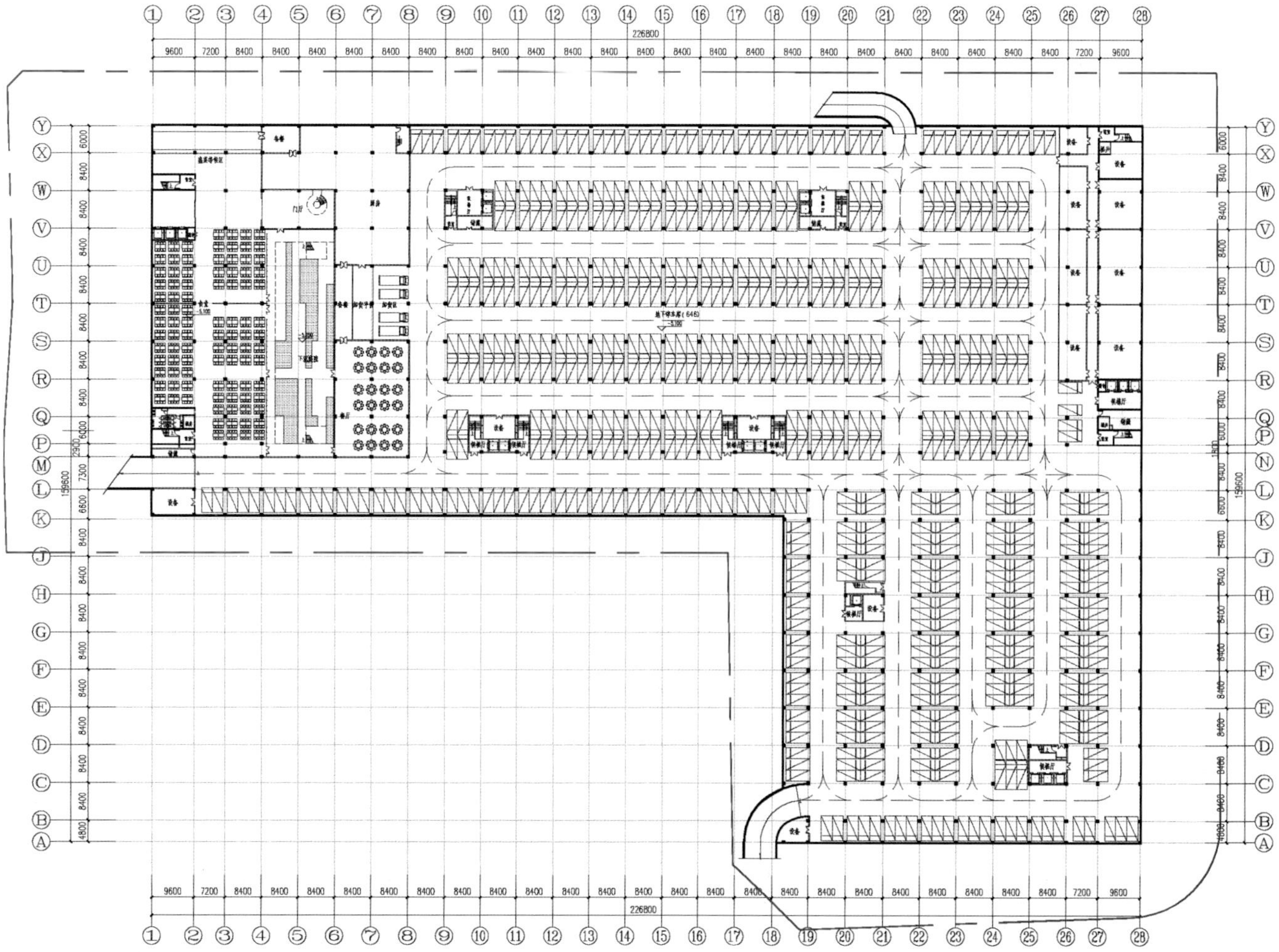

Plan of Basement Floor 地下一层平面图

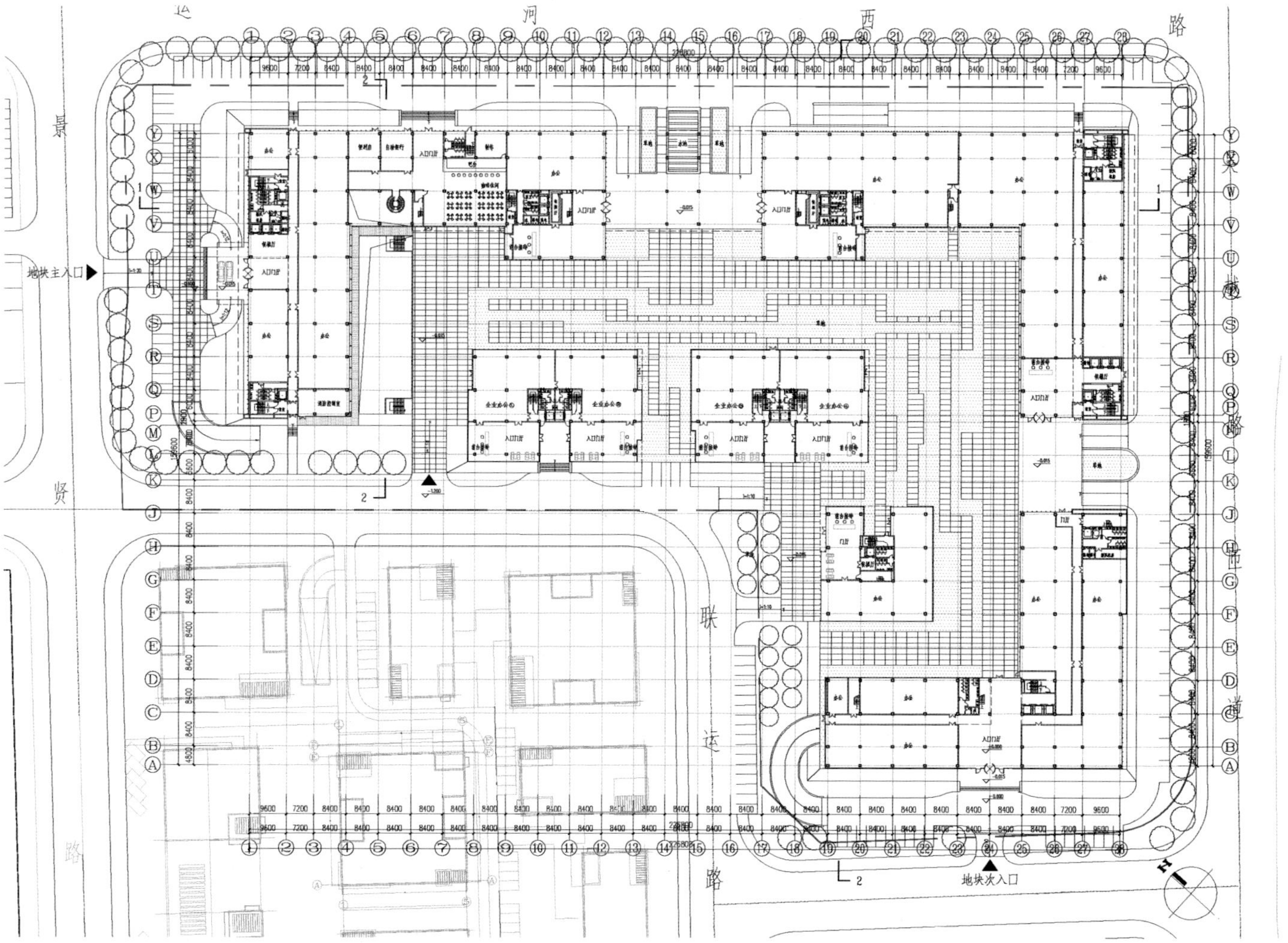

First Floor Plan 一层平面图

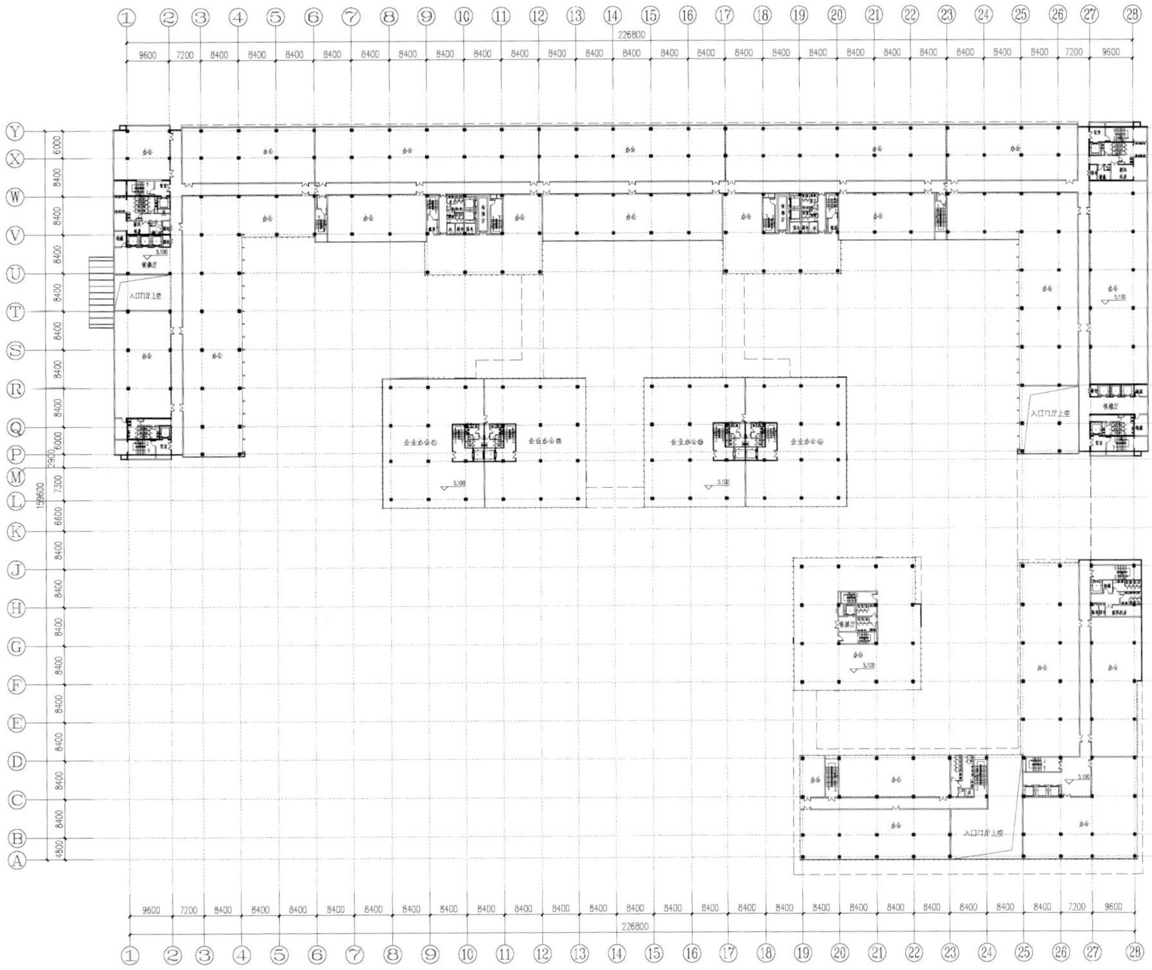

Second Floor Plan 二层平面图

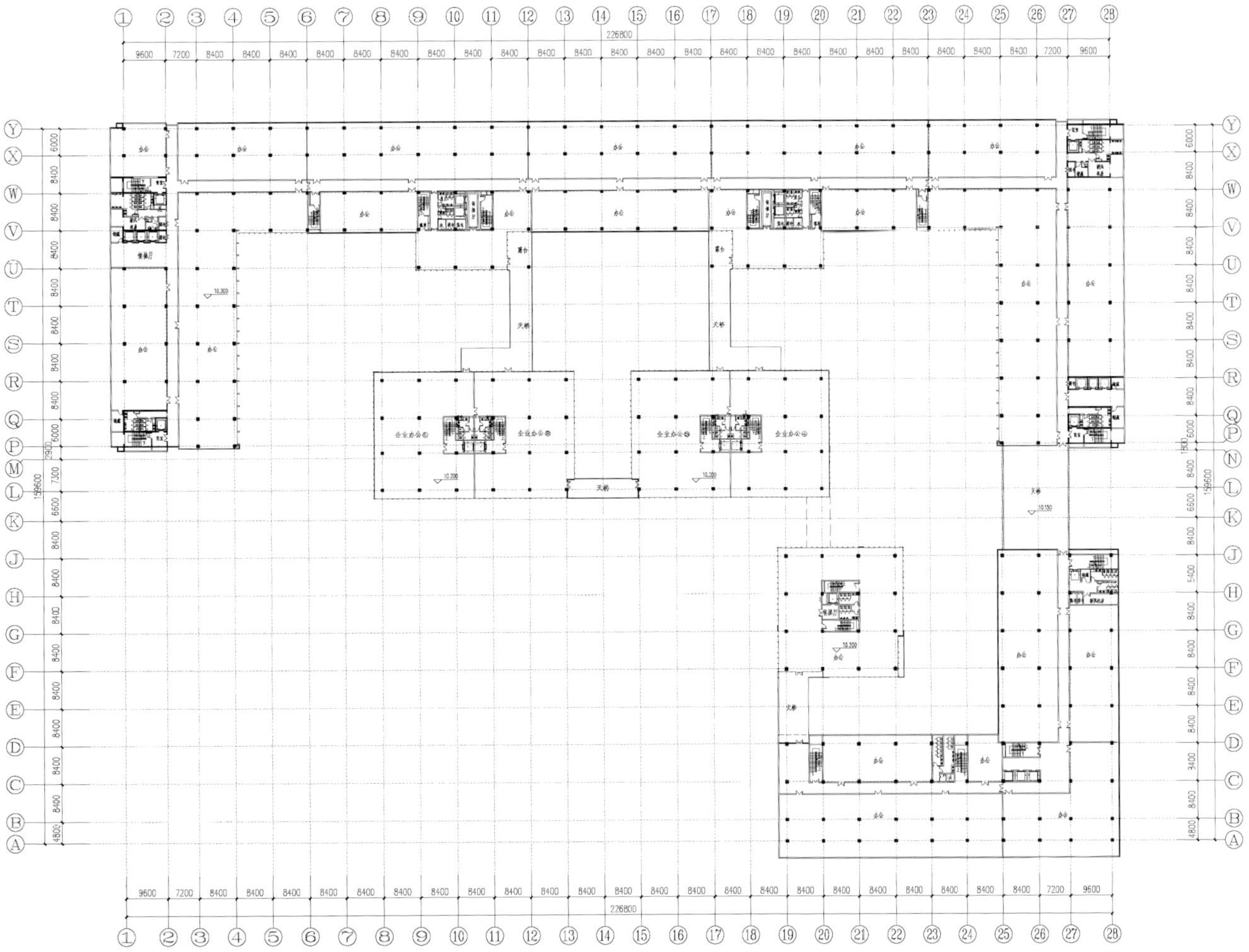

Third Floor Plan 三层平面图

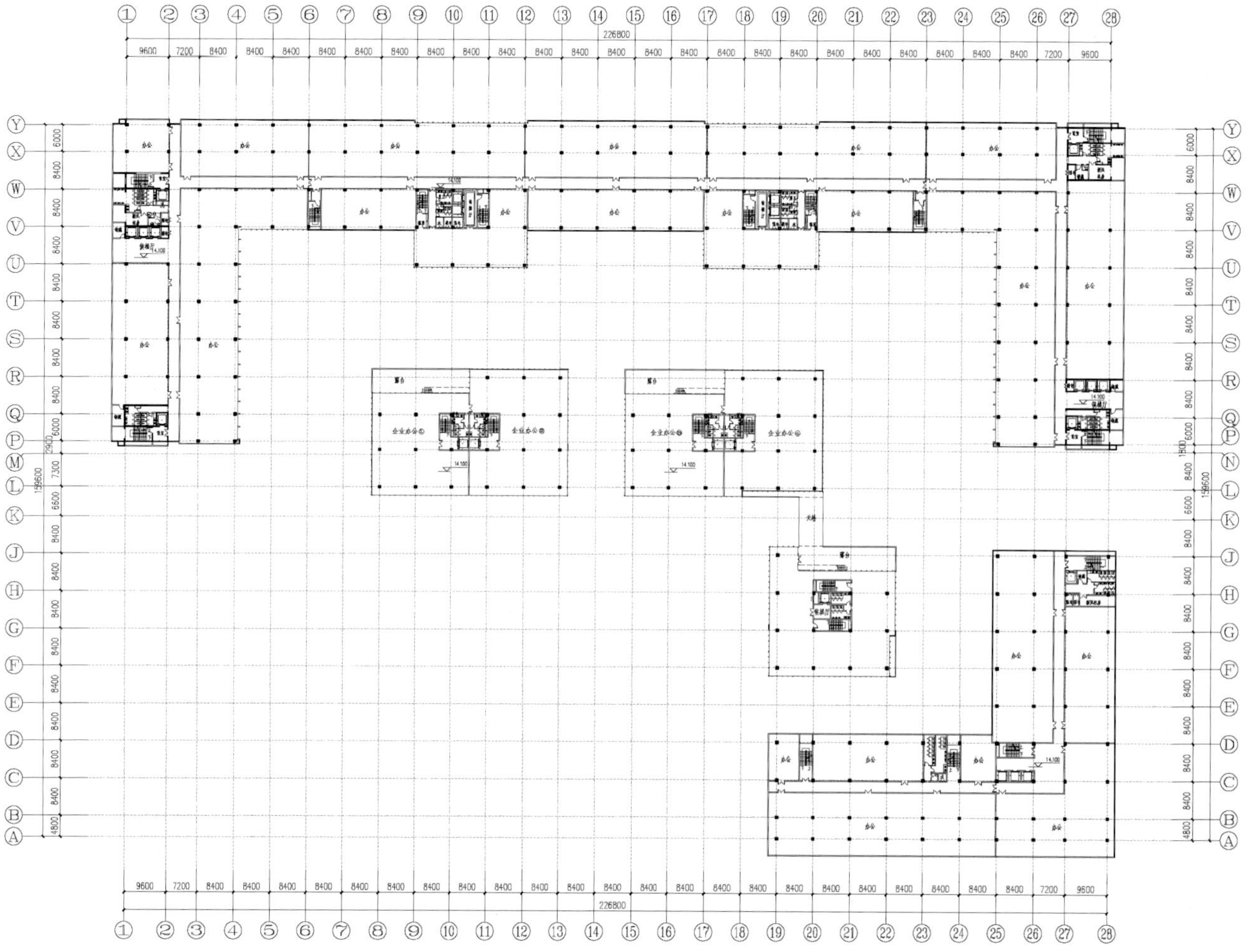

Fourth Floor Plan 四层平面图

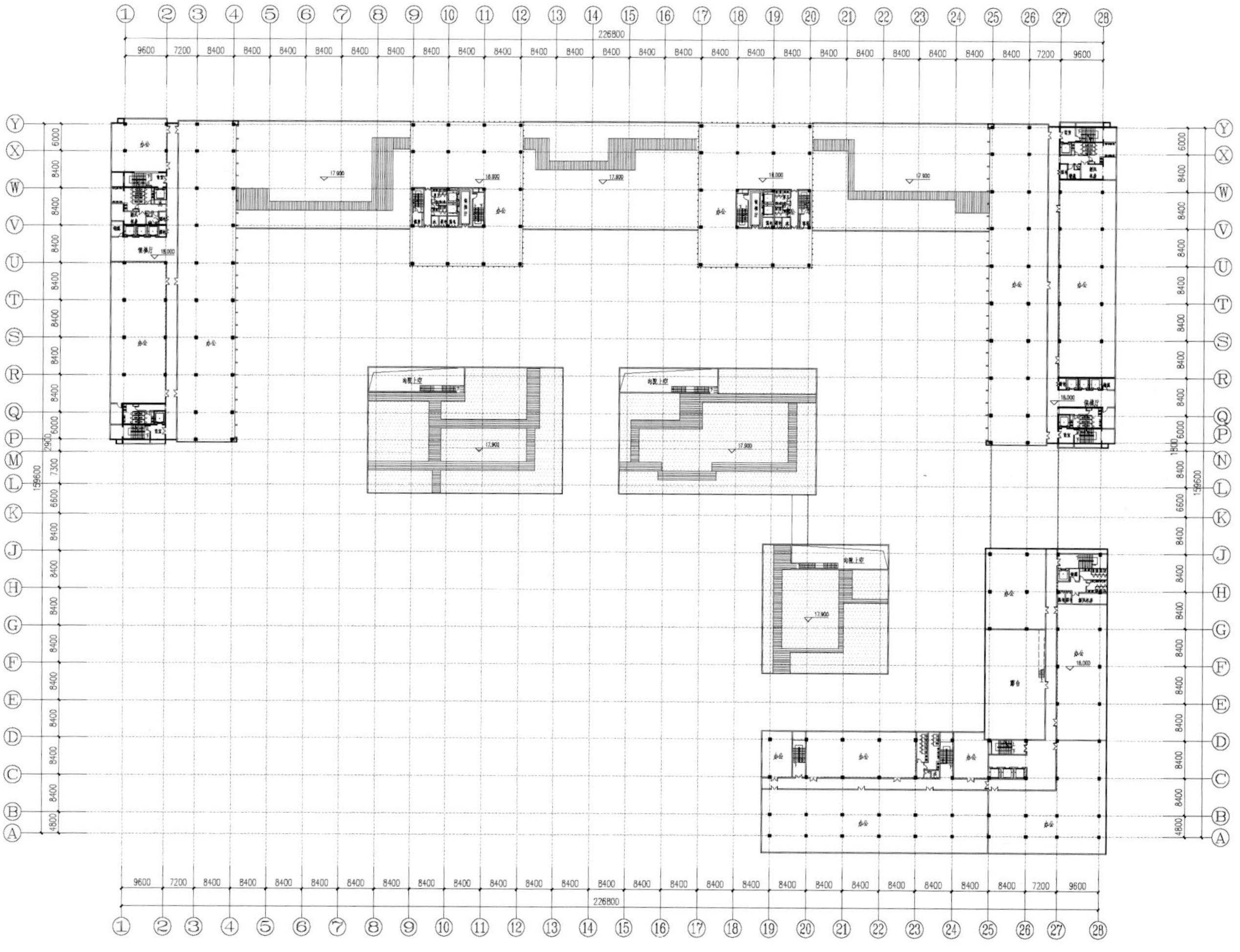

Fifth Floor Plan 五层平面图

■ Greening Landscape Design

The entire site is lifted by 1.2 m which is good for river sight seeing. The central section is designed open to the street; the bottom shifted floor, footways, shared spaces and roof activity plat forms all enjoy the greening; in combination with the courtyard, the roof greening follows the greening texture of the ground, providing perfect place for the people to appreciate the river landscape and creating micro-climate as well.

■ 绿化景观设计

整个地块抬高1.2 m，便于观赏运河景观。中心区面对街面作开敞设计，底层架空层、步行道、公共空间和屋顶活动平台，都能充分享受到园区绿化的盎然生机；结合庭院设置屋顶绿化，延续地面的绿化肌理，为人们观赏运河以及园区内风景提供了绝佳场所，并且创造园区小气候。

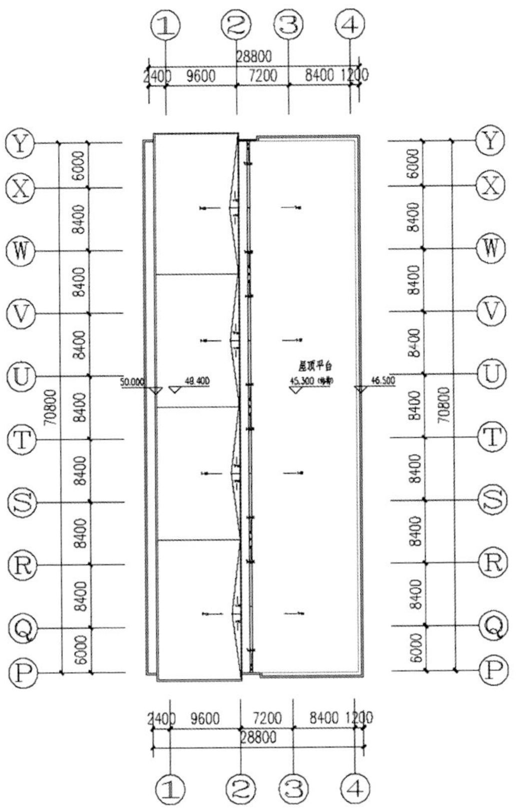

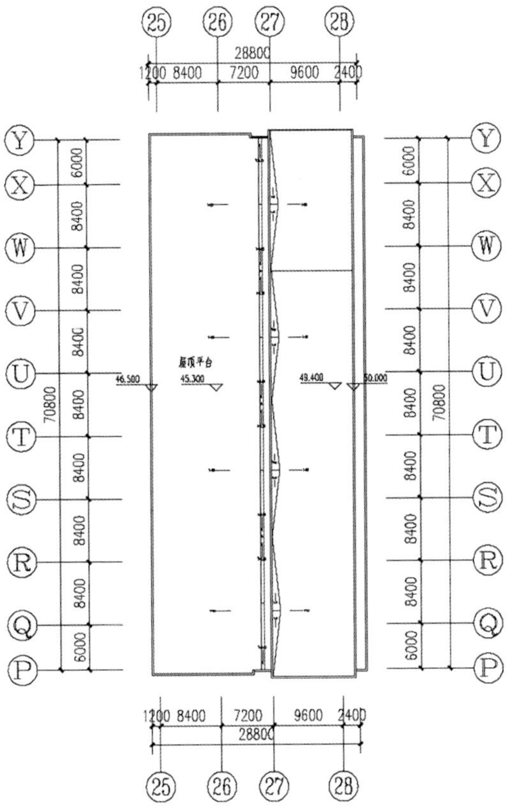

Roof Floor Plan 屋顶层平面图

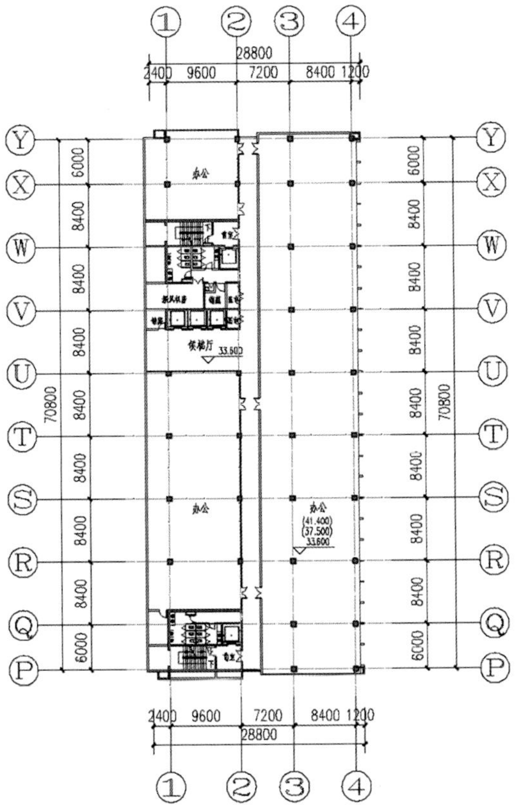

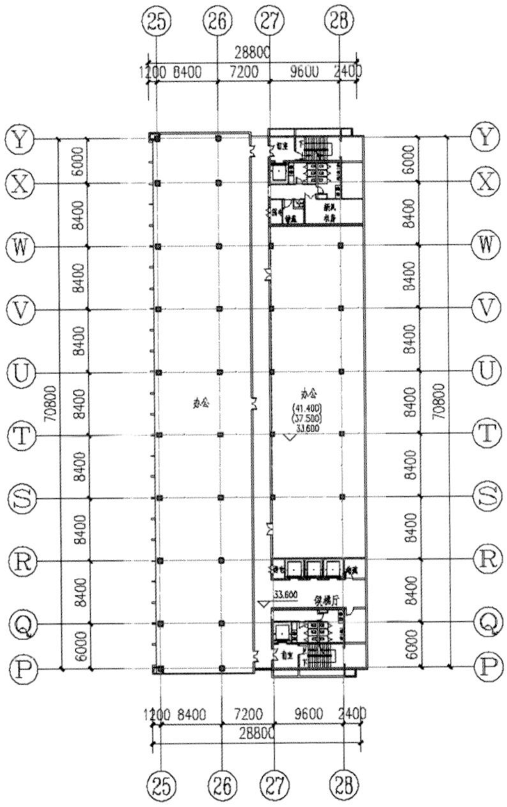

9~11 Floor Plan 9~11层平面图

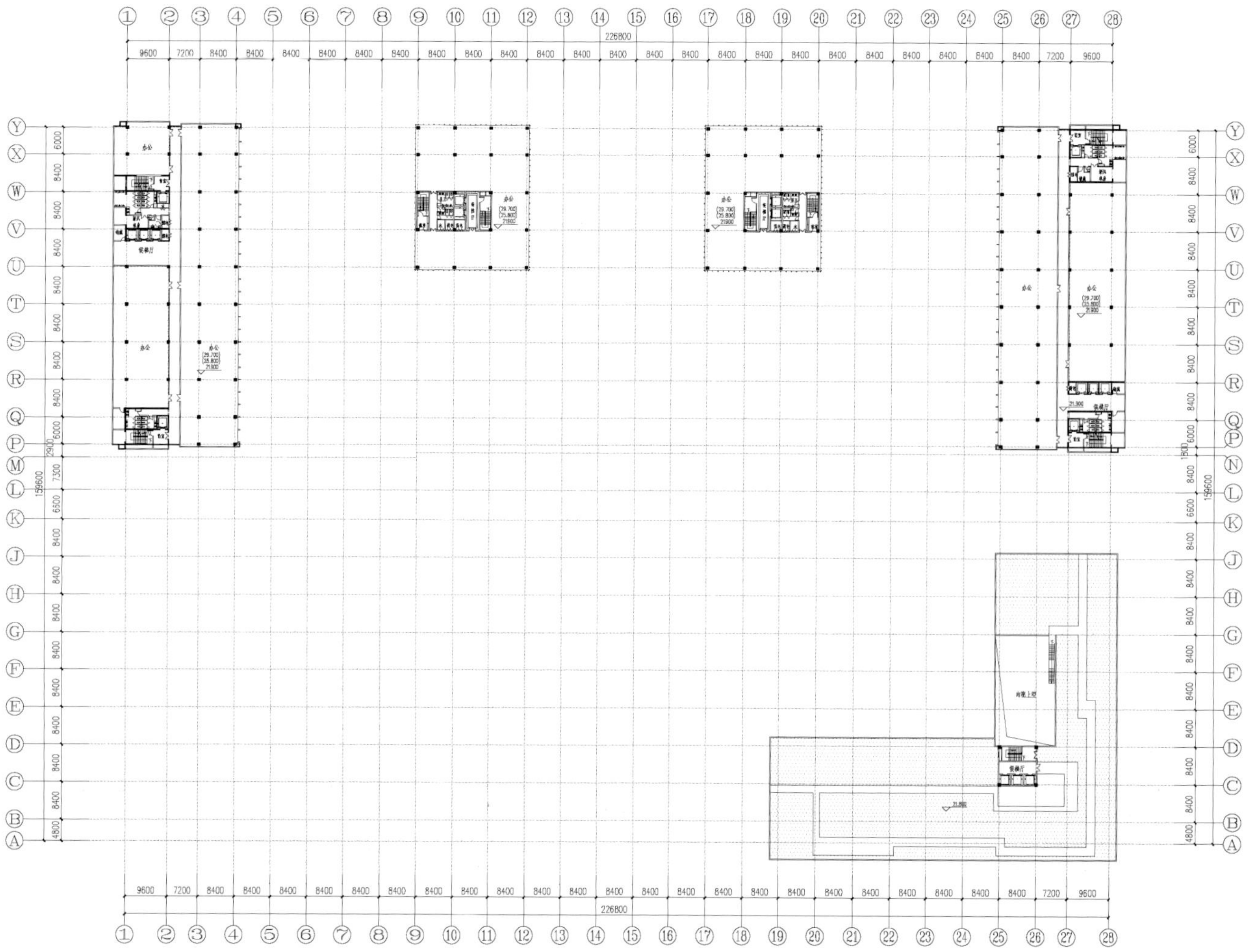

6~8 Floor Plan 6~8层平面图

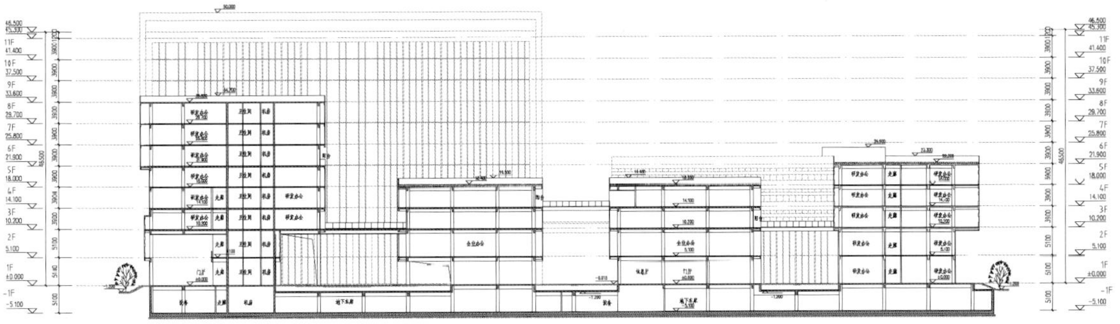

2-2 Section　2-2 剖面图

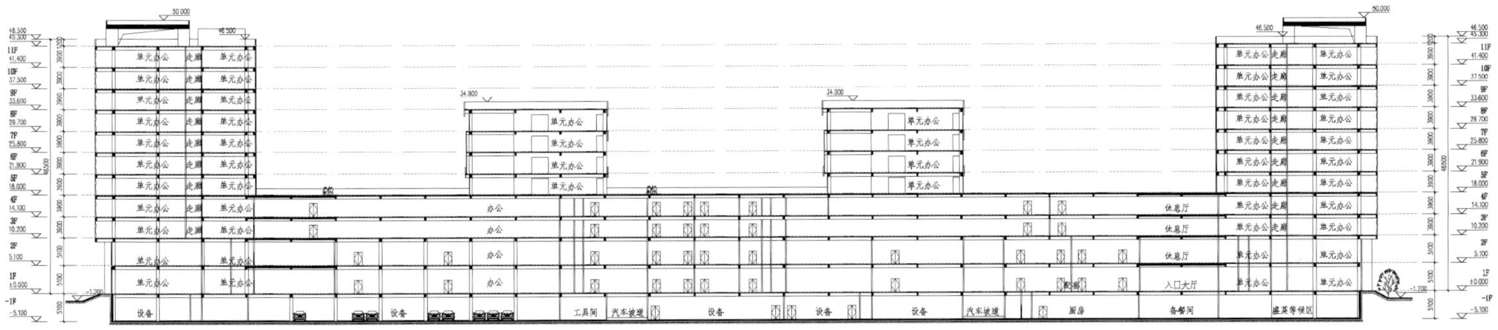

1-1 Section　1-1 剖面图

Xinhua Publishing Creative Center, Sichuan
四川新华出版创作中心

Keywords 关键词

Flexible Form
形体灵活

Architectural Language
建筑语言

Platform-style Layout
台地式布局

Features 项目亮点

Local materials such as massive ashlar wall and light bamboo shutter are used to create a flexible and free building form, which well matches the mountainous terrain and shapes the identity of the building.

建筑语言采用厚重的毛石墙面与轻盈的竹制百叶等具有地域特点的构造材料，建筑形体组合灵活自由，与山地地形融合，里面虚实有致，朴实厚重之中不乏轻巧。

Location: Dujiangyan, Sichuan, China
Layout Design: A&S International Architectural Design
Total Land Area: 48,288.3 m^2
Total Floor Area: 33,266.54 m^2

项目地点：中国四川省都江堰市
规划设计：北京翰时国际建筑设计咨询有限公司
总用地面积：48 288.3 m^2
总建筑面积：33 266.54 m^2

■ Overview

Located at Qingcheng Mountain Scenic Area, the project covers two plots over 48,288.3 m^2. It is west to Jianshe Road, south to Baihualing Road and east to Erwang Temple Hotel. The site is in a valley at the foot of the Qingcheng Mountain. The north plot is on the south slope of the valley that covers an area of 13,223.3 m^2, while the south plot is at the bottom of the valley that covers an area of 35,065 m^2. The site gives priority to slope which has large height difference with flourish mountain plantation around and favorable landscape environment.

■ 项目概况

四川新华出版创作中心位于都江堰市北部青城山景区内，分为两个地块，共占地48 288.3 m^2。地块东临建设路，北临百花岭路，西邻二王庙宾馆。基地位于青城山脚下一处山谷地形内，分为两个地块，北侧地块位于山谷南坡上，占地13 223.3 m^2，南侧地块位于谷底占地35 065 m^2。两个地块内地形以坡地为主，高差较大，周边山林植被茂盛，景观环境良好。

■ Planning

In terms of planning, centralized creative center is set in the flat part in the east side of the south lot, while the small-size single buildings such as creative community and fitness center are in the west side. Vertical platform type is used in the site. Entrance plaza at the east end is on the first level, main entrance of the creative center and the canteen are in the enclosed courtyard on the second level, fitness center and conference facility are on the third level. The north lot is for small buildings such as clubs and studios which scattered along the two sides of the road according to the terrain.

■ 规划布局

在规划上将南侧地块东侧较平坦部分布置为集中式综合创作中心，西侧部分布置创作群落、健身活动中心等小型单体建筑。地块采用台地式竖向布置，东端场地入口广场为第一级平台，综合创作中心的主入口以及其餐厅位于第二级平台上的围合庭院，康体中心、会议室布置于第三级平台，平台之间设室外步行路并可通向场地内部的各点式建筑。北侧地块为会所以及大师工作室等小型的单体建筑，结合地形分散布置于地块内道路两侧。

Tangshan Caofeidian Office

唐山曹妃甸办公楼

Keywords 关键词

Twist Shape
扭转造型

Towering
挺拔感

Open Space
开放空间

Location: Tangshan, Hebei, China
Architects: Synarchitects (Beijing)
Total Floor Area: 86,536 m^2

项目地点：中国河北省唐山市
建 筑 师：Synarchitects（北京）
总建筑面积：86 536 m^2

Features 项目亮点

The bottom of the building is parallel to the street to provide enough open space. With turning the top of the tower to the classical Chinese grid the architects twisted the whole building. It doesn't matter from which side you approach it, there will always be a new impression.

塔楼的底部与街道平行以留出更多的开放空间，通过扭转塔楼顶部至南北方向，整个大厦像是被扭弯了一般，不论从哪一个方向看，都给人耳目一新的感觉。

■ **Overview**

The development is located in the core area of Caofeidian Industrial Service Center, Hebei, offering the investors from Europe with high-end offices and residences.

■ **项目概况**

本工程位于河北省曹妃甸工业区综合服务区核心地段，主要是向欧洲来曹妃甸投资的商业人士提供高端的办公和居住场所。

■ **Architectural Design**

The building site is neighboring a river and is accessible from 4 sides through streets. On the clients recommendation the architects ensured to create two squares connecting the building to the north and to the south. According to local laws the clearance from the street gives the architects a centered building site receiving sufficient space to be able to satisfy the clients' request. The architects created a virtual building massing with the most possible floor area ratio and shaped it according to the building program with a tower facing eastwards to the river and a massive block to the west.

The city is an orthogonal grid aligned to east-west, which turned to north by 32.5 degree. All buildings are aligned to that grid, but this office building could attract interest because the architects have chosen a different one. The ideal Chinese city is based on an orthogonal grid, but facing north-south to harmonize between the forces of the cosmos, nature, and the humans that create the city. The tower is visible from afar, while the bottom of the building has to be parallel to the street to create enough open space. With turning the top of the tower to the classical Chinese grid the architects twist the whole building. It doesn' t matter from which side you approach it, there will always be a new impression. The extended lines of the towers body cut the rest of the volume, separate the functions and open the building' s plazas to the south and north. The new city plaza in the south will be partly a public park leading the people directly to the twisted tower gates.

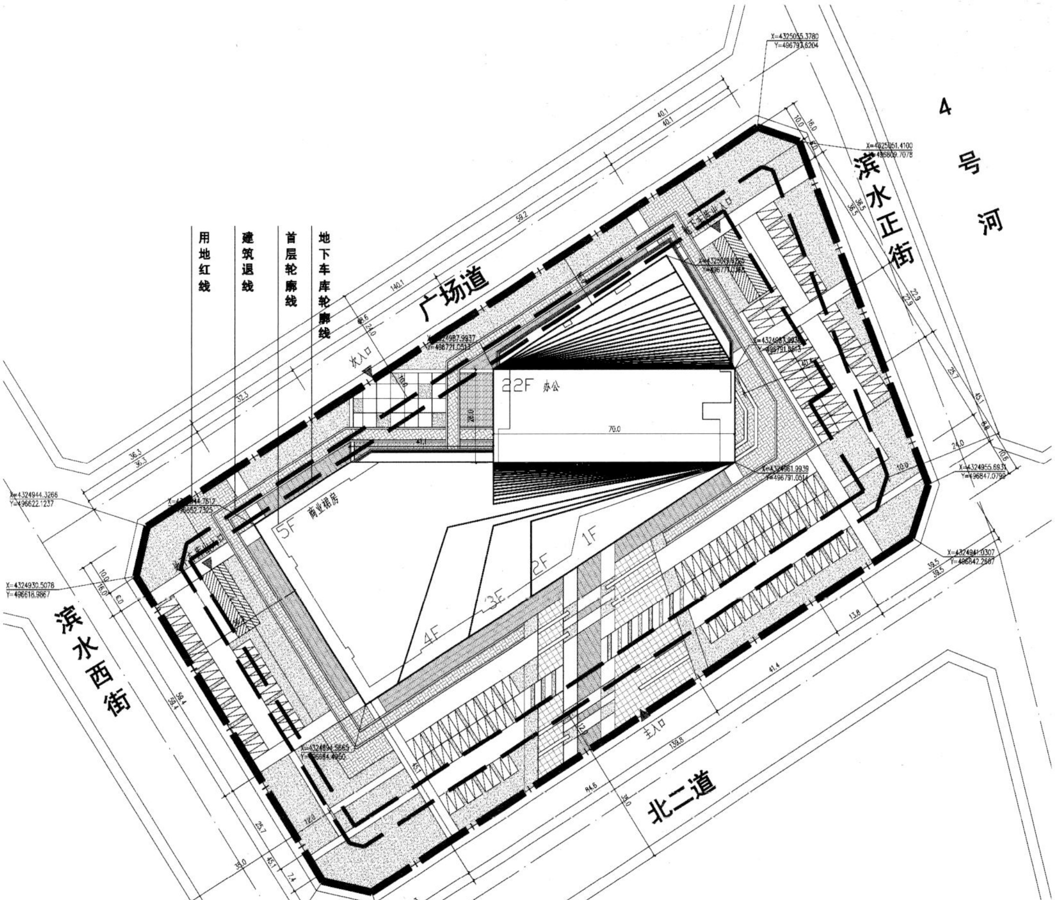

Site Plan 总平面图

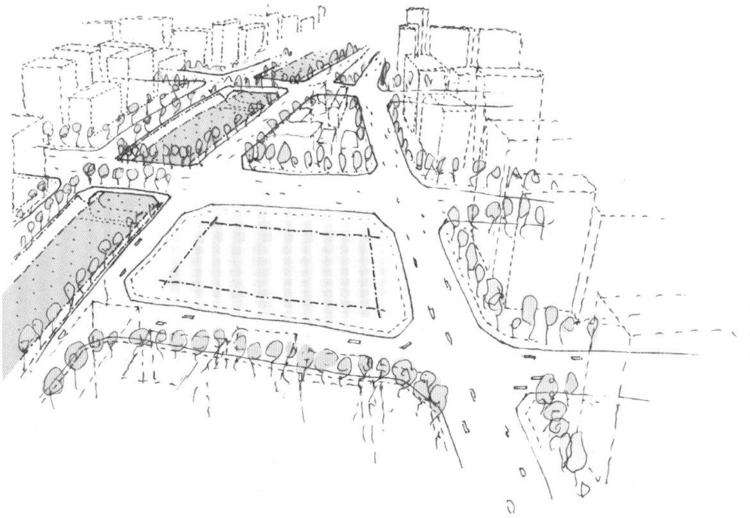

Sketch 1: Clearance 草图 1: 净空

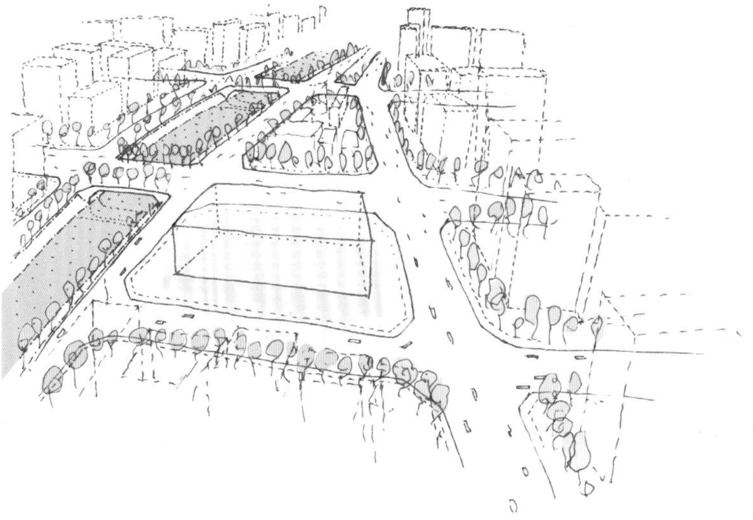

Sketch 2: Possible Volume 草图 2: 预估体量

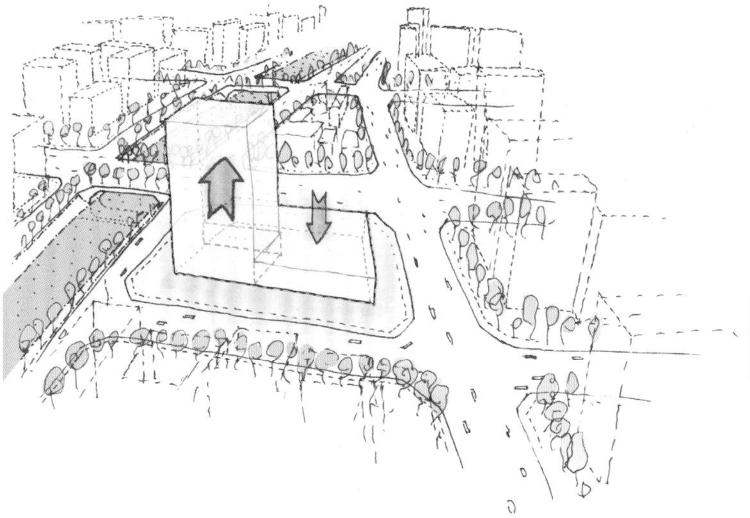

Sketch 3: Volumes According to Functions 草图 3: 功能体量

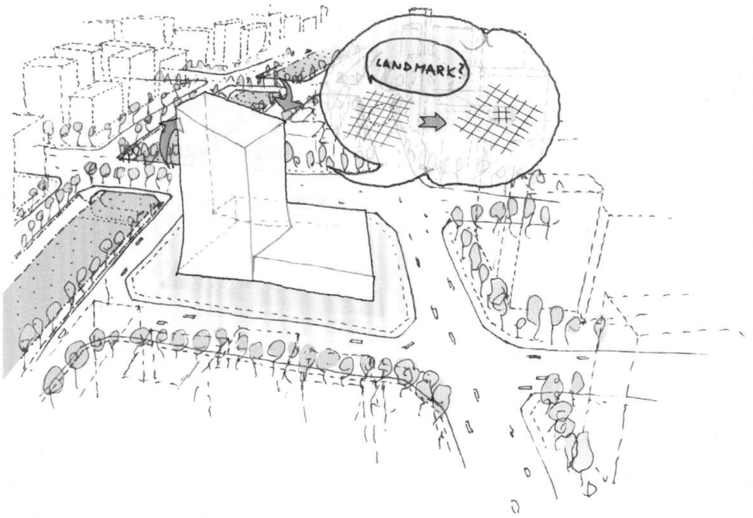

Sketch 4: North South 草图 4: 南北向

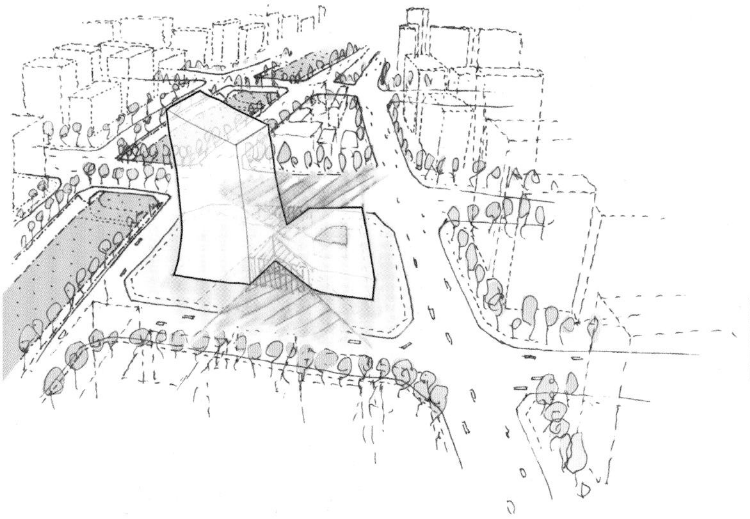

Sketch 5: North South 草图 5: 南北向

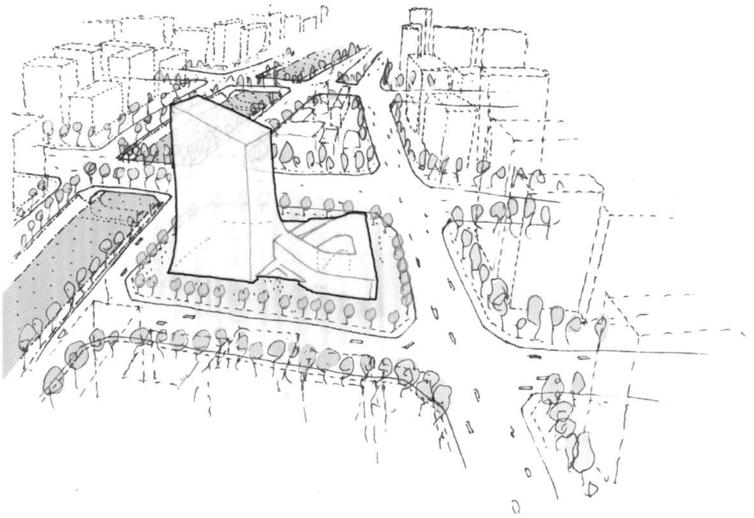

Sketch 6: Landmark 草图 5: 地标

■ 建筑设计

该项目濒临河流且四面都有街道直达，交通便利。遵循业主的建议，设计师在办公楼南北两侧分别建造一个广场与之相连。依据当地法律法规，两边街道的空隙使得办公楼独立居中并且有足够多的空间来满足广场所需。同时该项目还最大限度地利用了容积率，使塔楼东临河流，而主楼从最西端延伸过来，从而创造出一个与建筑方案相吻合的虚拟的建筑体态。

唐山市的交通线为向北偏移32.5度的东西向的正交网格状。在所有建筑物都是与该交通线路同向的情况下，该项目做出的这次尝试使其鹤立鸡群，引人注目。尽管唐山市以东西向正交网格线为基础，但坐北朝南使其达到宇宙、自然以及创造这个城市的人类之间的和谐共处，这才是理想的中国之城。该项目的塔楼能从很远的地方看到，而大厦的底部则与街道平行以留出更多的开放空间。通过扭转塔楼顶部至南北方向，整个大厦像是被扭弯了一般。不论从哪一个方向观看大厦，都给人耳目一新的感觉。从塔楼主体延伸出来的线条成了分割大厦功能区域以及南北两个广场的分割线，而这个全新的南广场在一定程度上将成为一个公园，并直接通往扭转塔楼的大门。

沿街商业　超市　金融　办公大堂　咖啡（办公配套）　交通/辅助/设备

First Floor Plan 一层平面图

餐饮　金融　咖啡简餐（办公配套）　交通/辅助/设备

Second Floor Plan 二层平面图

餐饮　办公　交通/辅助/设备

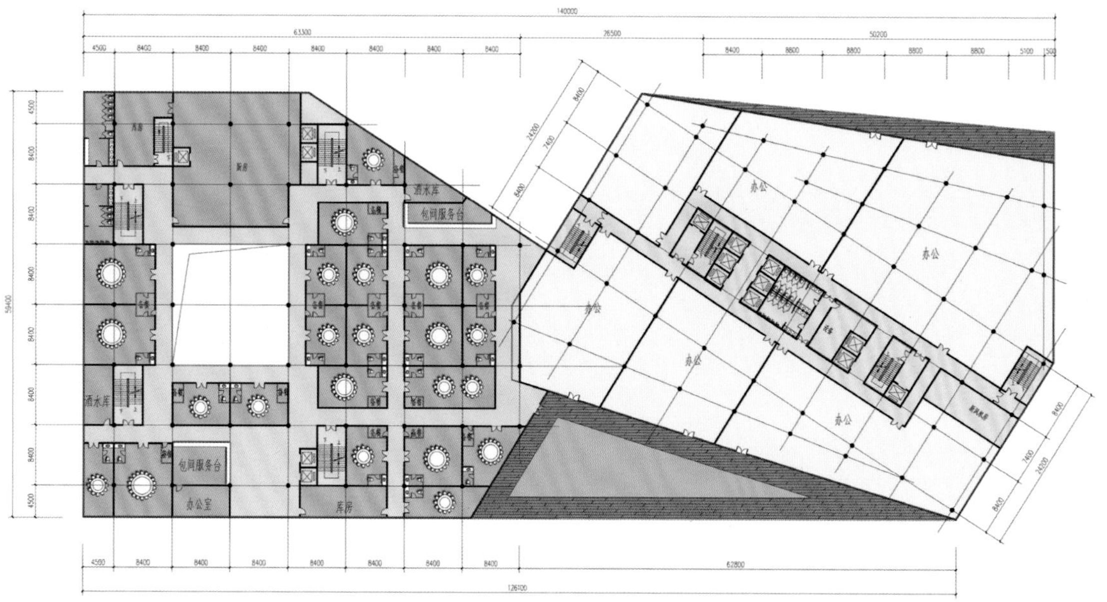

Third Floor Plan 三层平面图

会所部分（KTV） 办公 交通/辅助/设备

Fourth Floor Plan 四层平面图

会所部分（健身） 办公 交通/辅助/设备

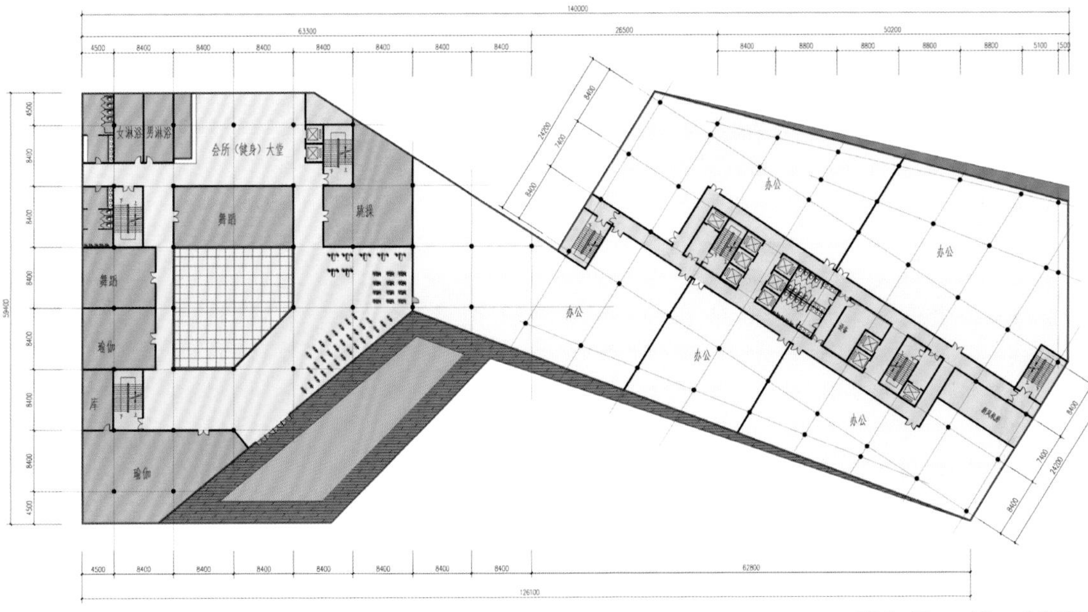

Fifth Floor Plan 五层平面图

办公室部分 开敞办公部分 交通/辅助/设备

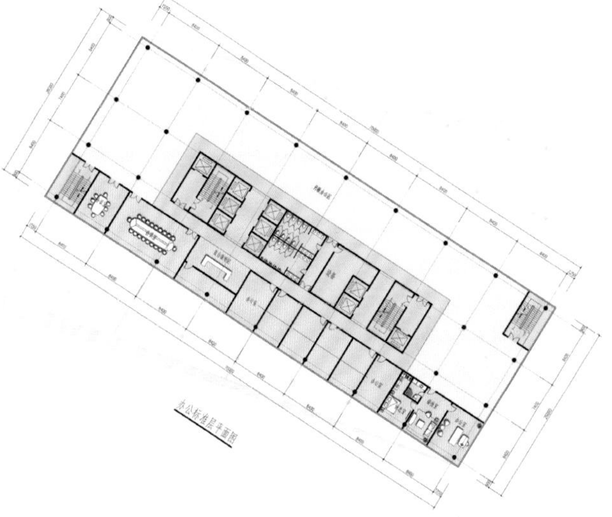

Plan for Standard Floor 标准层平面图

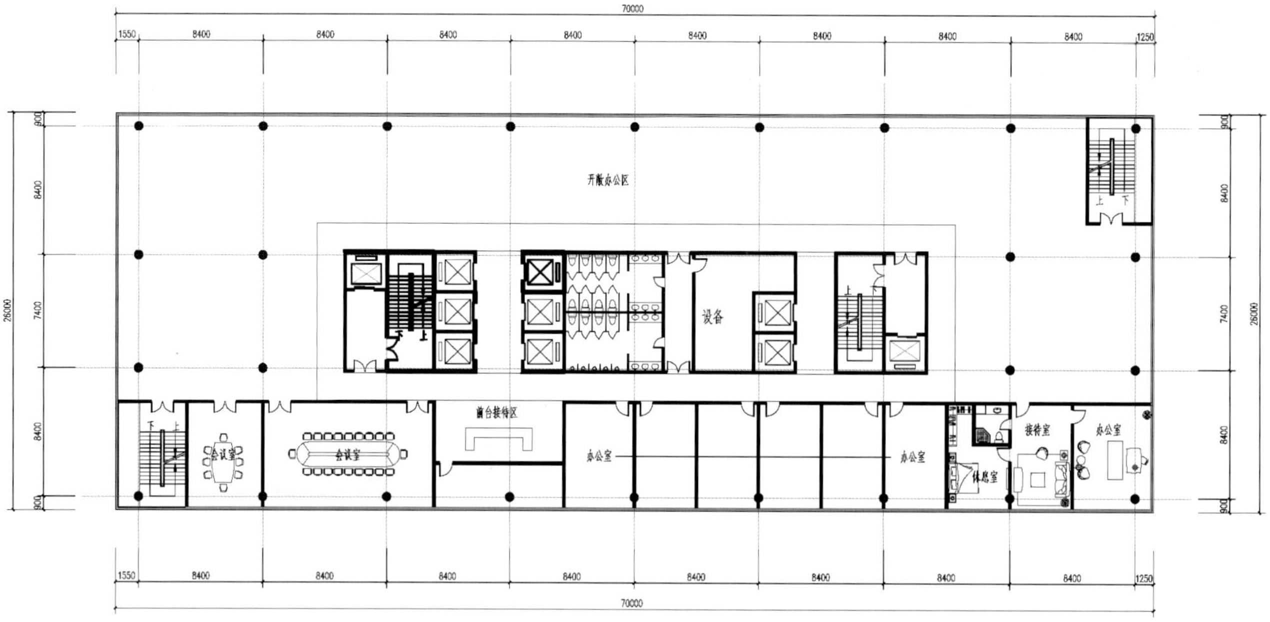

Standard Floor of Office Building　办公标准层平面图

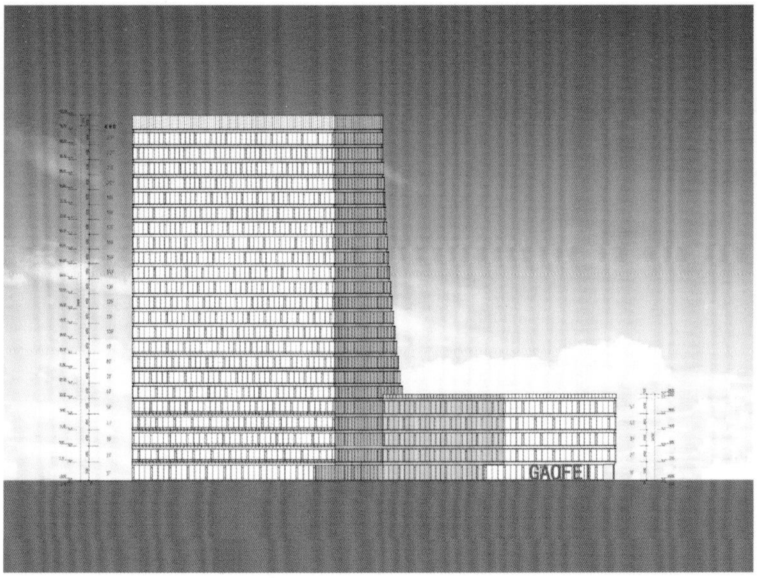

North Elevation 北立面图

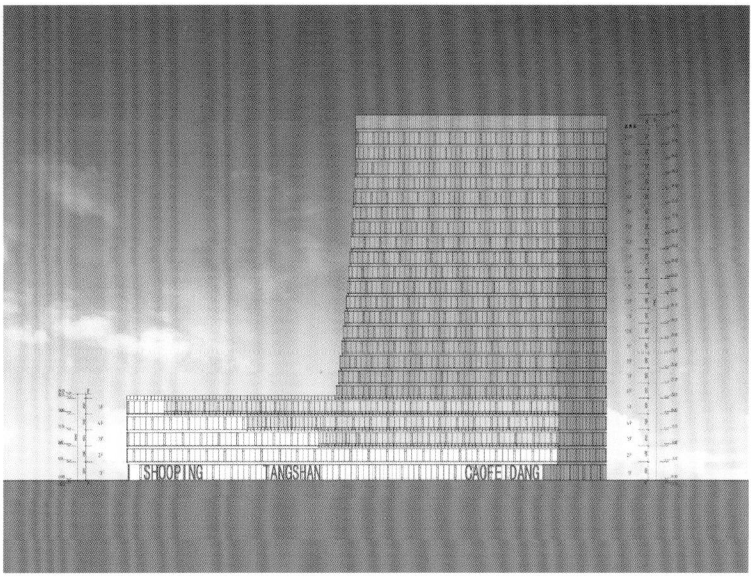

South Elevation 南立面图

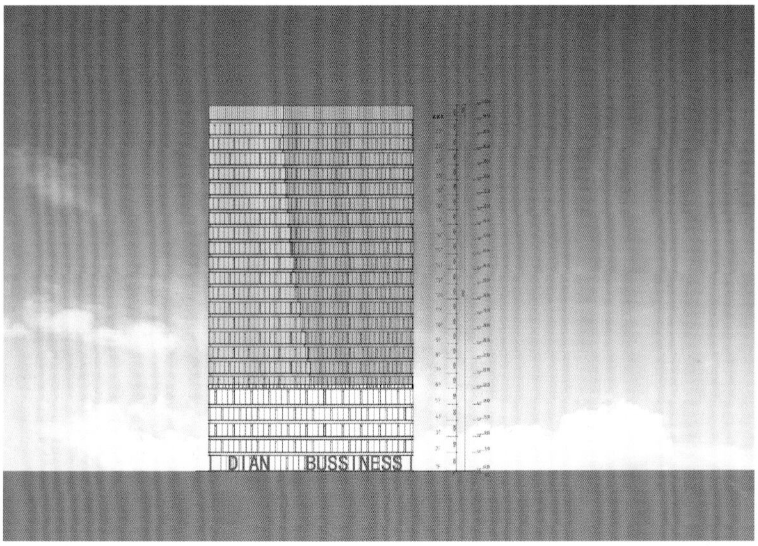

West Elevation 西立面图

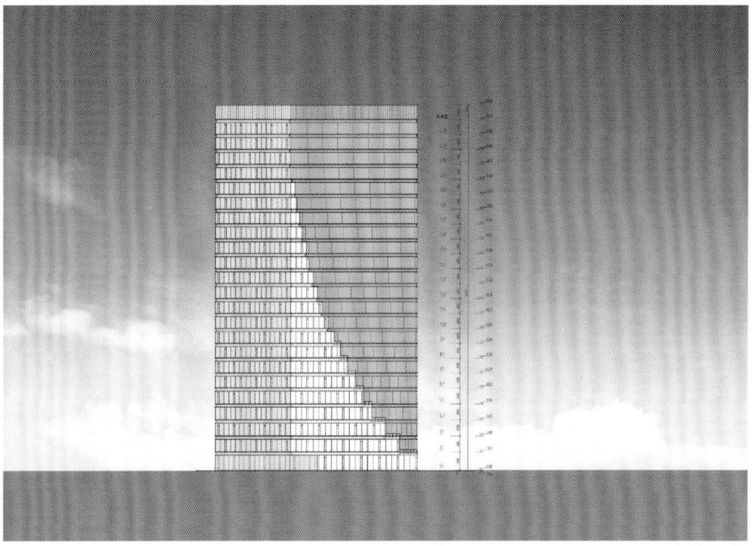

East Elevation 东立面图

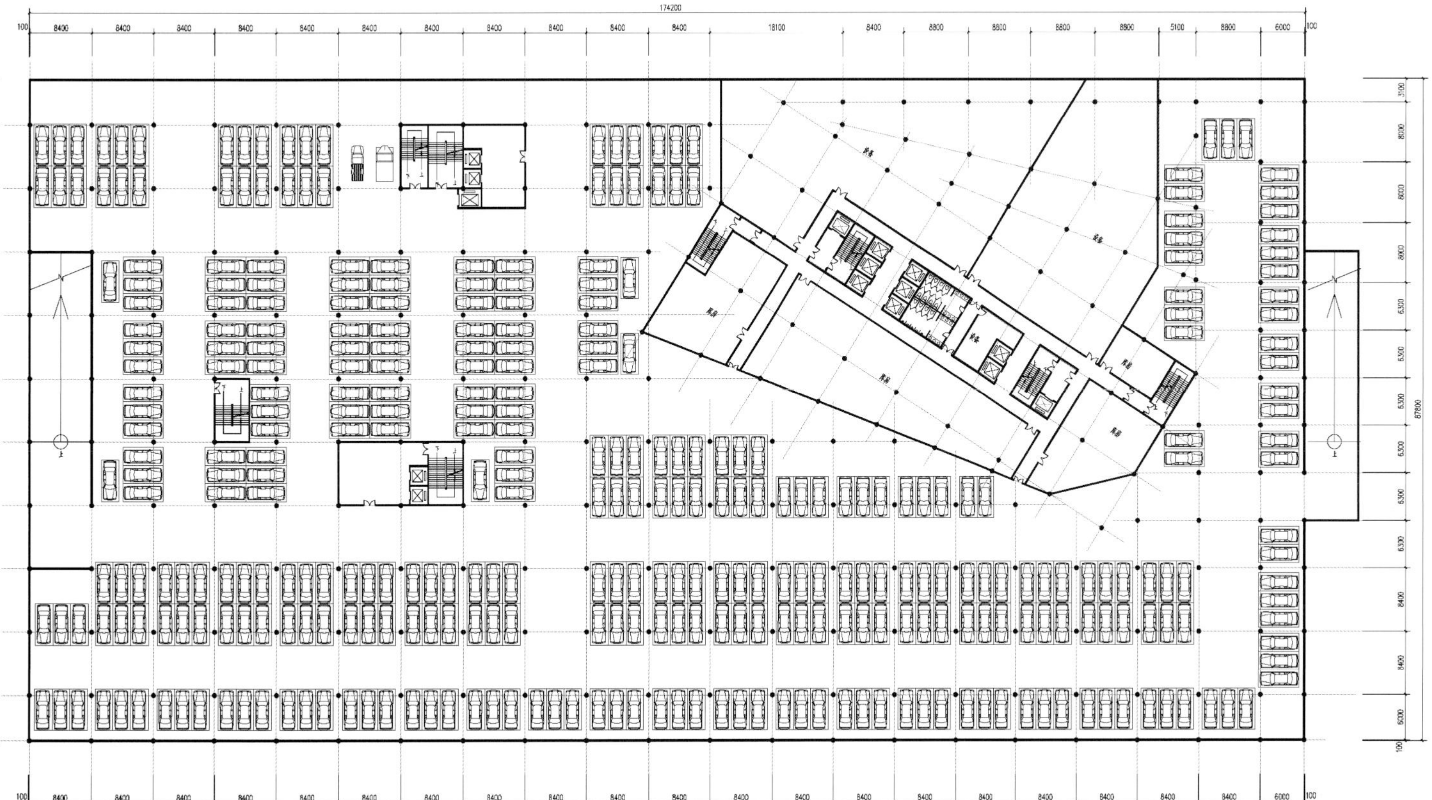

Salvatore Ferragamo
Dior
VISIBLE LIFT
rock

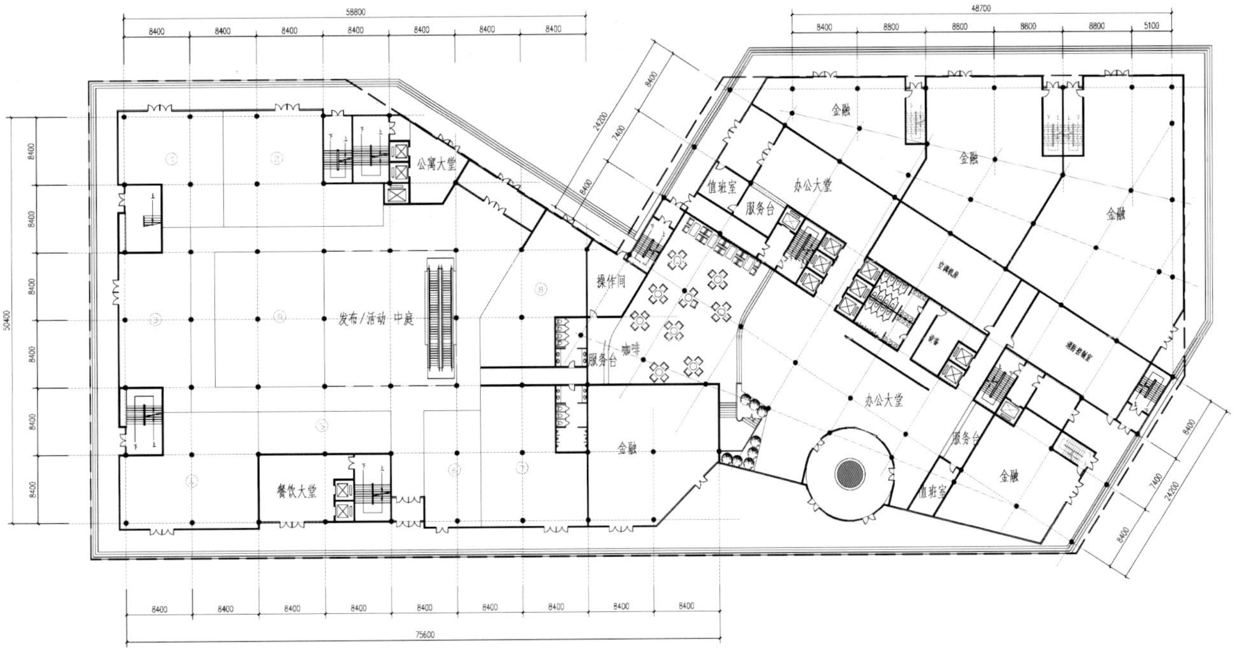
公寓大堂
发布／活动 中庭
餐饮大堂
操作间
服务台
咖啡
金融
值班室
办公大堂
金融
金融
金融
办公大堂
服务台
值班室
金融

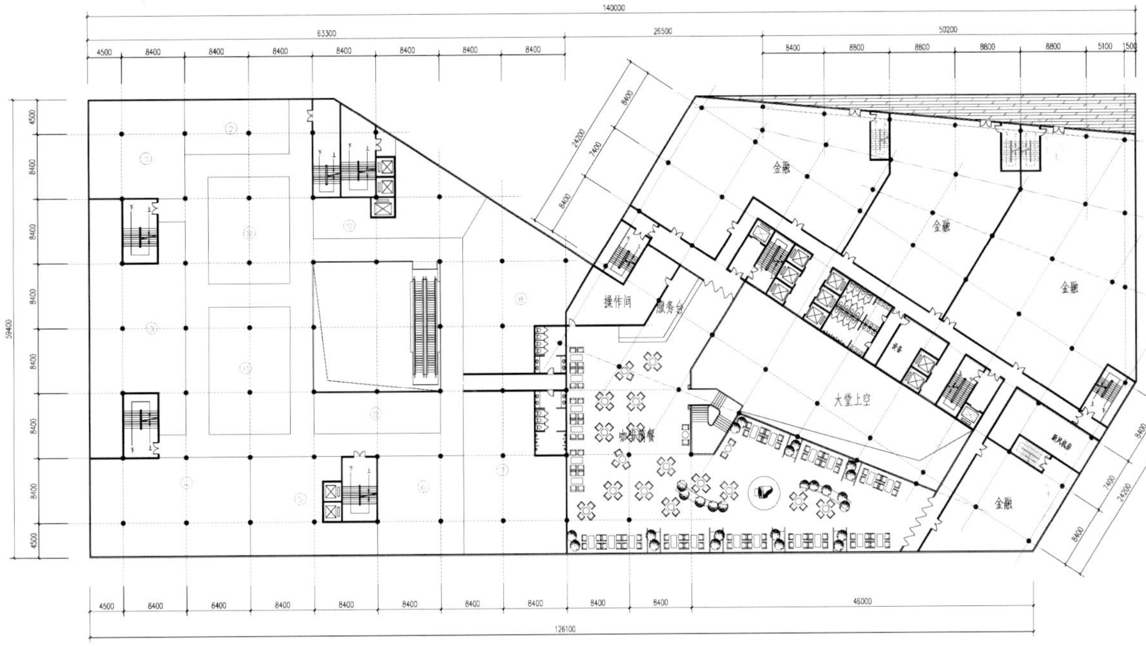
金融
金融
金融
操作间
服务台
大堂上空
金融

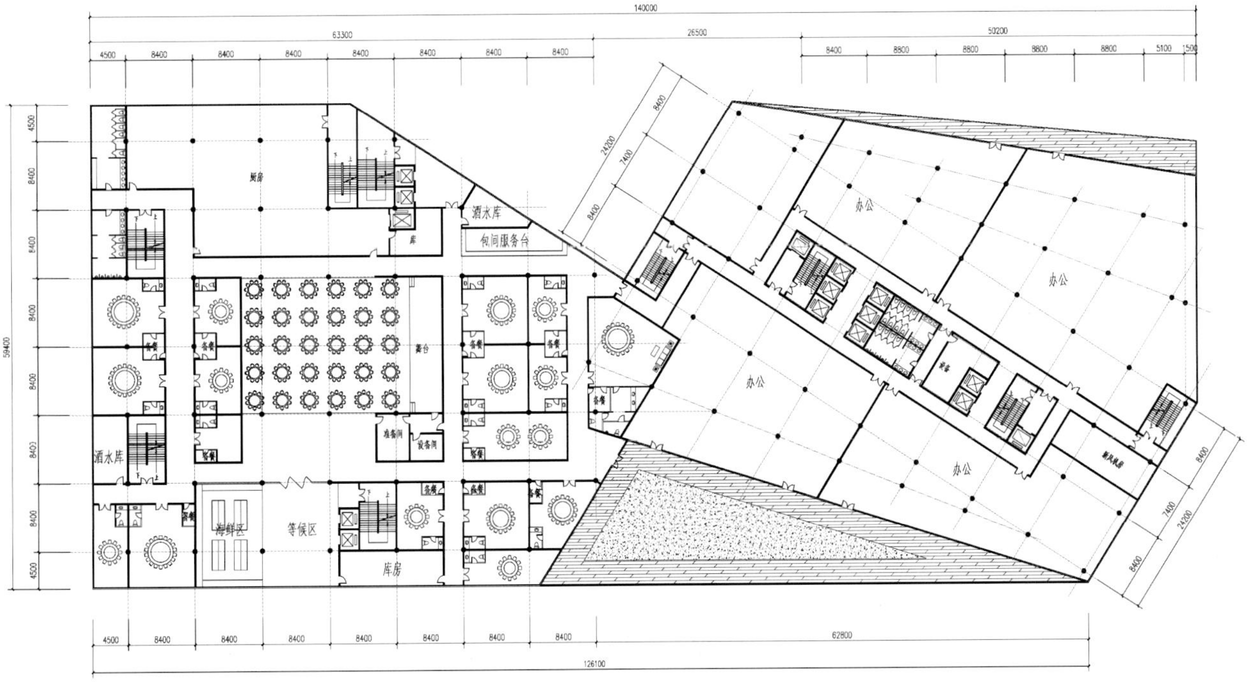
厨房
酒水库
包间服务台
舞台
酒水库
海鲜区
等候区
库房
办公
办公
办公
办公

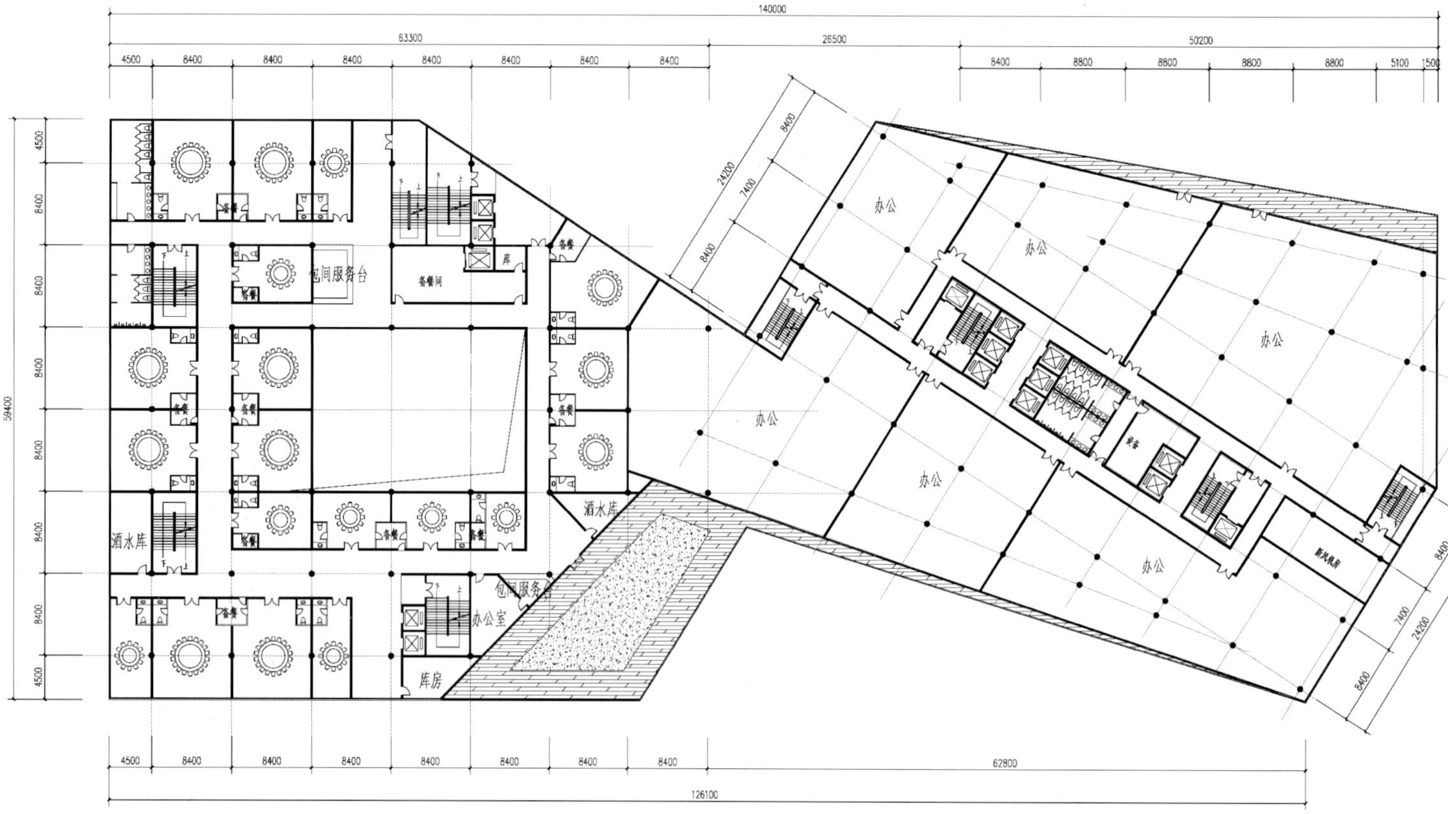
140000
63300
26500
50200
4500
8400
8800
5100
1500
59400
24200
7400
包间服务台
备餐间
酒水库
办公室
库房
库
备餐
办公
新风机房
126100
62800

机房层
23F
22F
21F
20F
19F
18F
17F
16F
15F
14F
13F
12F
11F
10F
9F
8F
7F
6F
5F
4F
3F
2F
1F
办公
咖啡简餐
办公大堂
金融

Section 剖面图

VERSACE
Dior
REVLON
DALIMIX

Transportation Building

交通建筑

Spatial Sequence
时空序列

Structural System
结构体系

Traffic Flow
交通流线

Visual Image
视觉形象

Zhuhai Shizimen Bridge
珠海十字门大桥

Keywords 关键词

Structure Design
结构设计

Wave Shape
波浪造型

Linear Road
线性路面

Features 项目亮点

The design was set out to create a simple and elegant structural solution for a bridge that would be a visual focal point both within the Shizimen District itself and along the Pearl River Delta coastline.

该设计方案以简明优雅的结构设计，旨在创造出十字门商务区及珠江三角洲沿岸的视觉焦点。

Location: Zhuhai, Guangdong, China
Client: Zhuhai Shizimen CBD Development Holdings CO., Ltd.
Architectural Design: 10 DESIGN
Designer: Gordon Affleck
Structural Engineering: Buro Happold
Bridge Length: 300 m
Completion: Expected 2014

项目地点：中国广东省珠海市
业　　主：珠海十字门中央商务区建设控股有限公司
建筑设计：拾稼设计
设 计 师：艾高登
结构工程设计：Buro Happold
桥　　长：300 m
完工时间：预计2014年

■ Overview

10 DESIGN and Buro Happold have won the design competition for the key signature gateway bridge for the new Shizimen Business District in Zhuhai, China. The feature bridge is the gateway entry to south China's new planned commercial hub and also marks the connection of the Shizimen Canal to the Pearl River Delta. The design set out to create a simple and elegant structural solution for a bridge that would be a visual focal point both within the Shizimen District itself and along the Pearl River Delta coastline.

■ 项目概况

10 DESIGN（拾稼设计）与Buro Happold结构顾问共同在中国珠海十字门商务区的门户大桥设计竞赛中获胜。这座特色大桥是通往中国南部新规划商务中心的门户，同时也是十字门商务区与珠江三角洲之间联通的标志。设计方案以简明优雅的结构设计，旨在创造出十字门商务区及珠江三角洲沿岸的视觉焦点。

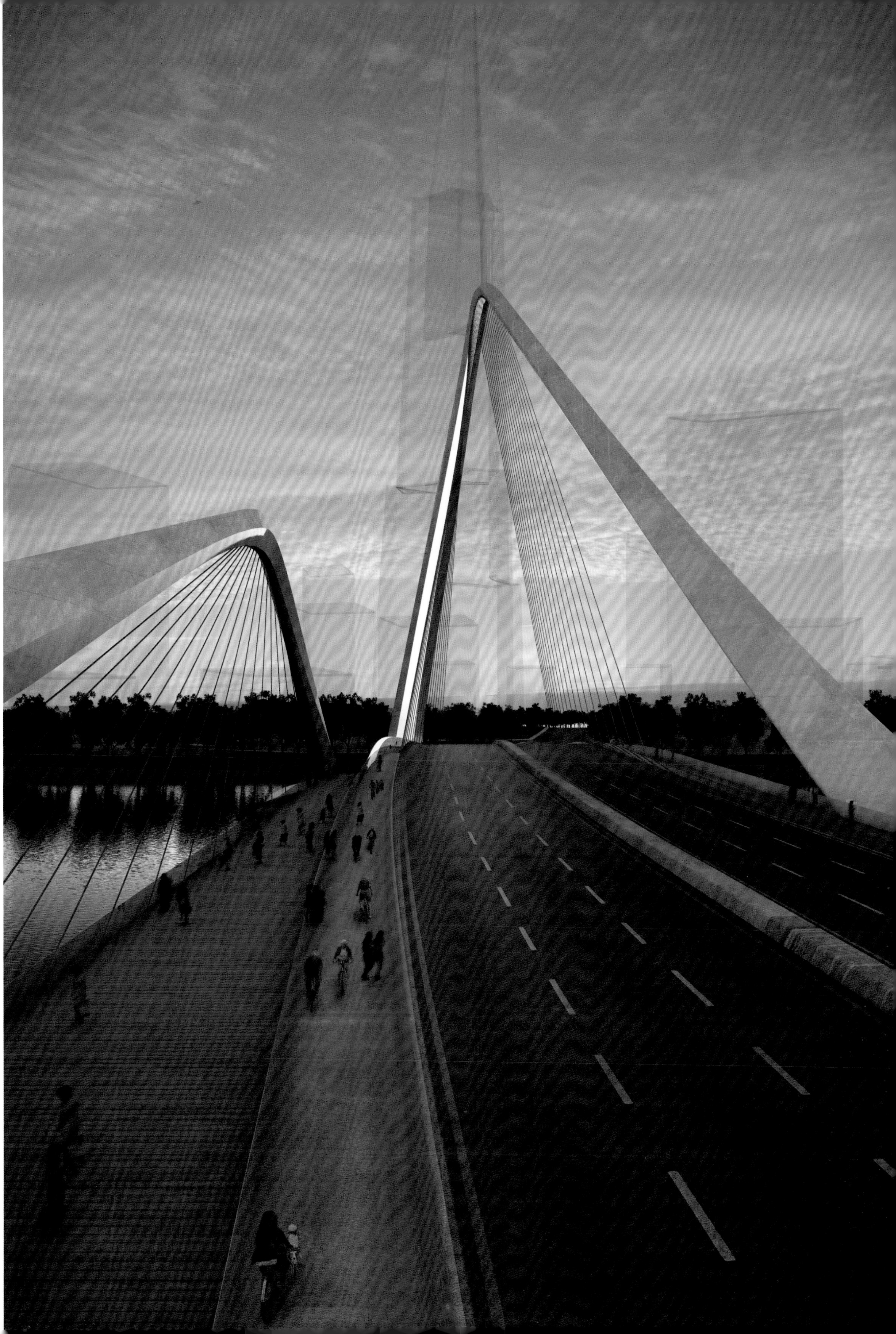

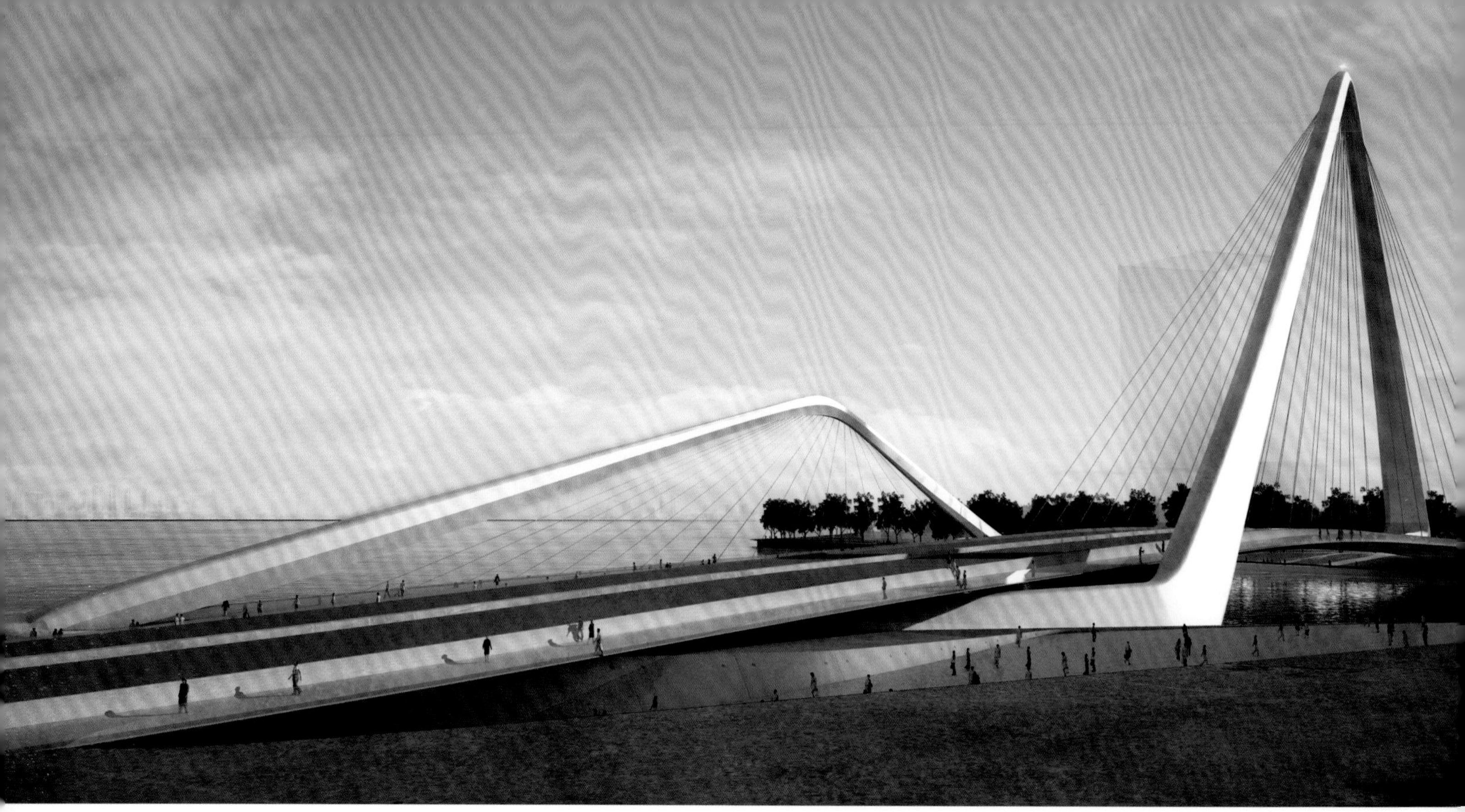

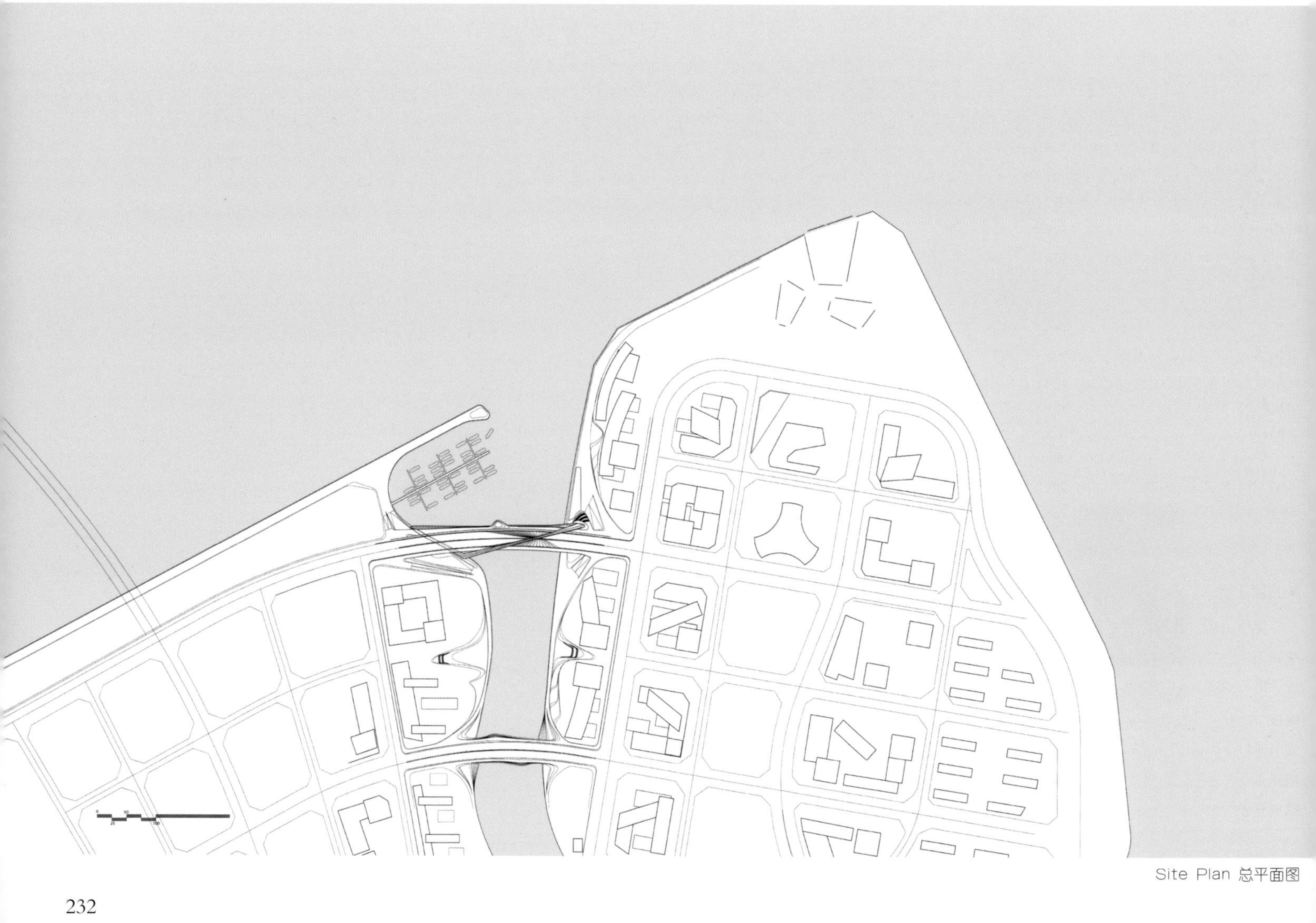

Site Plan 总平面图

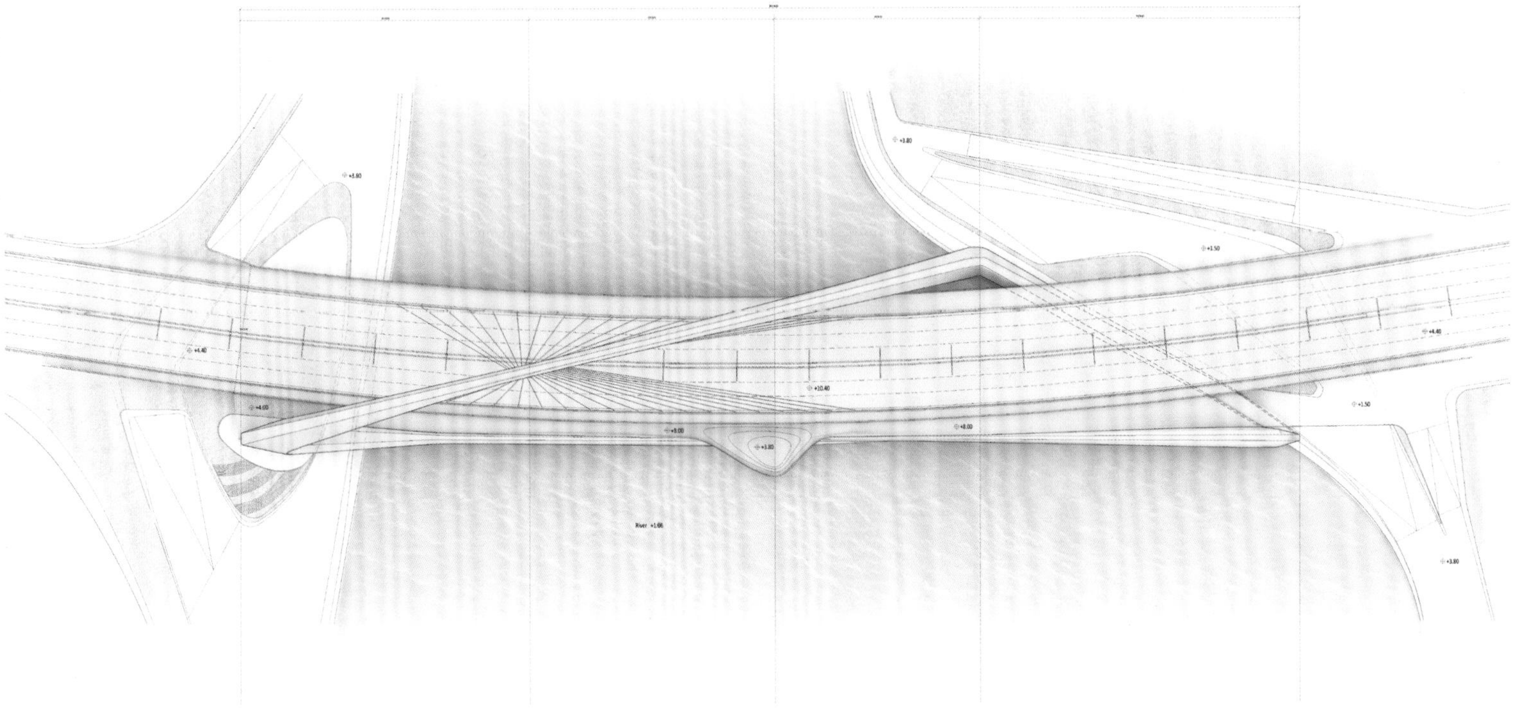

■ Architecture Design

The winning bridge scheme proposes an undulating ribbon of structure that wraps around the 6 lane road deck in a continuous sculptural loop form. At low levels the loop cradles the road deck, with the structural ribbon then soaring upwards to create 2 triangulated arch forms.

These triangular gateways, one primary and one secondary, then support the road deck over 2 separate distinct spans via suspension cables. The primary triangulated gate is about 90m in height and by diagonally crossing the deck of the bridge, celebrating not only the crossing of vehicular and pedestrian traffic, but also the meeting point of the canal with the Pearl River Delta. The continuity of the loop structure is visible from the opposite mainland shoreline to the west, and appears as a dramatic free standing infinity loop.

■ 方案设计

设计以波浪形丝带为结构造型，以线形回路样式环绕大桥的6条道路面。在低处，它像是呵护桥身的摇篮，向上却又摇身一变，高耸入云形成两道三角拱形结构。两道三角拱形结构一主一辅，通过缆索连接，支撑起两段跨度极大的桥面。主拱形结构高约90m，以对角方式斜跨整个桥面，不仅支撑了车辆与行人的交通要道，还成为了十字门商务区与珠江三角洲的汇合点。对面的大陆海岸依然可见延伸的回路型丝带结构，它就像飘逸的不断波动着的浪形丝带。

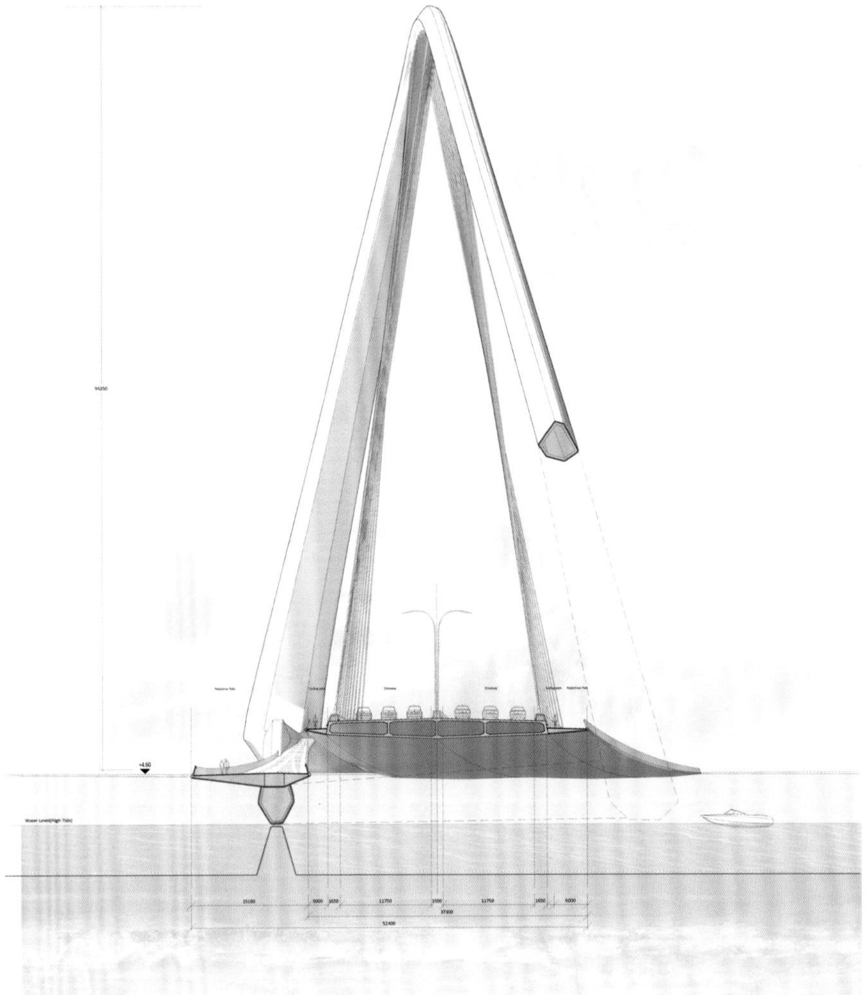

Section 剖面图

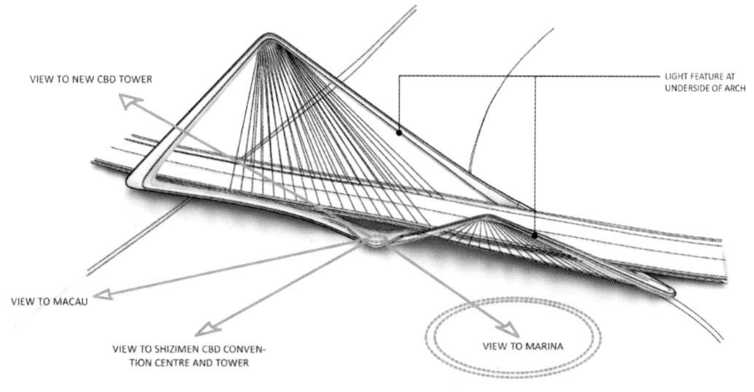
VIEW TO NEW CBD TOWER
LIGHT FEATURE AT UNDERSIDE OF ARCH
VIEW TO MACAU
VIEW TO SHIZIMEN CBD CONVEN-TION CENTRE AND TOWER
VIEW TO MARINA

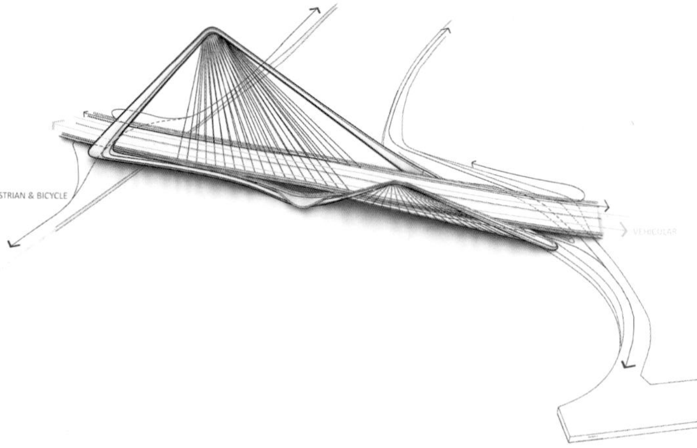
PEDESTRIAN & BICYCLE

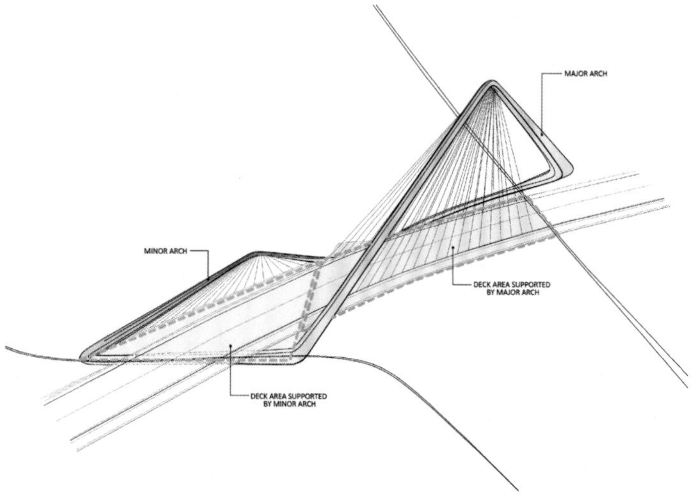
MAJOR ARCH
MINOR ARCH
DECK AREA SUPPORTED BY MAJOR ARCH
DECK AREA SUPPORTED BY MINOR ARCH

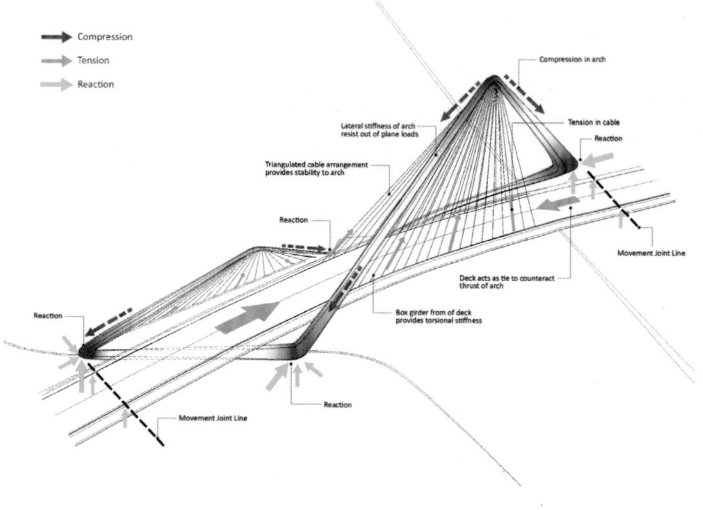
Compression
Tension
Reaction
Compression in arch
Tension in cable
Reaction
Lateral stiffness of arch resist out of plane loads
Triangulated cable arrangement provides stability to arch
Reaction
Movement Joint Line
Deck acts as tie to counteract thrust of arch
Box girder from of deck provides torsional stiffness
Reaction
Reaction
Movement Joint Line

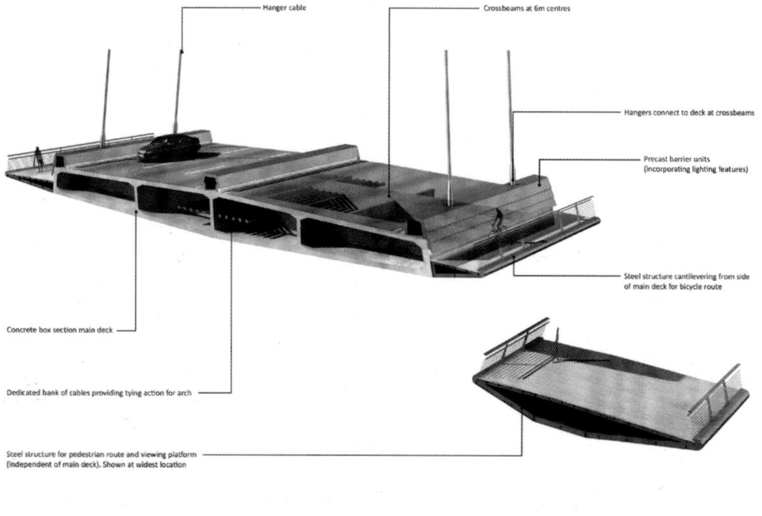
Hanger cable
Crossbeams at 6m centres
Hangers connect to deck at crossbeams
Precast barrier units (incorporating lighting features)
Steel structure cantilevering from side of main deck for bicycle route
Concrete box section main deck
Dedicated bank of cables providing tying action for arch
Steel structure for pedestrian route and viewing platform (independent of main deck). Shown at widest location

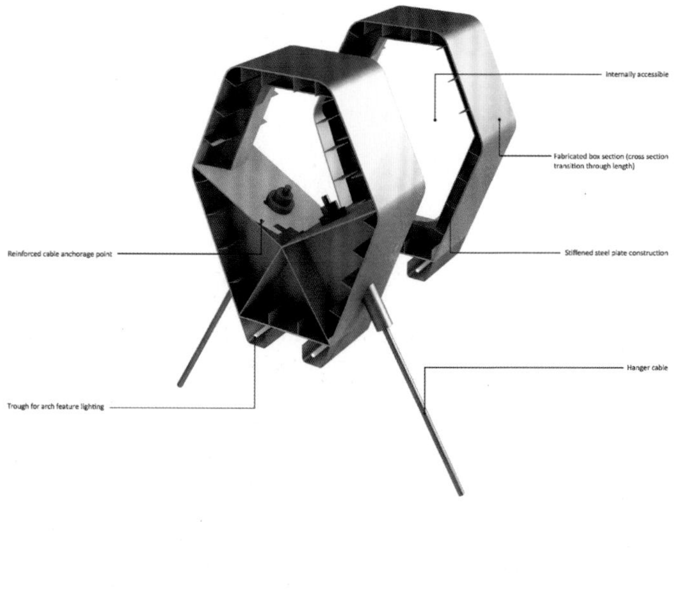
Internally accessible
Fabricated box section (cross section transition through length)
Stiffened steel plate construction
Reinforced cable anchorage point
Hanger cable
Trough for arch feature lighting

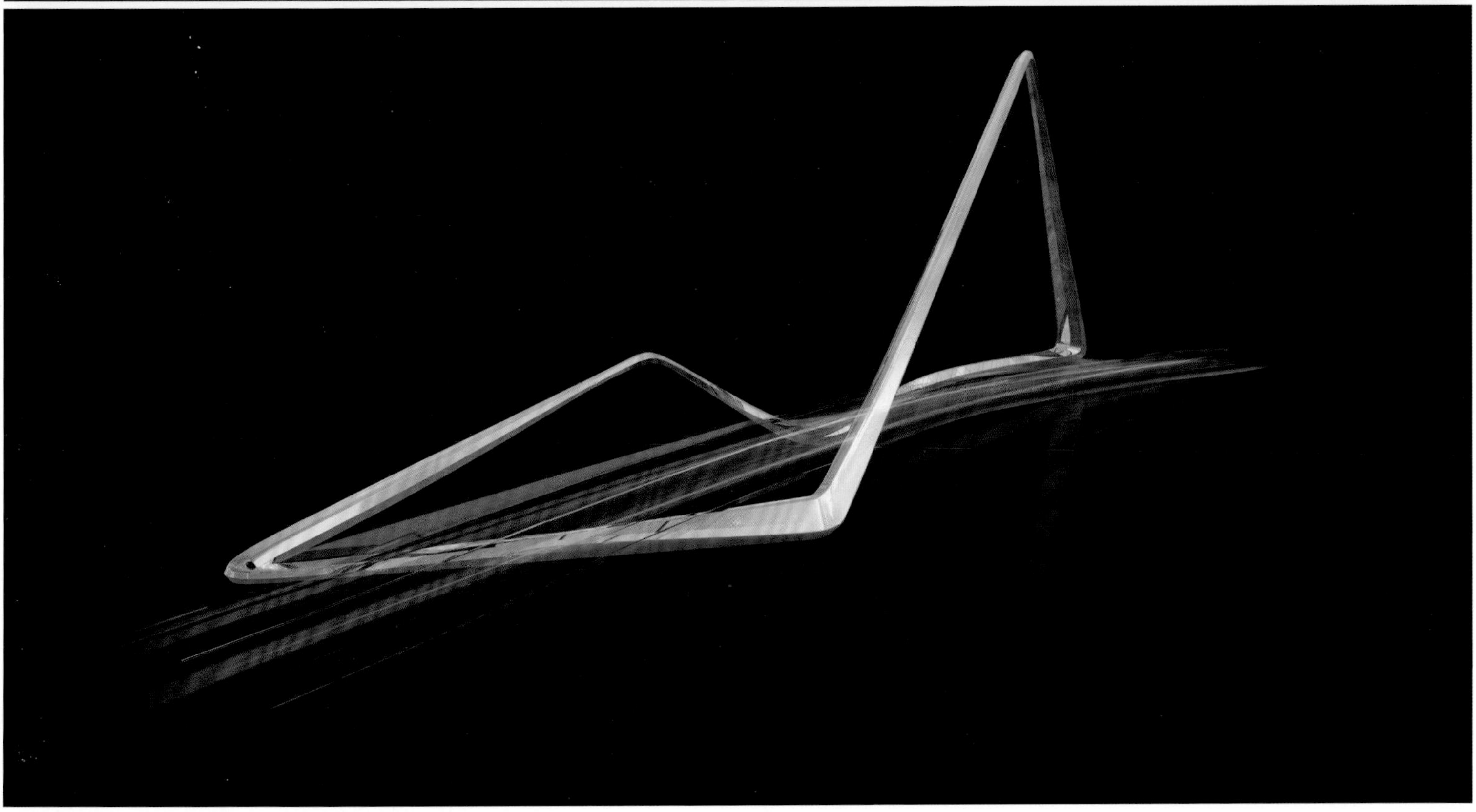

Passenger Station of Taibai Mountain Tourism Complex, Shanxi

陕西太白山旅游综合区客运站

Keywords 关键词

Architectural Form
建筑形式

Traffic Streamline
交通流线

Space System
空间体系

Features 项目亮点

The designers fully consider the relationship between the visitor center and the scenic area, and tries to make the main building body echo with Taibai Mountain, creating a portal image of overall sequence of time and space for the scenic spot.

设计充分考虑游客中心与景区之间的关系，使建筑主体与太白山相呼应，营造该景区总体时空序列的前端门户形象。

Location: Mei County, Shaanxi, China
Developer: Shaanxi Taibai Mountain Investment Group Co., Ltd.
Architectural Design: Xi'an Bai'an Architecture Design Co., Ltd.
Beijing Zhonghan International Architecture Design Co., Ltd.
Total Floor Area: 10,132 m^2

项目地点：中国陕西省眉县
开 发 商：陕西太白山投资集团有限公司
设计公司：西安百岸建筑设计有限公司
北京中翰国际建筑设计有限公司
总建筑面积：10 132 m^2

■ **Overview**

The project is located in Taibai Mountain Resort, Baoji, Shaanxi. Because of the project base facing Taibai Mountain in the south, the planning and layout fully consider the relationship between the visitor center and the scenic area, and tries to make the main building body echo with Taibai Mountain, creating a portal image of overall sequence of time and space for the scenic spot and travel route for the visitors.

■ **项目概况**

项目位于陕西省宝鸡市眉县太白山旅游度假区，由于项目基地南侧面向太白山，因此规划布局时充分考虑了游客中心与景区之间的关系，使建筑主体与山相呼应，呈现出对太白山的观望状，并由此成为该景区总体时空序列的前端门户形象，引导和起承游客的出行线路。

■ **Architectural Design**

The architectural form is the combination of the modern architecture scale, modern doors and windows forms, modern architectural surface and abstract mountain terrain with the traditional local-style residential building symbols of the central Shanxi Plain, natural mountain texture, traditional rhythm, traditional architecture and natural concept of the unity of heaven and earth, which can also be thought of as regional modern architecture.

The main building of visitor center is located in the central base, and achieves maximum developed surface in front of the former square, which facilitates people to quickly arrive in different function areas by the guide of the square. The main function of the comprehensive tourist center is constituted by 2-layer passenger transportation center, tourist service center and conversion platform, providing various services of travel demands. The relationship of the parking lot and the visitor center proceeds from most visitors' transfer (i.e. parking convenience) to carry on the reasonable design, and doesn't affect the integrity of the overall environmental space.

■ 建筑设计

建筑形式由现代建筑的尺度、门窗形式、建筑表皮和抽象的山势、关中民居建筑符号、自然的山地肌理，以及传统的韵律节奏、建筑与天地合一的自然理念相结合，也可以认为是现代建筑的地域化。

游客中心主体建筑位于基地中部，与前广场形成最大限度的展开面，便于人流通过广场的引导，快速抵达不同功能区域。综合游客中心主要由二层的客运中心和旅游服务中心以及转换平台组成，提供各种旅游需求的服务。停车场与游客中心的关系从方便大多数游客换乘（即停车便利性）的角度出发进行合理设计，并且不影响整体环境空间的完整性。

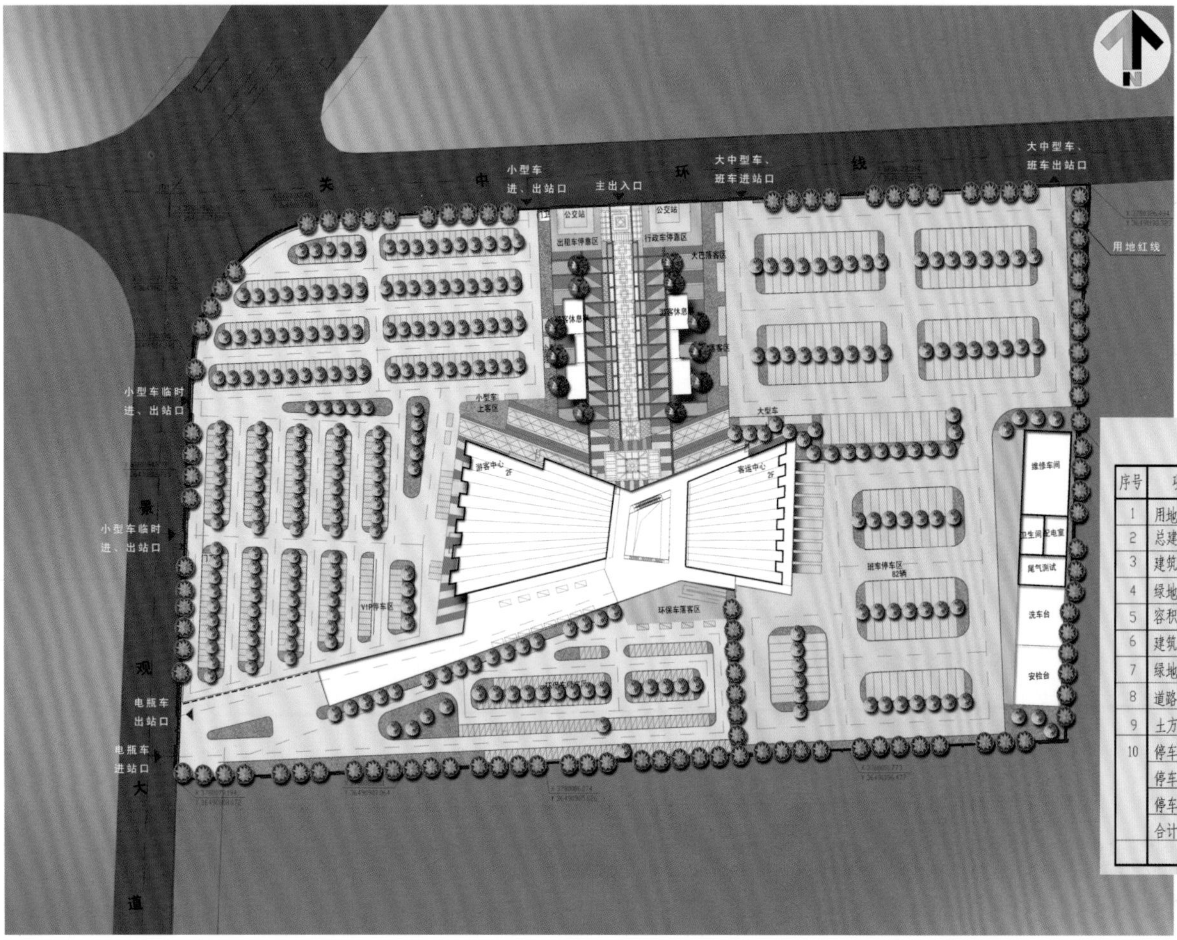

服务区主要技术经济指标

序号	项目	单位	数据	备注
1	用地面积	hm^2	10.9153	合163.73亩
2	总建筑面积	m^2	10132	
3	建筑基底总面积	m^2	7221	
4	绿地总面积	m^2	14000	
5	容积率		0.09	
6	建筑密度	%	6.62	
7	绿地率	%	12.83	
8	道路及停车场面积	m^2	58000	
9	土方工程量填/挖	m^3	125560/674630	
10	停车泊位 (3.5X18~22)	辆	88	
	停车泊位 (3.5X10~13)	辆	90	
	停车泊位 (3X6)	辆	55	
	合计	辆	233	

Site Plan 总平面图

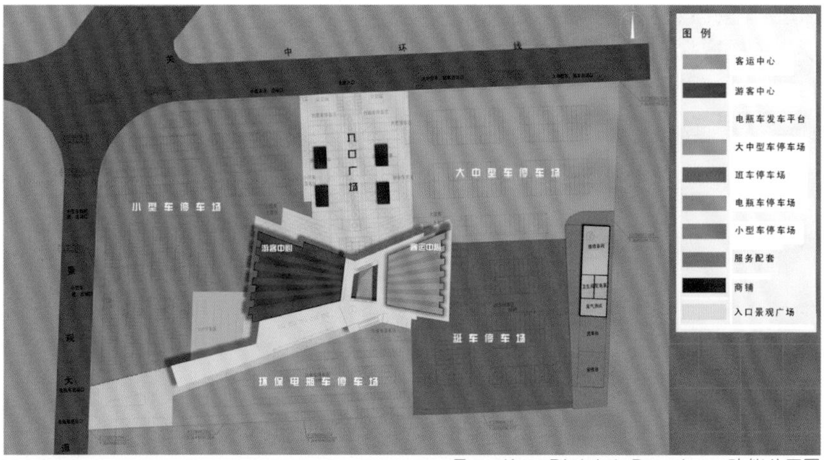

Function District Drawing 功能分区图

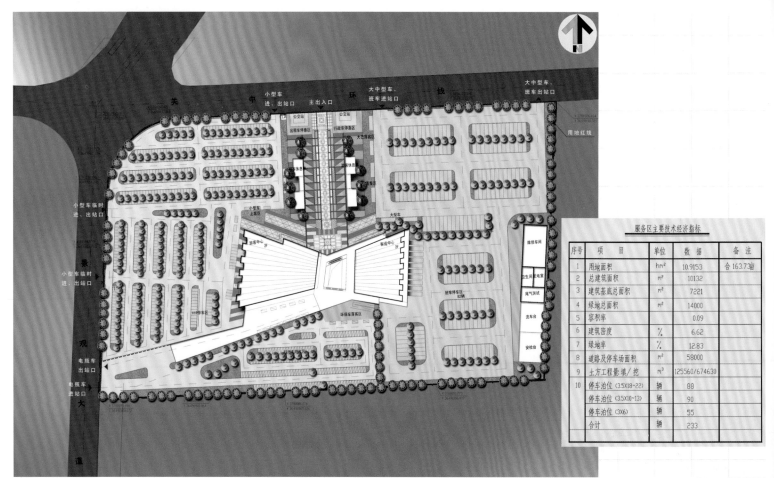

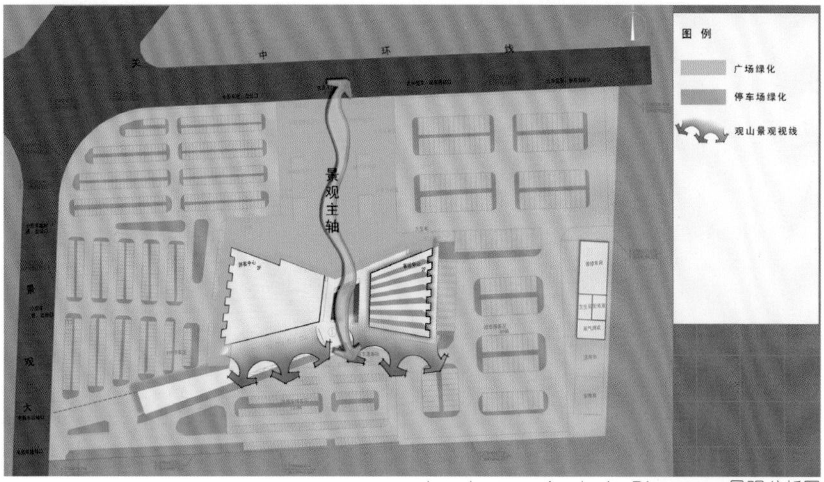

Landscape Analysis Diagram 景观分析图

■ Traffic Streamline Design

The convenient traffic environment brings main stream of people for the comprehensive tourist center; for the sake of the in-and-out of large stream of people, visitors' suspended stay and safety factor, the project sets the main entrances and exits on Guanzhong Loop Road in the north, and designs a large green square in the main entrance part. The square is not only the main pedestrian and traffic space, but also the defined and linking transition space of the each functional area of the tourist service center and other public space, and an important part of landscape space system of the visitor center.

The planning sets small vehicle parking lots on the west side of the visitor center, which form two secondary entrances and exits together with the storage battery car parking platform in Taibai landscape avenue, facilitating for the tourists to arrive or depart form Taibai Mountain Scenic Spot. Large and medium-sized passenger car and tour bus parking lots are on the east side of the visitor center with independent entrances and exits, and is quick and easy for the visitors to transfer and avoid the cross interference with the traffic streamline of the scenic spot.

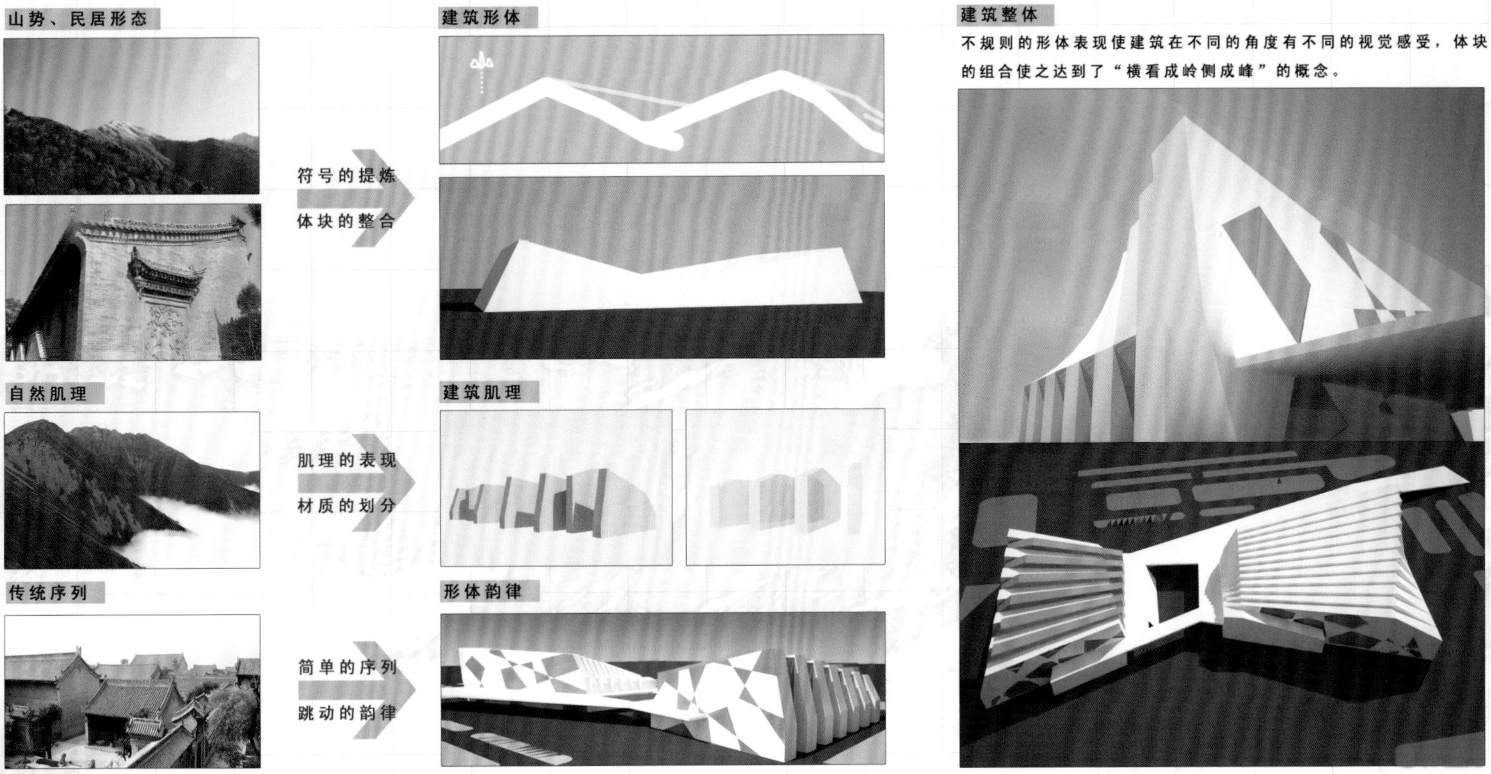

Design Concept 项目设计理念

■ 交通流线设计

基地北侧交通环境的便利为综合游客中心带来主要的人流，考虑到大量人流的出入和游客的暂缓停留行为以及安全因素，项目将主要出入口设置在北侧关中环线道路上，并在主入口部分设计了大型绿化广场。该广场不仅是主要的人流交通空间，还是对游客服务中心内部各功能领域与其他公共空间进行界定与衔接的过渡空间，同时也是游客中心景观空间体系的重要组成部分。

规划中将小型车辆的停车场布置在游客中心的西侧，与电瓶车的停车平台在太白景观大道上形成两个次要出入口，利于游客在太白山景点的往返出发。大中型客车与旅游班车停车场位于游客中心的东侧，设置独立出入口，易于游客快速便捷地换乘，避免与景区交通流线的交叉干扰。

New Loja High-speed Railway Station, Spain
西班牙洛哈新高铁站

Keywords 关键词

Large Deck
巨型平台

Cylindrical Supports
圆柱形支架

Large Facade Design
大立面设计

Location: Granada, Spain
Developer: ADIF (Administrador de Infraestructuras Ferroviarias)
Architedural Design: OPTA Arquitectos
Team: Juan Ramírez Sanz, Angel Sanz Rueda, Leticia Izquierdo Lahuerta, Daniel Vázquez Míguez, Belén Butragueño Díaz-Guerra
Total Floor Area: 20,116.67 m^2

项目地点：西班牙格拉纳达
开 发 商：ADIF (Administrador de Infraestructuras Ferroviarias)
建筑设计：西班牙OPTA建筑师事务所
团　　队：Juan Ramírez Sanz, Angel Sanz Rueda, Leticia Izquierdo Lahuerta, Daniel Vázquez Míguez, Belén Butragueño Díaz-Guerra
总建筑面积：20 116.67 m^2

Features 项目亮点

The protection requirements of traveler's flows through the platforms and parking canopies led to a global covering concept that unifies the complex image. Thus, the station extends its volume, creating an "apparent" large facade that helps to identify the station easily from afar.

为保护经过站台和停车场的客流，形成与建筑形象统一的遮盖物。车站看上去像是其体量延伸创造的一个“明显的”大立面，即使离车站很远，也能轻松识别。

■ Overview

The project consists of a high-speed railway station situated on the line Antequera to Granada, located near the town of Loja, in southern Spain.

The new railway line runs along the Sierra de Loja, mid slope, through pasture lands. Given a fixed position to the railways and the access road, the project had to deal with both boundaries. The station is located in parallel to the tracks, on a lower bound due to the difference in level between the existing natural ground and the new railway platform. The wild environment, clear and slightly away from the city center allows a gradual approach to the station from its link to the motorway (A-42) nearby.

■ 项目概况

该项目是一个位于安特克拉——格拉纳达线上的高铁车站，靠近西班牙南部小镇——洛哈。

新的铁路线沿洛哈山展开，缓坡中前行，穿过牧场。车站与轨道平行，鉴于地面与新站台的高度差，车站略低。车站远离市中心，在自然开阔的环境下以渐进的方式与附近的高速公路（A-42）联系起来。

■ Design Concept

Its condition of intermediate station, involved specific requirements, associated with a much more reduced functional programme than terminus stations. The proposal therefore considers two apparently opposite constraints:

– to ensure value for money, projecting a relatively modest and small passenger building.

– to achieve visibility and integration of the entire complex at the given location, in a wild landscape of great environmental value, with the hills skyline behind.

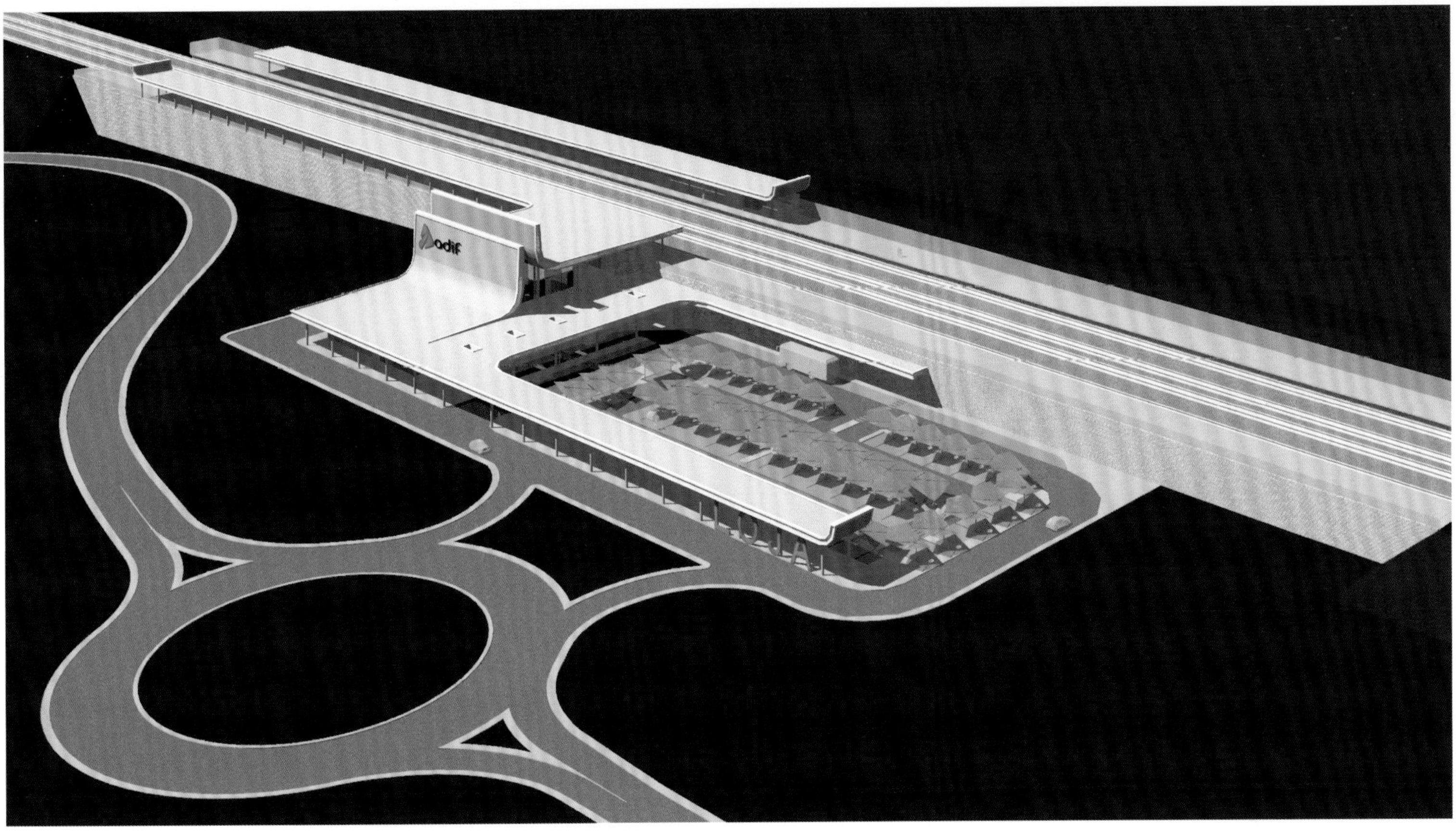

■ 设计理念

作为中间站，它涉及的具体要求与功能方案比总站要简洁。因此，该项目设计考虑了两个明显对立的约束条件：

——确保物有所值，展现一个相对稳重小体量的客运大楼；

——在给定的位置实现整个项目的可视性和一体化，充分利用周边自然景观价值与后方的群山天际线。

LOCATION Given a pre-fixed position to the Railway and the access road the project had to deal between both "boundaries"

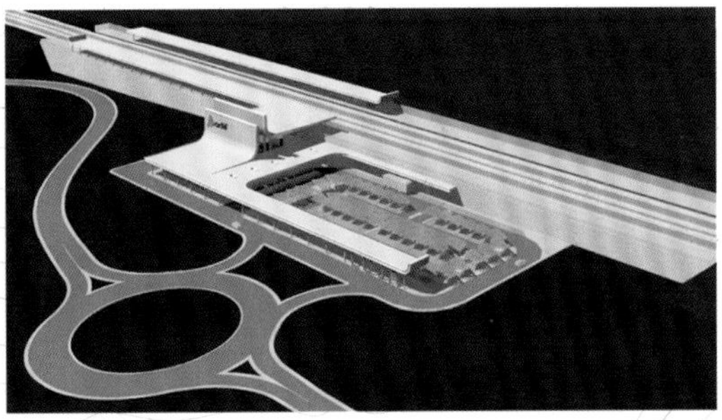

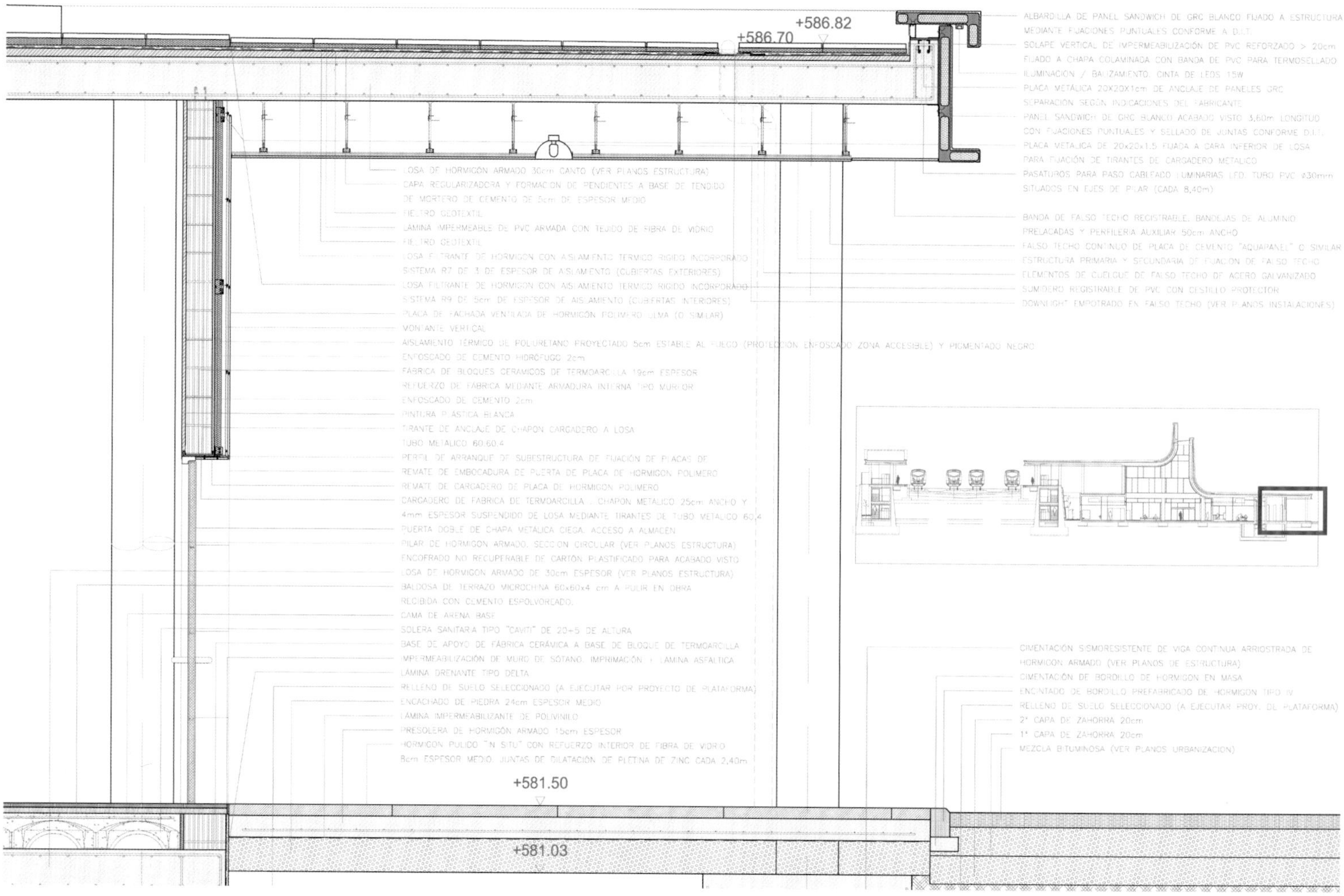
+586.82
+586.70
LOSA DE HORMIGÓN ARMADO 30cm CANTO (VER PLANOS ESTRUCTURA)
CAPA REGULARIZADORA Y FORMACION DE PENDIENTES A BASE DE TENDIDO
DE MORTERO DE CEMENTO DE 5cm DE ESPESOR MEDIO
FIELTRO GEOTEXTIL
LAMINA IMPERMEABLE DE PVC ARMADA CON TEJIDO DE FIBRA DE VIDRIO
FIELTRO GEOTEXTIL
LOSA FILTRANTE DE HORMIGON CON AISLAMIENTO TERMICO RIGIDO INCORPORADO
SISTEMA R7 DE 3 DE ESPESOR DE AISLAMIENTO (CUBIERTAS EXTERIORES)
LOSA FILTRANTE DE HORMIGON CON AISLAMIENTO TERMICO RIGIDO INCORPORADO
SISTEMA R9 DE 5cm DE ESPESOR DE AISLAMIENTO (CUBIERTAS INTERIORES)
PLACA DE FACHADA VENTILADA DE HORMIGÓN POLIMERO ULMA (O SIMILAR)
MONTANTE VERTICAL
AISLAMIENTO TÉRMICO DE POLIURETANO PROYECTADO 5cm ESTABLE AL FUEGO (PROTECCIÓN ENFOSCADO ZONA ACCESIBLE) Y PIGMENTADO NEGRO
ENFOSCADO DE CEMENTO HIDROFUGO 2cm
FÁBRICA DE BLOQUES CERAMICOS DE TERMOARCILLA 19cm ESPESOR
REFUERZO DE FÁBRICA MEDIANTE ARMADURA INTERNA TIPO MURFOR
ENFOSCADO DE CEMENTO 2cm
PINTURA PLÁSTICA BLANCA
TIRANTE DE ANCLAJE DE CHAPON CARGADERO A LOSA
TUBO METALICO 60.60.4
PERFIL DE ARRANQUE DE SUBESTRUCTURA DE FIJACIÓN DE PLACAS DE
REMATE DE EMBOCADURA DE PUERTA DE PLACA DE HORMIGON POLIMERO
REMATE DE CARGADERO DE PLACA DE HORMIGON POLIMERO
CARGADERO DE FABRICA DE TERMOARCILLA , CHAPON METALICO 25cm ANCHO Y
4mm ESPESOR SUSPENDIDO DE LOSA MEDIANTE TIRANTES DE TUBO METALICO 60.4
PUERTA DOBLE DE CHAPA METÁLICA CIEGA. ACCESO A ALMACEN
PILAR DE HORMIGON ARMADO. SECCION CIRCULAR (VER PLANOS ESTRUCTURA)
ENCOFRADO NO RECUPERABLE DE CARTÓN PLASTIFICADO PARA ACABADO VISTO
LOSA DE HORMIGON ARMADO DE 30cm ESPESOR (VER PLANOS ESTRUCTURA)
BALDOSA DE TERRAZO MICROCHINA 60x60x4 cm A PULIR EN OBRA
RECIBIDA CON CEMENTO ESPOLVOREADO.
CAMA DE ARENA BASE
SOLERA SANITARIA TIPO "CAVITI" DE 20+5 DE ALTURA
BASE DE APOYO DE FÁBRICA CERÁMICA A BASE DE BLOQUE DE TERMOARCILLA
IMPERMEABILIZACIÓN DE MURO DE SÓTANO. IMPRIMACIÓN + LAMINA ASFALTICA
LÁMINA DRENANTE TIPO DELTA
RELLENO DE SUELO SELECCIONADO (A EJECUTAR POR PROYECTO DE PLATAFORMA)
ENCACHADO DE PIEDRA 24cm ESPESOR MEDIO
LÁMINA IMPERMEABILIZANTE DE POLIVINILO
PRESOLERA DE HORMIGÓN ARMADO 15cm ESPESOR
HORMIGON PULIDO "IN SITU" CON REFUERZO INTERIOR DE FIBRA DE VIDRIO
8cm ESPESOR MEDIO. JUNTAS DE DILATACIÓN DE PLETINA DE ZINC CADA 2,40m
+581.50
+581.03
ALBARDILLA DE PANEL SANDWICH DE GRC BLANCO FIJADO A ESTRUCTURA
MEDIANTE FIJACIONES PUNTUALES CONFORME A D.I.T.
SOLAPE VERTICAL DE IMPERMEABILIZACIÓN DE PVC REFORZADO > 20cm
FIJADO A CHAPA COLAMINADA CON BANDA DE PVC PARA TERMOSELLADO
ILUMINACION / BALIZAMIENTO. CINTA DE LEDS 15W
PLACA METÁLICA 20X20X1cm DE ANCLAJE DE PANELES GRC
SEPARACION SEGÚN INDICACIONES DEL FABRICANTE
PANEL SANDWICH DE GRC BLANCO ACABADO VISTO 3,60m LONGITUD
CON FIJACIONES PUNTUALES Y SELLADO DE JUNTAS CONFORME D.I.T.
PLACA METALICA DE 20x20x1.5 FIJADA A CARA INFERIOR DE LOSA
PARA FIJACIÓN DE TIRANTES DE CARGADERO METALICO
PASATUBOS PARA PASO CABLEADO LUMINARIAS LED. TUBO PVC ø30mm
SITUADOS EN EJES DE PILAR (CADA 8,40m)
BANDA DE FALSO TECHO REGISTRABLE. BANDEJAS DE ALUMINIO
PRELACADAS Y PERFILERIA AUXILIAR 50cm ANCHO
FALSO TECHO CONTINUO DE PLACA DE CEMENTO "AQUAPANEL" O SIMILAR
ESTRUCTURA PRIMARIA Y SECUNDARIA DE FIJACION DE FALSO TECHO
ELEMENTOS DE CUELGUE DE FALSO TECHO DE ACERO GALVANIZADO
SUMIDERO REGISTRABLE DE PVC CON CESTILLO PROTECTOR
DOWNLIGHT EMPOTRADO EN FALSO TECHO (VER PLANOS INSTALACIONES)
CIMENTACIÓN SISMORESISTENTE DE VIGA CONTINUA ARRIOSTRADA DE
HORMIGON ARMADO (VER PLANOS DE ESTRUCTURA)
CIMENTACIÓN DE BORDILLO DE HORMIGON EN MASA
ENCINTADO DE BORDILLO PREFABRICADO DE HORMIGON TIPO IV
RELLENO DE SUELO SELECCIONADO (A EJECUTAR PROY. DE PLATAFORMA)
2ª CAPA DE ZAHORRA 20cm
1ª CAPA DE ZAHORRA 20cm
MEZCLA BITUMINOSA (VER PLANOS URBANIZACION)

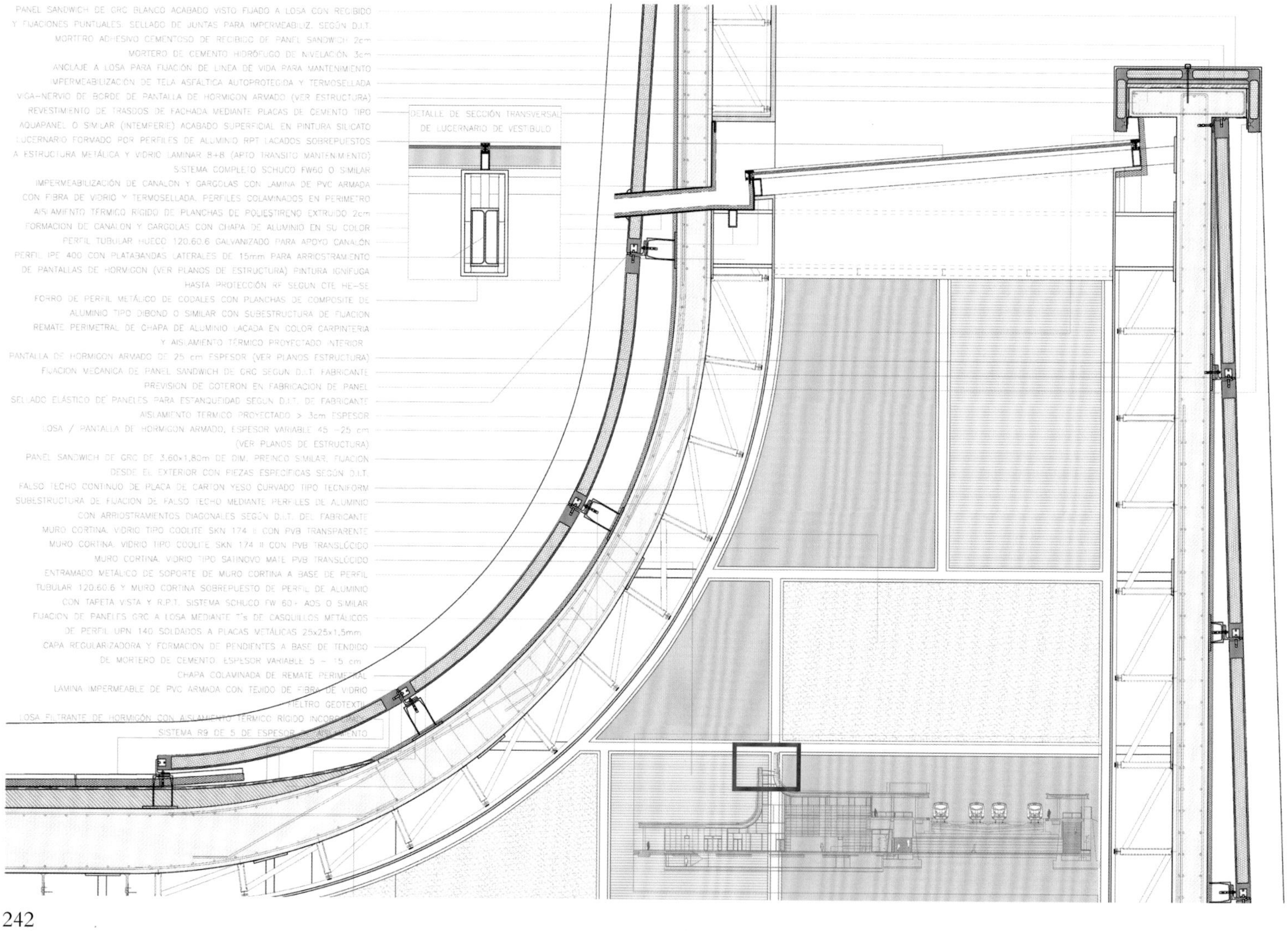
PANEL SANDWICH DE GRC BLANCO ACABADO VISTO FIJADO A LOSA CON RECIBIDO
Y FIJACIONES PUNTUALES. SELLADO DE JUNTAS PARA IMPERMEABILIZ. SEGÚN D.I.T.
MORTERO ADHESIVO CEMENTOSO DE RECIBIDO DE PANEL SANDWICH 2cm
MORTERO DE CEMENTO HIDRÓFUGO DE NIVELACIÓN 3cm
ANCLAJE A LOSA PARA FIJACIÓN DE LINEA DE VIDA PARA MANTENIMIENTO
IMPERMEABILIZACIÓN DE TELA ASFÁLTICA AUTOPROTEGIDA Y TERMOSELLADA
VIGA-NERVIO DE BORDE DE PANTALLA DE HORMIGON ARMADO (VER ESTRUCTURA)
REVESTIMIENTO DE TRASDOS DE FACHADA MEDIANTE PLACAS DE CEMENTO TIPO
AQUAPANEL O SIMILAR (INTEMPERIE) ACABADO SUPERFICIAL EN PINTURA SILICATO
LUCERNARIO FORMADO POR PERFILES DE ALUMINIO RPT LACADOS SOBREPUESTOS
A ESTRUCTURA METÁLICA Y VIDRIO LAMINAR 8+8 (APTO TRANSITO MANTENIMIENTO)
SISTEMA COMPLETO SCHUCO FW60 O SIMILAR
IMPERMEABILIZACIÓN DE CANALON Y GARGOLAS CON LAMINA DE PVC ARMADA
CON FIBRA DE VIDRIO Y TERMOSELLADA. PERFILES COLAMINADOS EN PERIMETRO
AISLAMIENTO TÉRMICO RÍGIDO DE PLANCHAS DE POLIESTIRENO EXTRUIDO 2cm
FORMACION DE CANALON Y GARGOLAS CON CHAPA DE ALUMINIO EN SU COLOR
PERFIL TUBULAR HUECO 120.60.6 GALVANIZADO PARA APOYO CANALÓN
PERFIL IPE 400 CON PLATABANDAS LATERALES DE 15mm PARA ARRIOSTRAMIENTO
DE PANTALLAS DE HORMIGON (VER PLANOS DE ESTRUCTURA) PINTURA IGNÍFUGA
HASTA PROTECCIÓN RF SEGUN CTE HE-SE
FORRO DE PERFIL METÁLICO DE CODALES CON PLANCHAS DE COMPOSITE DE
ALUMINIO TIPO DIBOND O SIMILAR CON SUBESTRUCTURA DE FIJACION
REMATE PERIMETRAL DE CHAPA DE ALUMINIO LACADA EN COLOR CARPINTERIA
Y AISLAMIENTO TÉRMICO PROYECTADO INTERIOR
PANTALLA DE HORMIGON ARMADO DE 25 cm ESPESOR (VER PLANOS ESTRUCTURA)
FIJACION MECANICA DE PANEL SANDWICH DE GRC SEGUN D.I.T. FABRICANTE
PREVISIÓN DE GOTERON EN FABRICACION DE PANEL
SELLADO ELÁSTICO DE PANELES PARA ESTANQUEIDAD SEGUN D.I.T. DE FABRICANTE
AISLAMIENTO TERMICO PROYECTADO > 3cm ESPESOR
LOSA / PANTALLA DE HORMIGON ARMADO, ESPESOR VARIABLE 45 -25 cm
(VER PLANOS DE ESTRUCTURA)
PANEL SANDWICH DE GRC DE 3,60x1,80m DE DIM. PREINCO SIMILAR. FIJACION
DESDE EL EXTERIOR CON PIEZAS ESPECÍFICAS SEGÚN D.I.T.
FALSO TECHO CONTINUO DE PLACA DE CARTON YESO CURVADO TIPO TECNIFORM
SUBESTRUCTURA DE FIJACION DE FALSO TECHO MEDIANTE PERFILES DE ALUMINIO
CON ARRIOSTRAMIENTOS DIAGONALES SEGÚN D.I.T. DEL FABRICANTE
MURO CORTINA. VIDRIO TIPO COOLITE SKN 174 II CON PVB TRANSPARENTE
MURO CORTINA. VIDRIO TIPO COOLITE SKN 174 II CON PVB TRANSLÚCIDO
MURO CORTINA. VIDRIO TIPO SATINOVO MATE PVB TRANSLÚCIDO
ENTRAMADO METÁLICO DE SOPORTE DE MURO CORTINA A BASE DE PERFIL
TUBULAR 120.60.6 Y MURO CORTINA SOBREPUESTO DE PERFIL DE ALUMINIO
CON TAPETA VISTA Y R.P.T. SISTEMA SCHUCO FW 60+ AOS O SIMILAR
FIJACION DE PANELES GRC A LOSA MEDIANTE T's DE CASQUILLOS METÁLICOS
DE PERFIL UPN 140 SOLDADOS A PLACAS METÁLICAS 25x25x1,5mm
CAPA REGULARIZADORA Y FORMACION DE PENDIENTES A BASE DE TENDIDO
DE MORTERO DE CEMENTO. ESPESOR VARIABLE 5 - 15 cm
CHAPA COLAMINADA DE REMATE PERIMETRAL
LAMINA IMPERMEABLE DE PVC ARMADA CON TEJIDO DE FIBRA DE VIDRIO
FIELTRO GEOTEXTIL
LOSA FILTRANTE DE HORMIGÓN CON AISLAMIENTO TÉRMICO RÍGIDO INCORPORADO
SISTEMA R9 DE 5 DE ESPESOR DE AISLAMIENTO
DETALLE DE SECCIÓN TRANSVERSAL
DE LUCERNARIO DE VESTIBULO

WINTER
1. Winter sun through glazings
2. Underfloor radiant heating
3. Air Treatment Unit (Ventilation) with heating recovery
4. A.C. heating reinforcement (extremely cold days)
5. Extra thermal isolation of the inhabitable space

SUMMER
1. Large shadowed areas, no direct sun through glazings
2. Underfloor radiant cooling
3. Air Treatment Unit (Ventilation) with cooling recovery
4. Night natural free-cooling in main lobby
5. A.C. cooling reinforcement (extremely hot days)
6. Extra thermal isolation of the inhabitable space

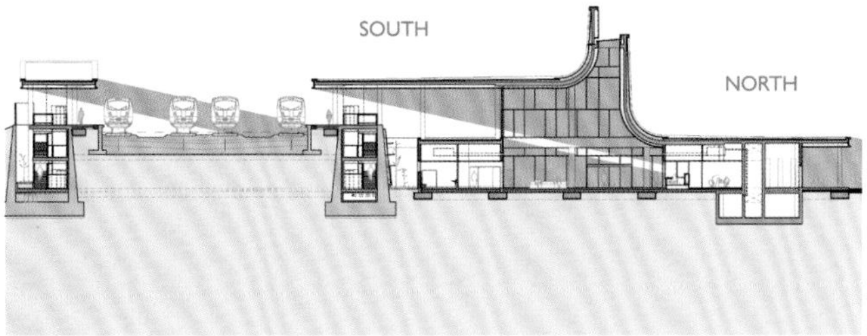

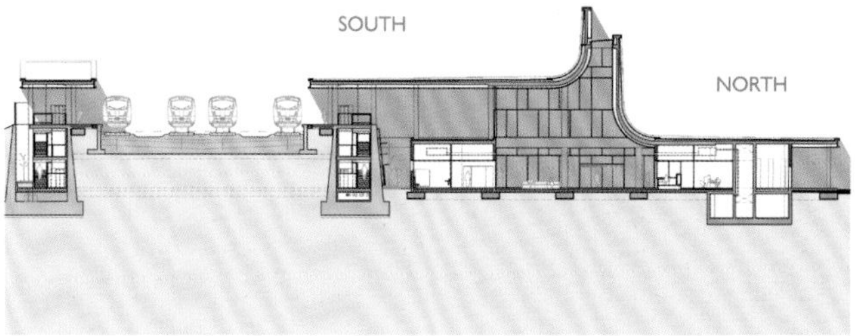

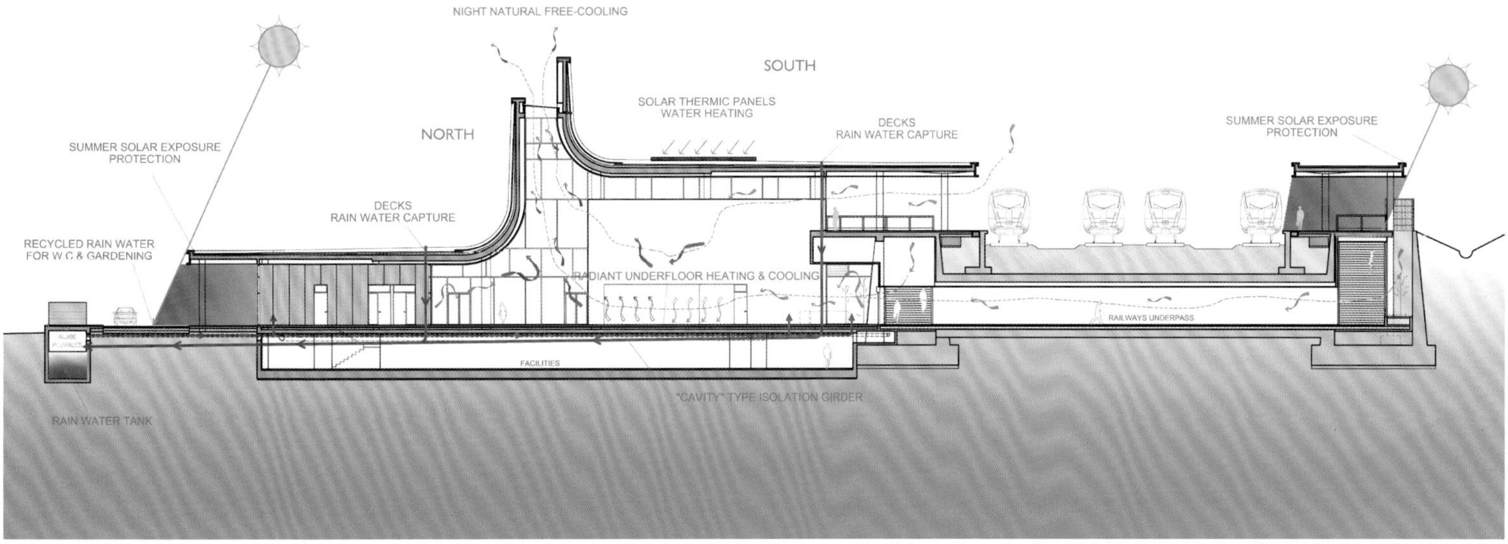

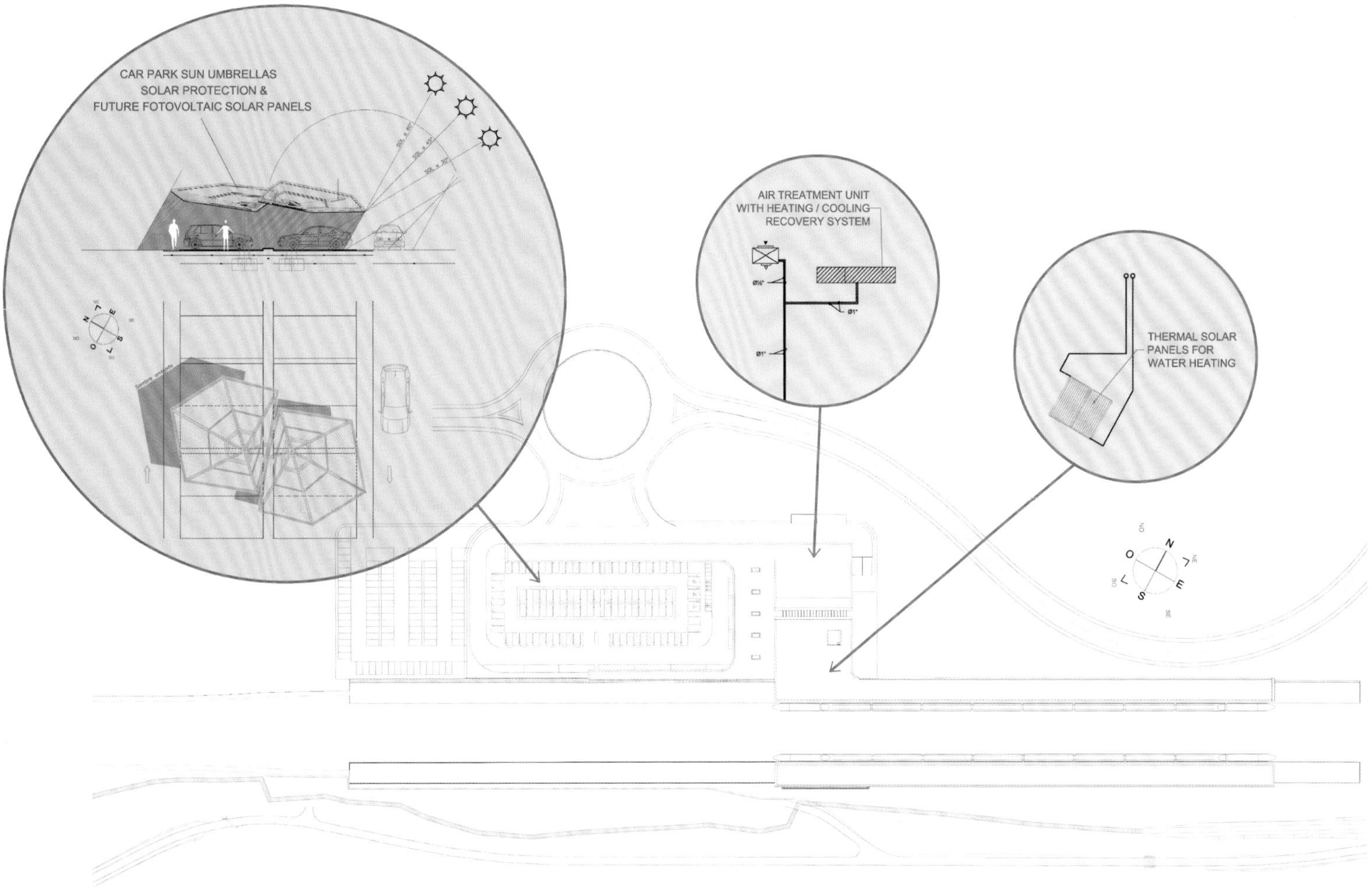

All facilities connections are made through an underground corridor, (accessible for maintenance) and vertical ducts to each area.

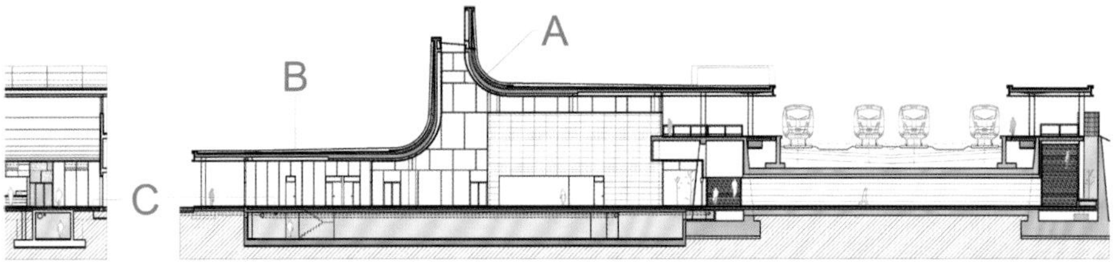

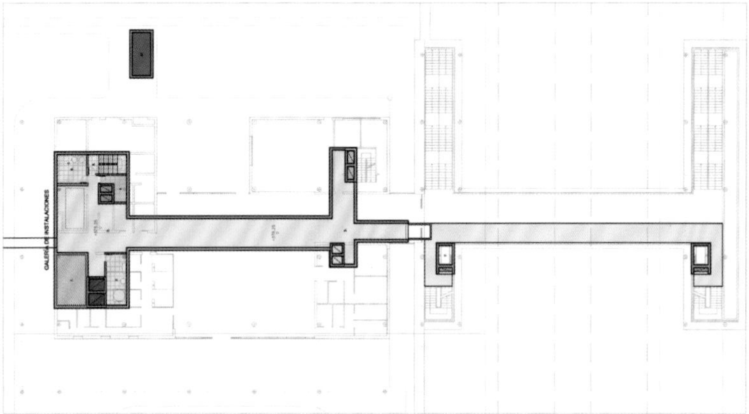

- Undergroung facilities corridor
- Water tanks
- Tickets office facilities connections
- Premises facilities connections
- Toilets facilities connections
- Ground level facilities connections
- Platform facilities connections
- Future needs
- Power supply

A- Curved deck with GRC panels, thermal isolation, waterproof sealing, concrete slab and indoor acoustic continuous ceiling

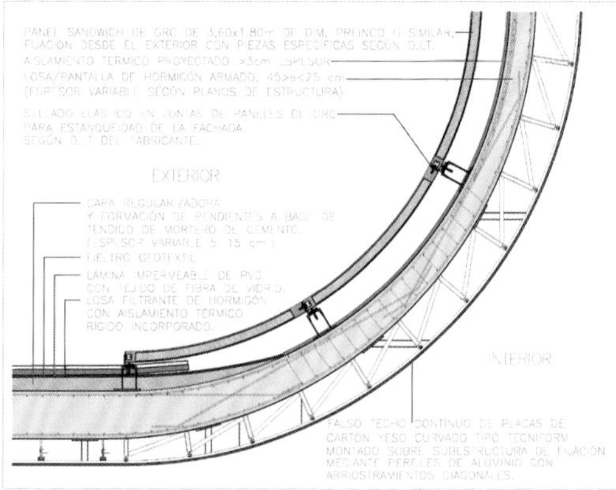

B- Flat deck with porous concrete tiles, thermal isolation, waterproof sealing, concrete slab and indoor acoustic continuous ceiling

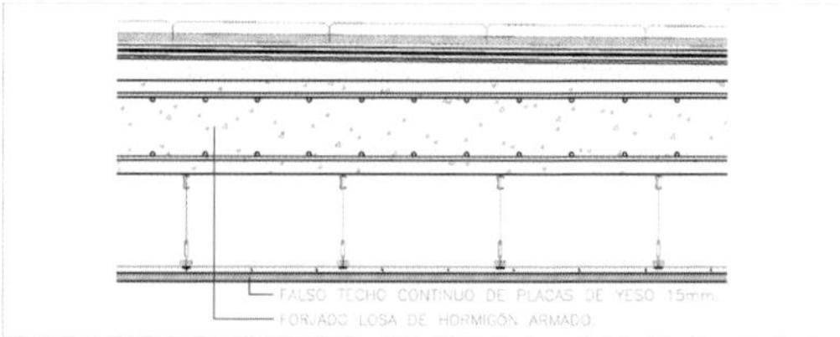

C- Cavity type girder with ceramic tiles, cement base, underfloor radiant system, thermal isolation, cavity system and girder

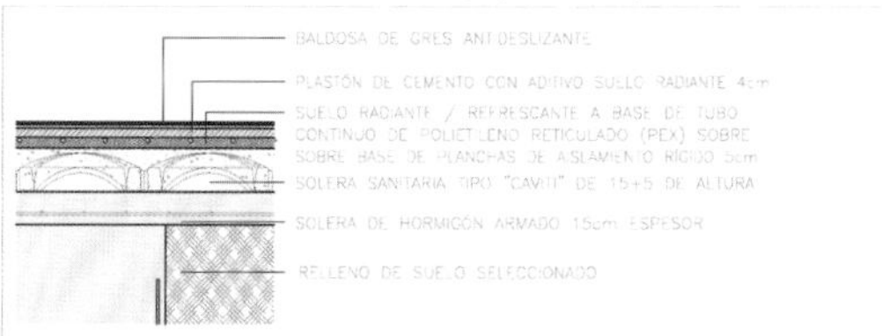

Seismic-resistant structure construction sequence.

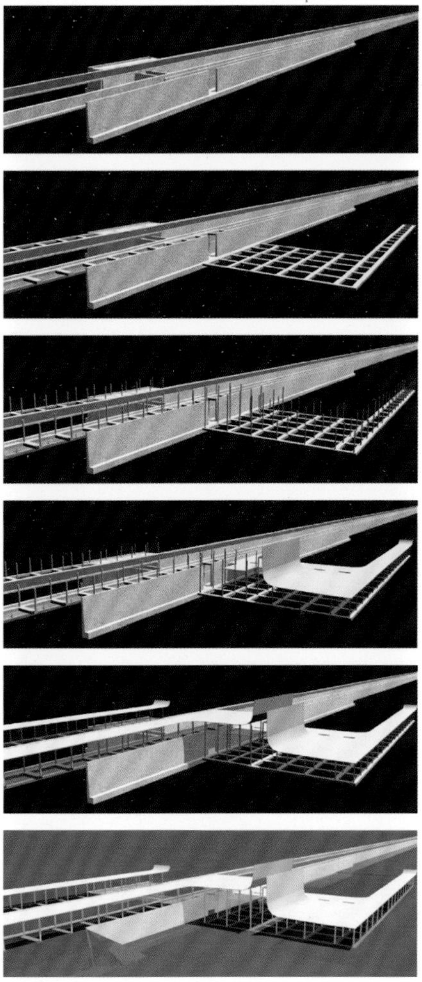

Loja (Granada) has the highest rate of Seismic activity in Spain.
This involved that the Station structure had to be designed and calculated to resist very severe seismic conditions.
We designed five especific structural elements:

1. Bidirectional foundations
2. Overdimensioned pilars
3. Anti-seismic "tubes"
4. Continous reinforced concrete slabs
5. Metalic reinforcements

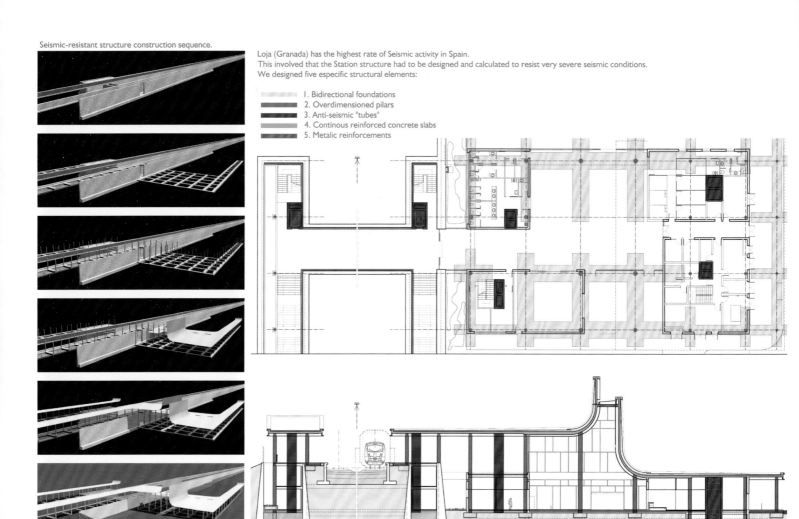

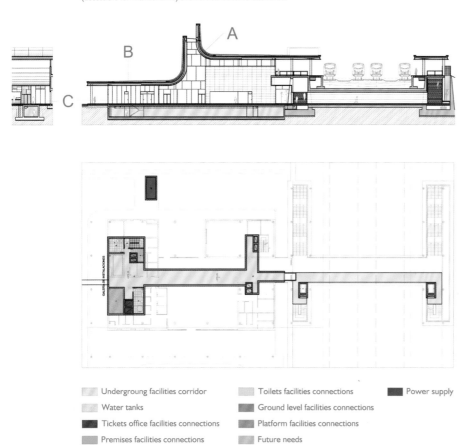

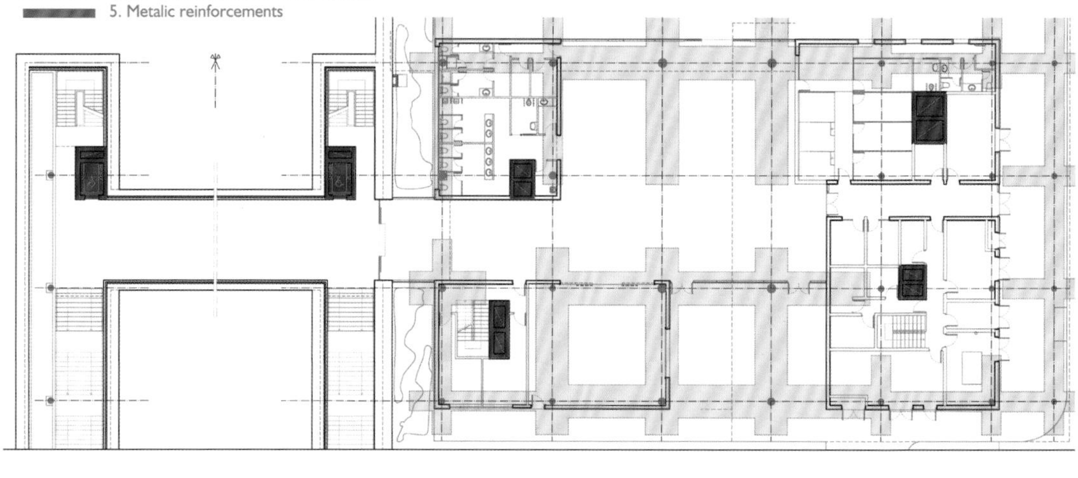

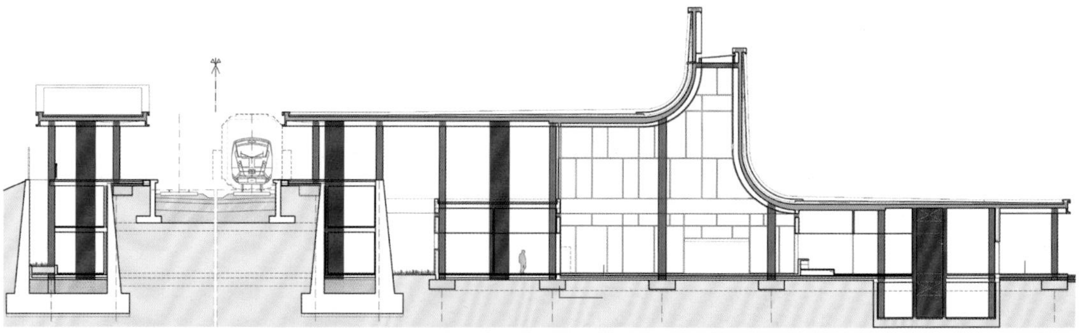

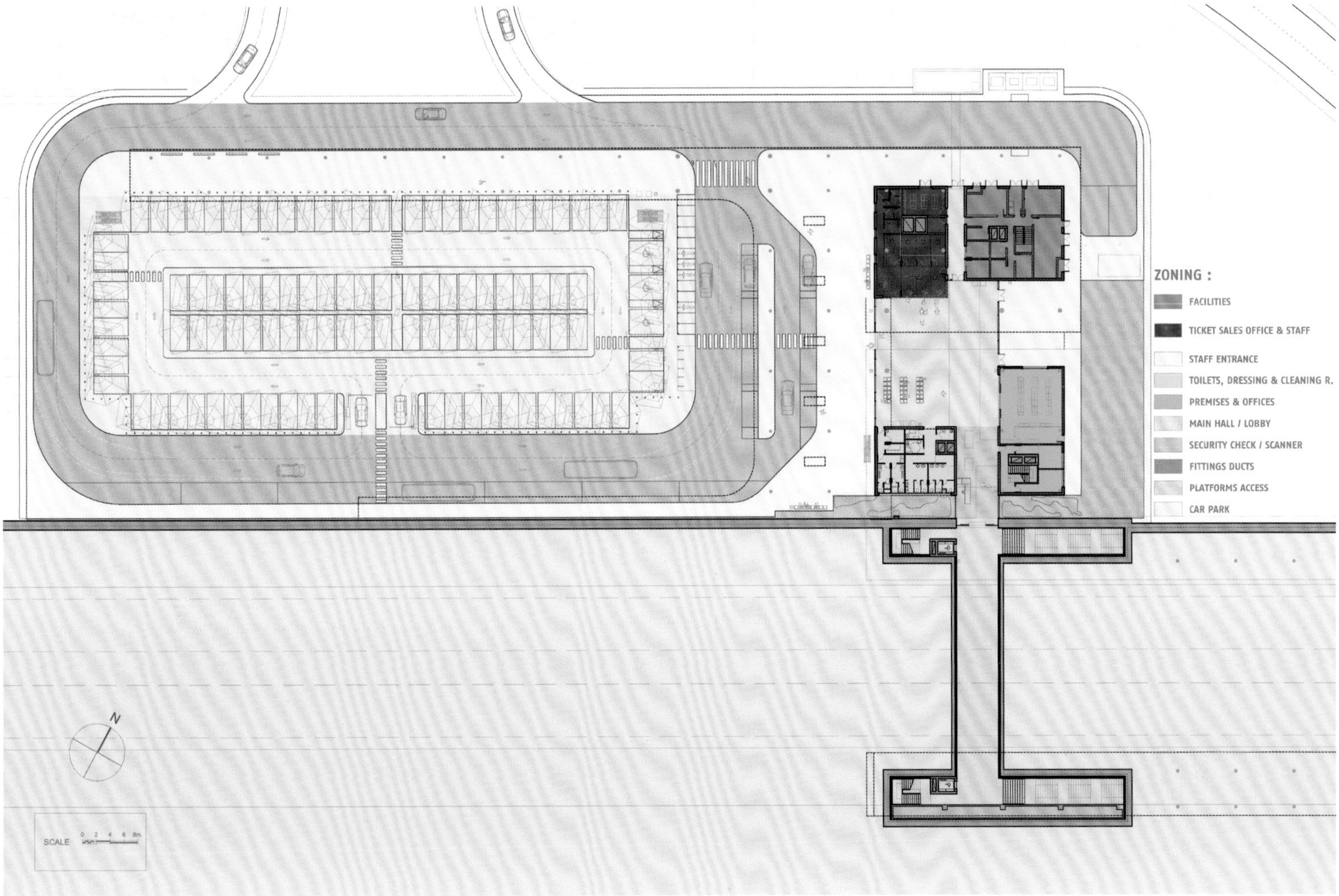
ZONING :
FACILITIES
TICKET SALES OFFICE & STAFF
STAFF ENTRANCE
TOILETS, DRESSING & CLEANING R.
PREMISES & OFFICES
MAIN HALL / LOBBY
SECURITY CHECK / SCANNER
FITTINGS DUCTS
PLATFORMS ACCESS
CAR PARK
N
SCALE

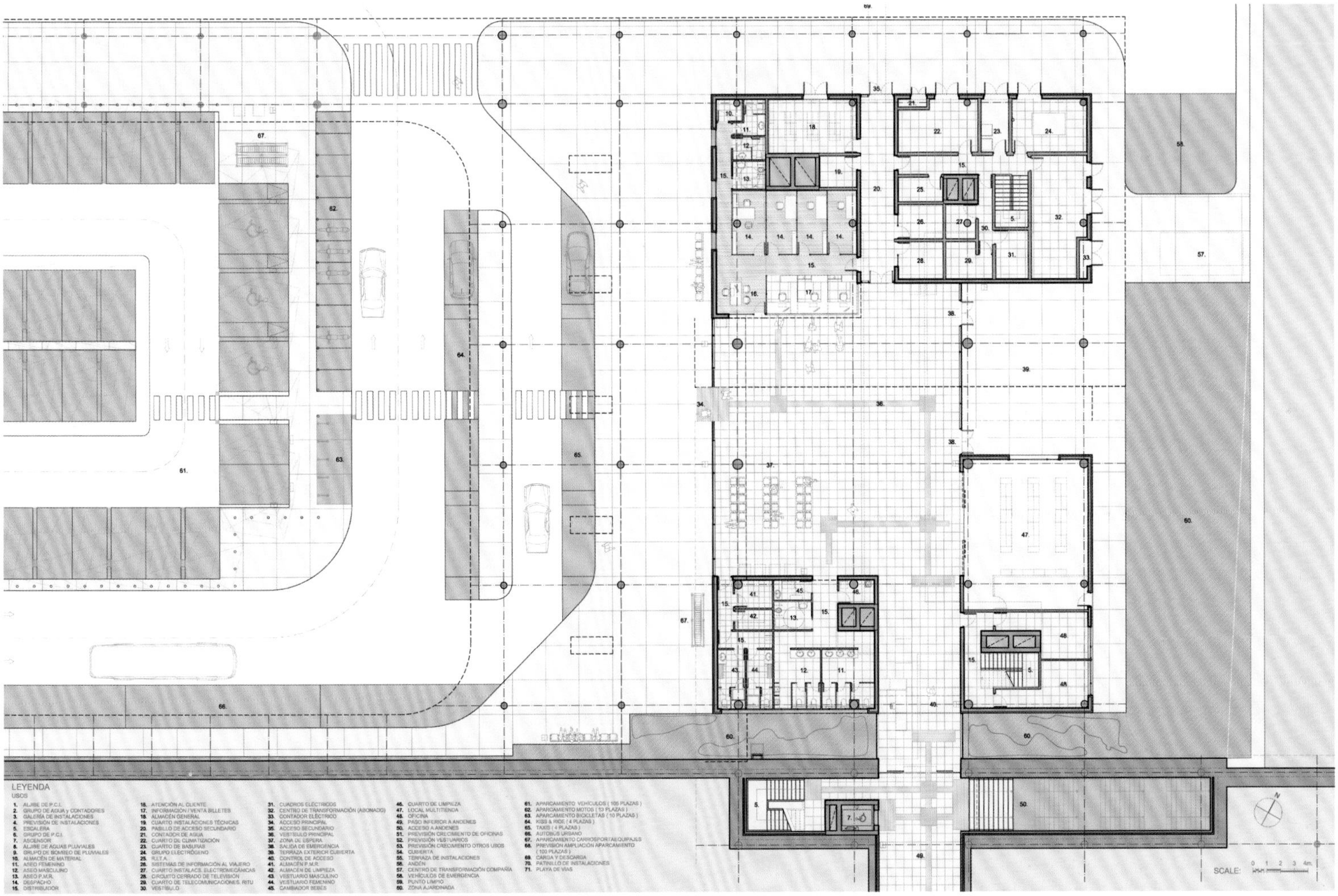
LEYENDA
USOS
1. ALJIBE DE P.C.I.
2. GRUPO DE AGUA y CONTADORES
3. GALERÍA DE INSTALACIONES
4. PREVISIÓN DE INSTALACIONES
5. ESCALERA
6. GRUPO DE P.C.I.
7. ASCENSOR
8. ALJIBE DE AGUAS PLUVIALES
9. GRUPO DE BOMBEO DE PLUVIALES
10. ALMACÉN DE MATERIAL
11. ASEO FEMENINO
12. ASEO MASCULINO
13. ASEO P.M.R.
14. DESPACHO
15. DISTRIBUIDOR
16. ATENCIÓN AL CLIENTE
17. INFORMACIÓN / VENTA BILLETES
18. ALMACÉN GENERAL
19. CUARTO INSTALACIONES TÉCNICAS
20. PASILLO DE ACCESO SECUNDARIO
21. CONTADOR DE AGUA
22. CUARTO DE CLIMATIZACIÓN
23. CUARTO DE BASURAS
24. GRUPO ELECTRÓGENO
25. R.I.T.A.
26. SISTEMAS DE INFORMACIÓN AL VIAJERO
27. CUARTO INSTALACS. ELECTROMECÁNICAS
28. CIRCUITO CERRADO DE TELEVISIÓN
29. CUARTO DE TELECOMUNICACIONES. RITU
30. VESTÍBULO
31. CUADROS ELÉCTRICOS
32. CENTRO DE TRANSFORMACIÓN (ABONADO)
33. CONTADOR ELÉCTRICO
34. ACCESO PRINCIPAL
35. ACCESO SECUNDARIO
36. VESTÍBULO PRINCIPAL
37. ZONA DE ESPERA
38. SALIDA DE EMERGENCIA
39. TERRAZA EXTERIOR CUBIERTA
40. CONTROL DE ACCESO
41. ALMACÉN P.M.R.
42. ALMACÉN DE LIMPIEZA
43. VESTUARIO MASCULINO
44. VESTUARIO FEMENINO
45. CAMBIADOR BEBÉS
46. CUARTO DE LIMPIEZA
47. LOCAL MULTITIENDA
48. OFICINA
49. PASO INFERIOR A ANDENES
50. ACCESO A ANDENES
51. PREVISIÓN CRECIMIENTO DE OFICINAS
52. PREVISIÓN VESTUARIOS
53. PREVISIÓN CRECIMIENTO OTROS USOS
54. CUBIERTA
55. TERRAZA DE INSTALACIONES
56. ANDÉN
57. CENTRO DE TRANSFORMACIÓN COMPAÑÍA
58. VEHÍCULOS DE EMERGENCIA
59. PUNTO LIMPIO
60. ZONA AJARDINADA
61. APARCAMIENTO VEHÍCULOS (105 PLAZAS)
62. APARCAMIENTO MOTOS (13 PLAZAS)
63. APARCAMIENTO BICICLETAS (10 PLAZAS)
64. KISS & RIDE (4 PLAZAS)
65. TAXIS (4 PLAZAS)
66. AUTOBÚS URBANO
67. APARCAMIENTO CARROSPORTAEQUIPAJES
68. PREVISIÓN AMPLIACIÓN APARCAMIENTO (100 PLAZAS)
69. CARGA Y DESCARGA
70. PATINILLO DE INSTALACIONES
71. PLAYA DE VIAS
N
SCALE

adif

adif
LOJA

adif

■ Architectural Design

The protection requirements of traveler's flows through the platforms and parking canopies led to a global covering concept that unifies the complex image. Thus, the station extends its volume, creating an "apparent" large facade that helps to identify the station easily from afar. The result is a large "active" deck that provides large areas of shade and protection.

This roof that embraces the complex folds at its ends as a leaf blowing in the wind or a flying carpet, with luminescent edges that can be seen from a distance, offering a clear vision of unity, both day and night, as an interlaced calligraphic gesture in the landscape. The main building is located next to the tracks underpass, as a nexus between the two levels of passenger's traffic: top (trains), bottom (vehicles). One single main lobby serves four "boxes" where the main services are grouped together in categories. The two deck levels encounter at the lobby in a vertical extension that generates a skylight to bring in natural northern light into the station. These vertical folds serve as a distant facade of the station and structurally act as large beams that free the lobby from pillars. The rest of the structure is solved with a regular grid of cylindrical supports and anti-seismic "tubes" hidden in the interior of the station that also fulfill the function of facilities holders, which are accessible via an underground tunnel. Loja has the highest rate of seismic risk in Spain.

The large deck is also considered a "Haima" that protects the station from extreme solar exposure of southern Spanish sun. Therefore, the exterior walls of the station are set back, so that the roof cantilevered perimeter ensures proper summer shading in order to avoid inside overheating. This design concept guarantees a lower energy demand to the building's air conditioning system, which is the main power consumption equipment in this station. An open air car park, with sun umbrellas designed to hold photovoltaic panels in the future, is placed besides the main entrance. Also several platforms and lanes for public transport provide fast connections for passengers. The entire complex is fully accessible to handicapped.

■ 建筑设计

为保护经过站台和停车场的客流，形成与建筑形象统一的遮盖物。车站看上去像是其体量延伸创造的一个“明显的”大立面，即使离车站很远，也能轻松识别。总之，它是一个“活动的”平台，可提供大面积遮阳和防护。

建筑的屋顶抱住体量，形态如和风绿叶或飞毯，发光的边缘从远处就能看到。主楼位于地下铁轨旁，连接两个层次的旅客运输：顶部（火车）、底部（车辆）。一个单一的主大厅供应四个“盒子”——主要服务集中归类的地方。双层平台引至大堂，略下垂形成南向天窗，将自然光引入车站。这些褶皱作为建筑外观一部分，可远观，还能起到梁、桁作用，免去了柱状物的使用。该结构剩余部分使用常用圆柱形支架，抗震的“管子”隐藏在车站内部，通过地下隧道可到达。洛哈是西班牙地震风险最高的地方。

巨型平台可以阻挡烈阳。车站外墙退后设置，悬挑的屋面也可适当地遮挡夏阳，以避免内部过热。这样的设计降低了车站主要电力消耗设备——空气调节系统的能源需求。露天停车场布置在正门旁，上盖覆有光伏面板的“阳伞”，也设计了几个平台和通道方便公共交通快速连接。此外，整个建筑设计也体现了对残疾人的关爱。

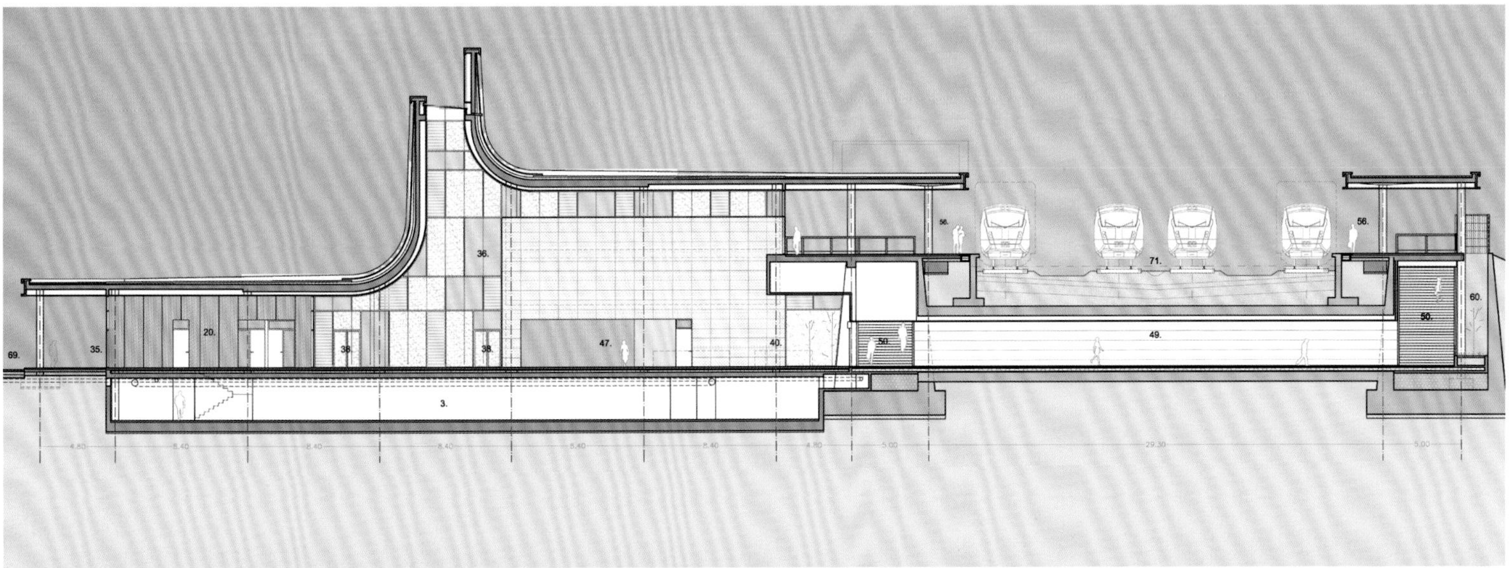

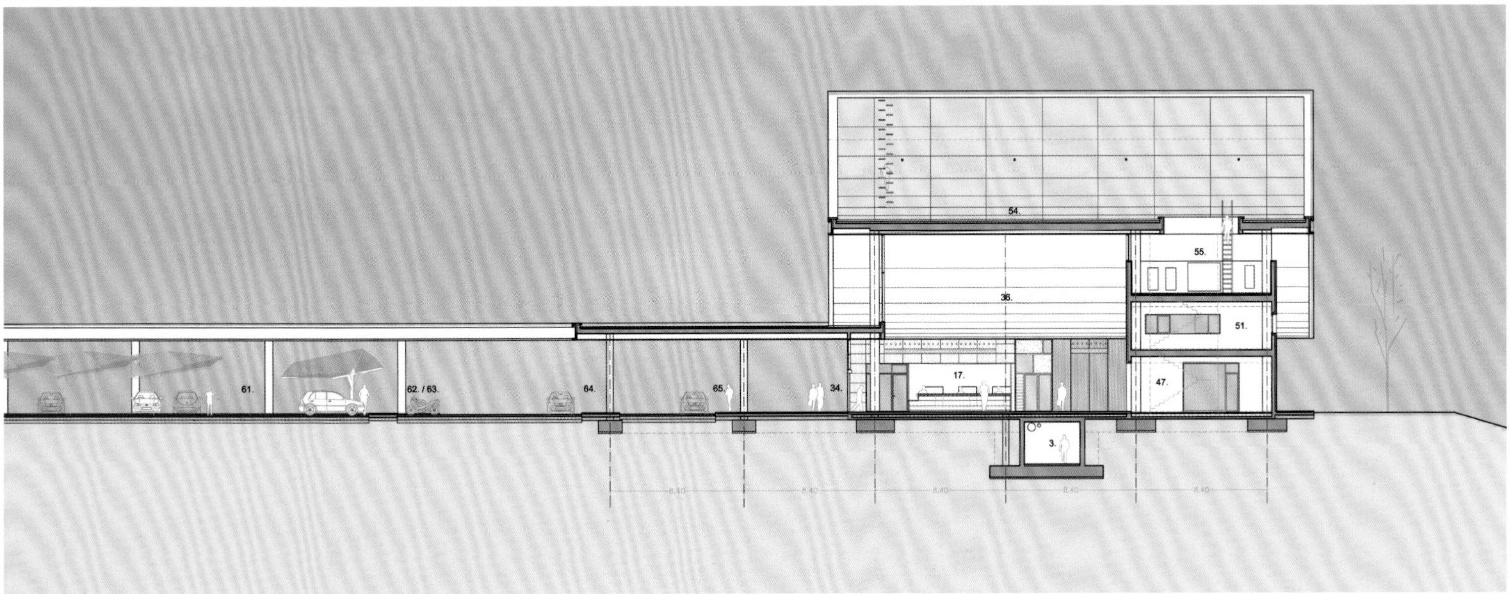

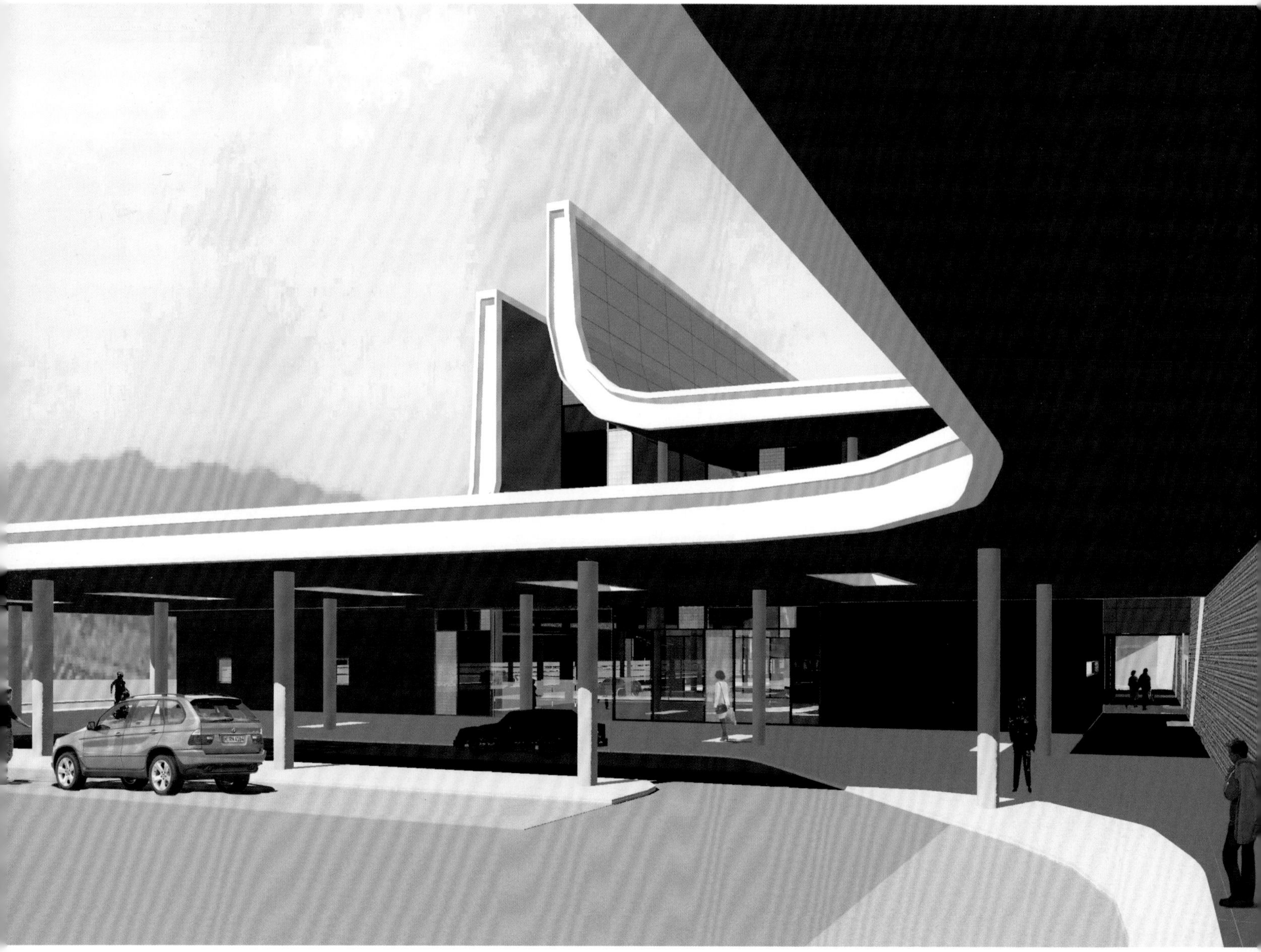

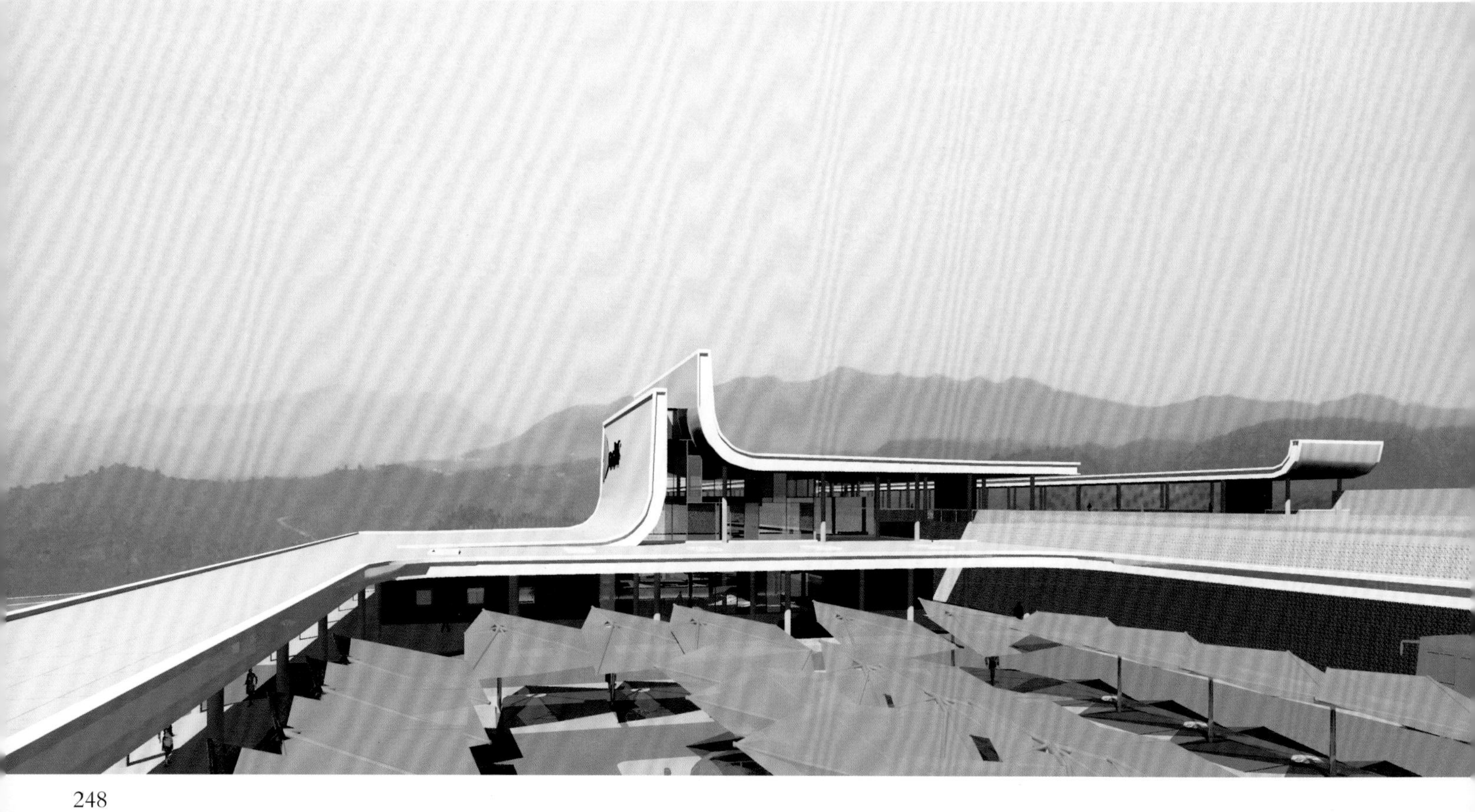

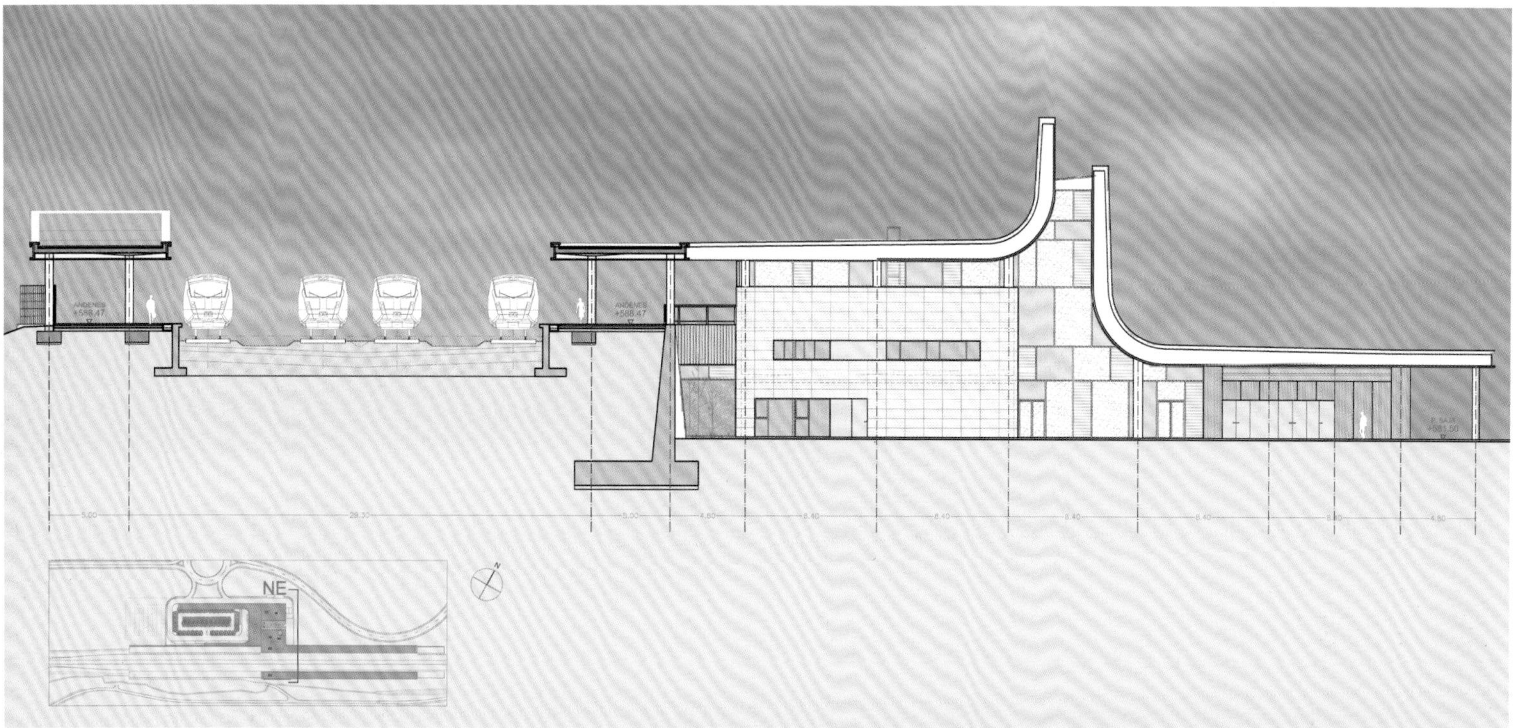
NE

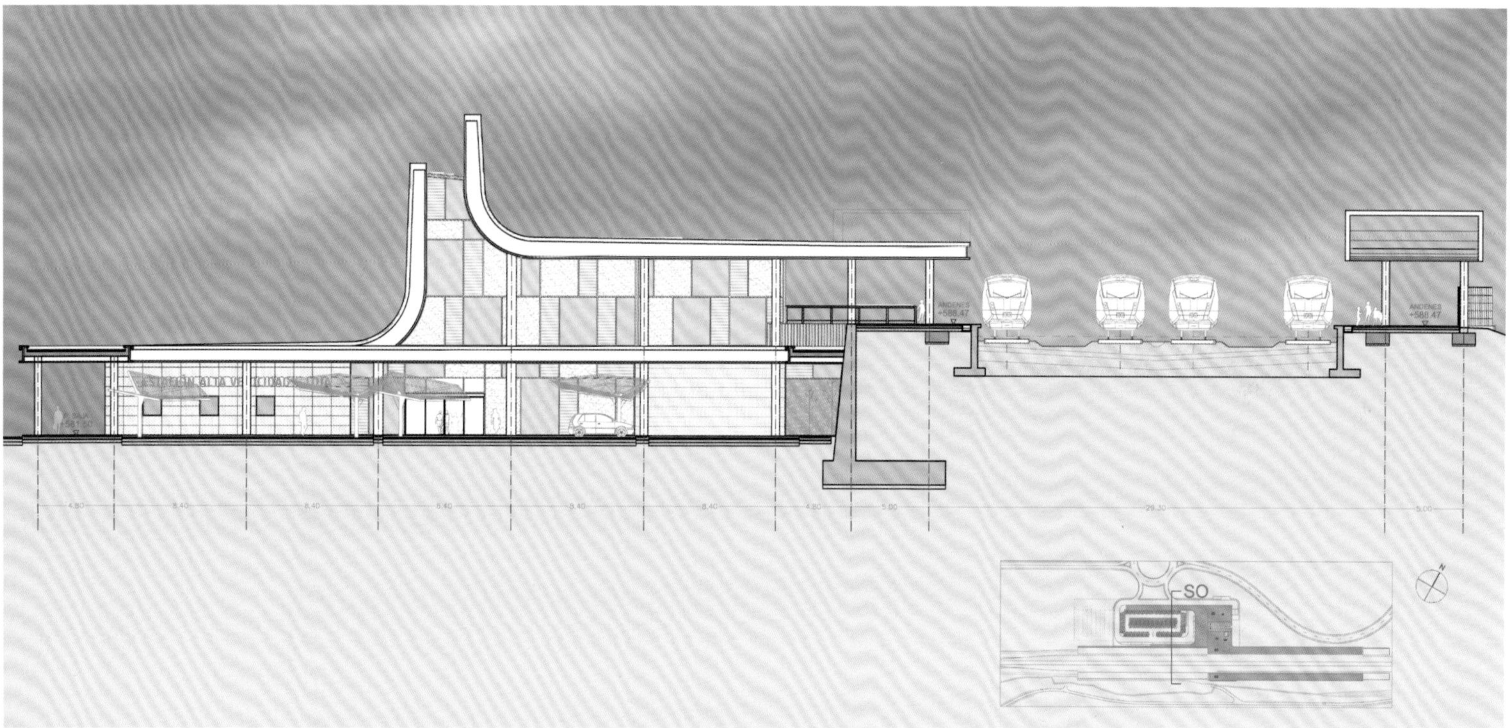
SO